AF661464

MICROBIAL BIOTECHNOLOGY

Dr. H.A Modi is M.Sc., M.Phil. and Ph.D. in Microbiology. Presently, he is a Reader (Microbiology) at Department of Life Sciences, Gujarat University, Ahmedabad. He is visiting Professor at P.G. Departments of North Gujarat University, Saurastra University, Gujarat Vidyapeeth, South Gujarat University and other Universities of Gujarat. He is a recognized Ph.D. guiding teacher in Microbiology at Gujarat University and North Gujarat University, supervising Ph.D. students in the fields of Enzyme biotechnology, Bioconversion technology, Mushroom Technology, Biocontrol and Food microbiology. He has presented several scientific presentations at National as well as International Conferences, seminars and symposia. His research papers have been published in journals of National and International repute. He is a member of many prestigious academic and professional organizations. He has published two volumes of Elementary Microbiology which are approved as text books at many University curricula. He has also authored books namely, "Industrial Microbiology" and "Microbiology for Nursing and Home-Science" published by University Book Production Board, Ahmedabad. He is a co-author of "Biofertilizers and Organic Farming." Aavishkar Publishers & Distributors (Jaipur) has published his book—"Introductory Food Microbiology."

Dr. Modi has widely traveled in countries like Brazil, Germany, England etc. for academic purposes under Group Study Exchange Programmes. He has successfully carried out research projects funded by UGC, DBT, GUJCOST etc.

FERMENTATION TECHNOLOGY SERIES

MICROBIAL BIOTECHNOLOGY

Dr. H.A. Modi

Microbiology Laboratory
Department of Life Sciences
University School of Sciences
Gujarat University
Ahmedabad-380 009 (Guj.)

Pointer Publishers

Jaipur 302 003 (Raj) India

Distributed by
Prem C. Bakliwal for
Aavishkar Publishers, Distributors
807, Vyas Building, Chaura Rasta
Jaipur 302 003 (Raj) India
Phone : 0141-2578159
e-mail : aavishkarbooks@hotmail.com

© H.A. Modi

ISBN 978-81-7132-591-7

First Published in 2009 by
Mrs. Shashi Jain for
Pointer Publishers
Vyas Building, S.M.S. Highway
Jaipur 302 003 (Raj) India
Phones : 0141-2578159, 2708286
e-mail : pointerpub@hotmail.com
web site : www.pointerpublishers.com

All rights reserved. No part of this publication may be reproduced or copied for any purpose by any means, manual, mechanical or electronic, without prior and written permission of the copyright owner and the Publishers.

Printed at
Sheetal Printers
Jaipur 302 003 (Raj) India

ACKNOWLEDGEMENTS

No book can be completed without the dedicated help of many individuals. I appreciate the help of all faculty members of Microbiology and Biotechnology subject, spreaded all over India, who have given me insightful suggestion, comments and ideas.

I would like to express my eternal gratitude to Principal (Retd.) B.N. Gandhi for his inspiration, moral support, and valuable help. I also express my feeling of thanks to Prof. N.K. Jain for his continuous encouragement.

I would like to thank all members of my family (especially my father and mother, my wife—Dharmishta, my son—Niraj, my daughter—Ami, my son-in-law Mr. Chetankumar and my dearest grand-son—"Krish") for "putting up" with me during the preparation of this book. Those hours that I spent working on the book were hours during which I neglected them.

I am grateful to Pointer Publishers, Jaipur for their concern, efforts and encouragement, especially for their excellent cooperation in the task of preparing and publishing this book in best possible form.

—H.A. Modi

PREFACE

Louis Pasteur established the biological concept of **"Fermentation Technology"** which is now flourished as an applied science known as—**"Industrial Microbiology"** or **"Microbial Biotechnology"**. This is an important branch of Microbiology and Biotechnology, where economically valuable products are produced from cheaper raw materials (e.g. agrowastes/industrial wastes containing organic substrates, etc.) by using mass culture of microorganisms. This concept is seedling for the birth of this book.

Chapter-1 introduces the concept and range of fermentation processes and products developed by fermentation technology. The chronological development of Industrial Fermentation Technology is also highlighted in the same chapter. Chapter-2 explains kinetics of microbial growth and fermentation through which various economically important products are formed. Chapter-3 focuses on the concept of microbial screening, development of primary screening techniques, significance of secondary screening studies as well as preservation of industrially important microbial cultures.

Molecular genetic techniques can be used to manipulate genes in order to alter the expression and production of microbial products, including the expression of novel recombinant products. The discovery of recombinant DNA technology (In vitro gene manipulation techniques) has dramatically broadened the spectrum of microbial genetic manipulations and subsequently formation of fermentation products. Chapter-4 describes various gene manipulation techniques adopted for gene cloning in Genetic Engineering at great depth. The brief introduction of mutations is given in chapter-5 and role of mutation in classical strain improvement programme is discussed in later part of the chapter. Classical strain improvement by various recombination methods (including fungal parasexuality and protoplast fusion) is reviewed in chapter-6.

Chapter-7 introduces the principles of Medium Engineering and Metabolic Engineering, which are very important in fermentation processes. Chapter-8 explains the concept of formulation for industrial fermentation media as well as medium optimization using statistical experimental designs. The choice of a good fermentation medium and raw materials is virtually as important to the success of an industrial fermentation as is the selection of a fermentative organism. Essential raw materials used for formulating industrial fermentation media are elaborately discussed in chapter-9.

Sterilization of fermentation media, air and fermenter (bioreactor) is of supreme importance in a number of microbial fermentation industries. For better understanding of industrial sterilization, fundamentals of microbial control and principles of thermal kill are introduced in chapter-10. Industrial sterilization of fermentation media and fermenter are described in chapter-11, while air sterilization is elaborately explained in chapter-12. Inoculum preparation is a critical stage in an industrial fermentation process. Various aspects of inoculum development, by taking suitable examples, are discussed in chapter-13.

A fermenter (or Bioreactor), a heart of fermentation industry, is a system which provide controlled environmental conditions for the growth of microbes (and/or production of specific metabolites) in culture whilst preventing entry and growth of contaminating microbes from the outside environment. A detailed design of industrial fermenter (as well as additional accessories) is explained in chapter-14. Alternative fermenter vessel designs are also described in the same chapter. For any fermentation in a fermenter, measurement and control of various types of bioprocess parameters are essential to take care of smooth product formation and so these important bioprocess parameters are discussed in chapter-15 by the help of many schematic figures. Aeration and agitation, most essential processes for fermentation industries are especially discussed in detail in chapter-16.

"Fluid Rheology" (chapter-17) and "Form Formation and its Control" (chapter-18) topics are elaborately discussed in this book by assigning separate status of chapter.

The marriage between fermentation technology and computer science is a fortuitous one, likely to stimulate, nurture and expand both areas of human knowledge. Without computers, fermentation technology would be processing slowly and painfully, rather than exploding as it is today. Computer applications in fermentation technology and industry are reviewed in chapter-19. The conversion of laboratory procedure to an industrial process is termed "Scale-up". Various aspects and significance of "Scale-up" and "Scale-down" are explained in chapter-20.

A fully referenced text is not, in my view, appropriate to a book of this type aimed at readers with little if any knowledge of the subject. However, a list of selected bibliography is included, to assist students in expanding their knowledge on the subject. The materials in each chapter are arranged in logical, systematic and concise sequences. Tables, schematic figures, photographs and data have been largely mentioned when their presentation is necessary for better understanding.

The attempt is done that book should prove to be an useful source of information for anyone with an interest in fermentation technology for both undergraduate as well as postgraduate courses of microbiology. The contents of this book will also be useful to the students of Biotechnology, Industrial Microbiology, Pharmaceutical sciences and technology, Food science and technology, Agricultural Microbiology, Dairy sciences and technology, Public health sciences, Biochemical Engineering and other fermentation-related commerical and vocational courses. This book will also be proved as an useful reference to get first hand information of fermentation technology and industrial microbiology for the people who are working in fermentation-based industries.

—H.A. Modi

CONTENTS

1

Concept and Range of Fermentation Technology

1.1 INTRODUCTION

Microorganisms are capable of growing on a wide range of substrates and can produce a remarkable spectrum of products. The relatively recent advent of in vitro genetic manipulation has extended the range of products that may be produced by microorganisms and has provided new methods for increasing the yields of existing ones. The commercial exploitation of biochemical diversity of the fermentation industry and the techniques of genetic manipulation has given this well-established industry the opportunity to develop new processes and to improve existing ones.

1.2 CONCEPT OF FERMENTATION TECHNOLOGY

During the last century, the word **"Fermentation"** has gained new meanings and applications have been extended to various fields. Following points explain the concept of fermentation:

(1) Originally, the term "fermentation" was derived from the Latin word *"fervere"* (= to boil), thus referring to the bubbling observed when sugar-starchy-fruit-malted grain materials undergo a transform to yield alcohol by yeasts. The boiling or bubbling appearance is developed due to production of CO_2.

(2) Later, the term "fermentation" was applied to the process of producing alcohol which was formed from the sugar, regardless of whether the causative agent was or was not biological.

(3) Louis Pasteur defined "fermentation" to any anaerobic process through which microorganisms obtain energy for growth in absence of O_2.

(4) Today, "fermentation" has much broader meaning and it applies to both aerobic as well as anaerobic metabolic activities of microorganisms in which specific chemical changes are brought about in any organic substrate.

However, fermentation is interpreted differently by biochemists and microbiologists.

(5) To a biochemist, in strict biochemical sense, the word "fermentation" means any energy-generating process in which organic compounds act as both electron donors and acceptors, that is, an anaerobic process where energy is produced without the participation of oxygen or other inorganic electron acceptors.

(6) Industrial microbiologists have extended the meaning of "Fermentation technology" to describe any microbial process or technology for the production of product having commercial value, by the mass culture of microorganisms. This is the most acceptable definition. In this book fermentation is used in its broader, microbiological context.

Louis Pasteur established the biological concept of **"Fermentation technology"** which is now flourished as a science known as—**"Industrial Microbiology"** or **"Microbial Biotechnology"**. So, Industrial Microbiology or Microbial Biotechnology is entirely based on the concept of fermentation technology. This is an important branch of Microbiology, where economically valuable products are produced from cheaper raw materials (e.g. agrowastes/industrial wastes containing organic substrates, etc.) by using mass culture of microorganisms.

1.3 RANGE OF FERMENTATION PROCESSES AND PRODUCTS

Microbial fermentation processes and products may be classified into the following major groups:

(1) Those that produce microbial cells as the product **[Microbial Biomass]**

(2) Those that produce **microbial metabolites**,

- (a) **Primary Metabolites**
- (b) **Secondary Metaboiltes**

(3) Those that produce **"Microbial Enzymes"**

(4) Those that modify a compound which is added to the fermentation—the transformation processes **[Biotransformation Products]**

(5) Those that produce **"Recombinant Products"**

(6) Food industry products—**"Fermented Foods"**

(7) Elicitors-induced products.

1.3.1 Microbial Biomass

Microbial biomass involves the production of microbial cells as major fermentation end products. It has two major applications

(a) as a source of feed protein in form of SCP

(b) as microbial inoculant needed in fermented foods, agriculture, waste treatment as well as vaccine preparation.

(a) SCP Production—Two major porcesses applied for the commercial production of yeast biomass involve production of baker's yeast and yeast—SCP [Single Cell Protein].

Instead of using brewer's or distiller's yeast obtained from brewery or distillery; at present baking industry has large scale usage of product called Baker's yeast. It involves the large scale production of selected yeast *Saccharomyces cerevisiae* culture by fed-batch fermentation process on molasses medium. Fed-batch culture involves maximum production efficiency with no concomitant formation of alcohol.

During World War I, Baker's yeast grown on molasses at a large scale, had been used as a protein supplement for human consumption and animal feed.

Later on, Germeny during World War II applied *Candida utilis,* cultivated on sulfite waste liquor from paper pulp industry as a single cell protein product for humans and animals. This process had since been used for large production of yeast biomass to get SCP in U.S.A.; U.S.S.R.; Switzerland and Taiwan.

In 1981, Imperial Chemical Industries in England, scaled up SCP production on methanol. Using aerobic bacterium *Methylophilus methylotrophus* cultivated on methanol, the SCP product *"pruteen"* for animal feed was obtained. The patent is still in application.

The first novel *'Mycoprotein'* obtained from *Fusarium graminearum* has also received the Government approval for human use in England. Its filamentous texture confers on it a natural appearance similar to meat enabling the product to command a realistic market price and gain consumer appeal.

(b) Microbial inoculants—

(i) Food applications—Various microbial inoculants are used in a variety of food applications. They have characteristic biochemical features that result into changes in food properties like texture, modification, presentation, flavour development and nutritional improvement.

Commercial yeast starter cultures for use in alcoholic beverage processes; cheese starter cultures in dairy industry and various lactic-starter cultures for different types of fermented milk products other than cheese are few examples.

(ii) Non-food application—Microbial inoculants are also required in a variety of non-food applications. It includes use of insecticidal bacteria for plant disease biocontrol; legume inoculants for use as biofertilizers, Methanogens for use in waste treatment processes and sulphur oxidizers for mineral recycling.

Microbial biomass production is also required in bacterial or viral vaccines, used in prophylactic treatment against infectious diseases.

Various "Microbial Biomass" products are summarized in Table 1.1.

Table 1.1 : Microbial biomass

	Fermentative microbe	Product
I.	**SCP production**	
1.	*Saccharomyces cerevisiae*	Baker's yeast
2.	*Candida utilis*	Single cell protein on hydrocarbon
3.	*Methylophilus methylotrophus*	SCP—Pruteen on methanol

contd. ...

... contd.

	Fermentative microbe	Product
4.	*Fusarium graminearum*	Mycoprotein
II.	**Microbial Inoculants—Food applications**	
5.	*Streptococcus cremoris*	Cheese
	+ *Streptococcous lactis*	Starter culture
6.	*Lactobacillus acidophilus*	'Probiotic' acidophilus milk
	+ *Streptococcus thermophilus*	
7.	*Lactobacillus bulgaricus*	Starter culture for yoghurt
III.	**Microbial Inoculants—Non-food application**	
8.	*Bacillus thuringiensis*	Insecticides
9.	*Thiobacillus ferreoxidans*	Metal bioleaching
10.	*Azotobacter agilis*	Biofertilizer
11.	*Rhizobium leguminosorum*	Legume inoculant
12.	*Methanogens*	Waste treatment

1.3.2 Microbial Metabolities

When a microbial culture is grown in a nutrient medium, it utilizes the nutrients to synthesize macromolecular components of cellular structure and also many low molecular weight compounds required for cellular activity.

The cell passes through various phases of growth. The initial time of adoptation is called lag phase during which growth does not occur.

1.3.2.1 (a) Primary metabolites

Following lag phase the growth rate increases exponentially with time at a constant rate. This is logarithmic phase (also known as Trophophase). During this phase, energy-producing or Catabolic pathways generate ATP and coenyzme molecules. Biosynthetic/ anabolic reactions start and various products of metabolism essential to the growth of cells are synthesized. It includes aminoacids, nucleotides, proteins, carbohydrates and Lipids. These products are referred to as the primary products of metabolism [Primary Metabolites].

Many products of primary metabolism have economic importance (Table : 1.2).

E.g. hexoses are converted to pyruvate via EMP pathway. This pyruvate is involved in anaerobic production of ethanol by Yeast, L-Lactic acid by *Lactobacillus* and to acetone-butanol by *Clostridium acetobutylicum.*

In aerobic metabolism, it enters TCA cycle which gives intermediates like citric acid, itaconic acid etc. by fungi. It also produces amino acids like—L-glutamate or L-lysine by *Corynebacterium* and *Brevibacterium* Spp.

In general, the synthesis of primary metabolites by wild type microbial strains is always in sufficient quantity to meet the cellular requirement for growth. For specific industrial fermentation to occur, industrial micobiologist should modify the wild-type strain. For overproduction of a given primary metabolite having industrial potential, cultural conditions are modified. Strains having improved productivity are selected for industrial processes.

Some primary products of microbial metabolism and their commercial application have been listed in table : 1.2.

Table 1.2 : Primary metabolites

	Product	Fermentative microbe	Application
1.	Ethanol	*Saccharomyces cerevisiae*	Gasohol fuel, solvent. Active ingredient of alcoholic beverage
2.	Citric acid	*Aspergillus niger*	Soft drink prep. Blood anticoagulant
3.	Lactic acid	*Lactobacillus* Spp.	Plastic prep.
4.	Acetone-butanol	*Clostridium acetobutylicum*	Solvents
5.	L-Lysine	*Corynebacterium glutamicum*	Feed supplement
6.	L-glutamate	*Brevibacterium flavum*	Flavour enhancer
7.	Cyanocobalmin (Vit. B_{12})	*Propionibacterium shermanii*	Pharmaceutical digestive aid
8.	Riboflavin (Vit. B_2)	*Ashbya gosypii*	Feed supplements
9.	Xanthan gum	*Xanthomonas campestris*	Oil recovery

1.3.2.2 (b) Secondary metabolites

Bu'Lock *et al.* suggested the term *'idiophase'* to describe the phase of a culture during which products other than primary metabolites are synthesized, products which do not have an obvious role in cell metabolism. The metabolites produced during the idiophase are referred to as secondary metabolites. The interrelationships between primary and secondary metabolism are illustrated in Figure 1.1, from which it may be seen that secondary metabolites tend to be synthesized from the intermediates and end-products of primary metabolism. Although the primary metabolic routes shown in Figure 1.1 are common to the vast majority of microorganisms, each secondary metabolite would be synthesized by very few microbial taxa. Also, not all microbial taxa undergo secondary metabolism; it is a common feature of the filamentous fungi and bacteria and the sporing bacteria but it is not, for example, a feature of the *Enterobacteriaceae*. Thus, although the taxonomic distribution of secondary metabolism is far more limited than that of primary metabolism, the range of secondary products produced is enormous. It is sometimes difficult to categorize a product as primary or secondary, and the kinetics of production of certain compounds may change, depending on the growth conditions employed.

At first sight it may seem anomalous that microorganisms produce compounds which do not appear to have any metabolic function and are certainly not by-products of catabolism as are, for example, ethanol and acetone. However, many secondary metabolites exhibit antimicrobial properties and, therefore, may be involved in competition in the natural environment, others have, since their discovery in idiophase cultures, been deomonstrated to be produced during the trophophase where, it has been claimed, they act in some form of metabolic control. Although the physiological role of secondary metabolism continues to be the subject of considerable debate its relevance to the fermentation industry is the commercial significance of the secondary metabolites. Table : 1.3 summarizes some of the industrially important secondary metabolites.

Table 1.3 : Commercially produced secondary metabolite products

Product	Activity	Source (producing microorganism)
Avermectin	Antihelminthic	*Streptomyces avermitilis*
Bestatin	Antitumour	*Streptomyces olivoreticuli*
1, 3-Diphenethylurea	Antidepressant	*Streptomyces* sp.
Detoxin	Detoxicant	*S. caespitosus*
Herbicidin	Herbicide	*S. sagamonensis*
Cyclosporin	Immunosupperssor	*Trichoderma polysporum*
Piercidin	Insecticide	*S. mobaraensis*
Tetranactin	Miticide	*S. aureus*
Gibberellic acid	Plant growth regulator	*Gibberella fujikuroi*
Slaframine	Salivation inducer	*Rhizoctonia leguminicola*
Azalomycin	Antiprotozoal	*S. hygroscopicus*
Aabomycin A	Antiviral	*S. hygroscopicus*
Penicillin	Antibacterial	*Penicillium chrysogenium*
Streptomycin	Antimicrobial	*Streptomyces griseus*
Griseofulvin	Antifungal	*Penicillium griseofulvum*

The production of microbial metabolites may be achieved in continuous, as well as batch, systems. The chronological separation of trophophase and idiophase in batch culture may be studied in continuous culture in terms of dilution rate. Secondary metabolism will occur at relatively low dilution rates (growth rates) and, therefore, it should be remembered that secondary metabolism is a property of slow-growing, as well as stationary, cells. The fact that secondary metabolites are produced by slow-growing organisms in continuous culture indicates that primary metabolism is continuing in idiophase-type cells. Thus, secondary metabolism is not switched on to remove an accumulation of metabolites synthesized entirely in a different phase; synthesis of the primary metabolic precursors continues through the period of secondary biosynthesis.

The control of the onset of secondary metabolism has been studied extensively in batch culture and, to a lesser extent, in continuous culture. The outcome of this work is that a considerable amount of information is available on the interrelationships between the changes occuring in the medium and the cells at the onset of secondary metabolism and the control of the process. Primary metabolic precursors of secondary metabolites have been demonstrated to induce secondary metabolism, for example, tryptophan in alkaloid biosynthesis and methionine in cephalosporin biosynthesis. On the other hand, medium components have been demonstrated to repress secondary metabolism, the earliest observation being that of Saltero and Johnson in 1953 of the repressing effect of glucose on benzyl penicillin formation. Carbon sources that support high growth rates tend to support poor secondary metabolism and Table 1.4 cites some example of this situation. Phosphate sources have also been implicated in the repression of secondary metabolism, as exemplified in Table 1.4.

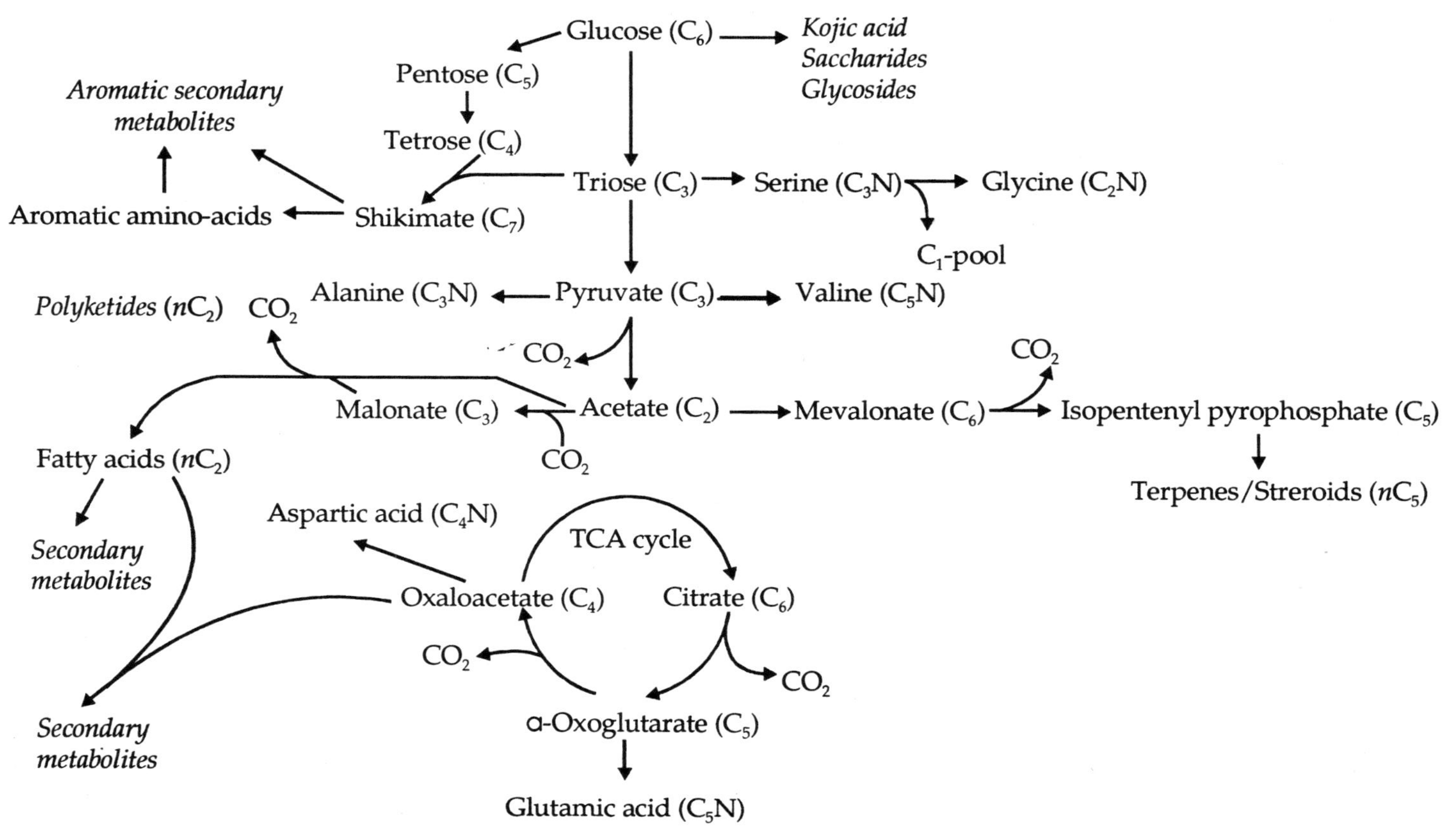

Fig. 1.1 : The interrelationships between primary and secondary metabolism. Primary catabolic routes are shown in heavy lines and secondary products are italicized (Turner, 1971).

Table 1.4 : Some examples of the repression of secondary metabolism by medium components

Medium component	Repressed secondary metabolite
Glucose	Penicillin
Glucose	Actinomycin
Glucose	Neomycin
Glucose	Streptomycin
Phosphate	Candicidin
Phosphate	Streptomycin
Phosphate	Tetracycline

Therefore, it is essential that repressing nutrients should be avoided in media to be used for the industrial production of secondary metabolites or that the mode of operation of the fermentation maintains the potentially repressing components at sub-repressing levels.

1.3.3 Microbial Enzymes

Enzymes are biological catalyst of high molecular weight proteinic compounds. They change during the reaction, but on completion of the reaction reappear in their original form.

Commerical production of enzymes from microbial sources have many advantages over plants and animal sources.

(i) Microbial enzymes can be produced in very large quantity by established fermentation technique.

(ii) Improvement in productivity and quality of microbial enzyme is possible by use of gene biotechnology.

(iii) It is possible to control the enzyme production in microbes by induction, overcome of feedback repression, mutation as well as fermentation paramenters control.

Commercial exploitation of isolated microbial enzymes dates back to Jokichi Takamine. He in 1894 patented a method for diastatic enzyme production from mould under the trade name Takadiastase. For this he cultivated the fungus on wheat bran medium by solid state fermentation. Later on Boidin and Effront (1917) respectively from France and Germany developed patent for amylases by cultivation of *Bacillus subtilis* in surface culture. Following use of submerged fermentation technique, many enzymes were produced at very large scale.

Introduction of microbial enzymes lay foundation for the development of enzyme application technology in particular areas of food processing. It includes use of amylase for preparation of mash in brewing and for reduction of dough viscocity in baking industry, desizing fabrics in textile industry and preparation of hydrolysed starch syrup in sugar industry. Pectinase is used for preparation of fruit juices. At present preparation of High Fructose Corn Syrup (HFCS) from corn starch, employ uses of amylase and glucose isomerase enzymes. HFCS serves as caloric sweetner and is used in place of sucrose in preparation of syrup base for pharmaceuticals, alcoholic beverages and soft drinks. Few microbial enzymes like alkaline protease has application in detergent industry. Proteases are also used in dehairing and bating of leather and tenderization of meat. Few Enzymes serve as digestive aids in pharmaceutical preparation. Streptokinase has medicinal application for dissolving blood clots. Glucose oxidase is applied for removal of glucose in preparation of dried eggs. Use of *Mucor miehei* rennet in cheese manufacture and application of commercial cellulases produced by *Trichoderma* in food processing, alcohol production and cellulosic waste treatment are very common. Penicillin acylase is used to hydrolyze penicillin side-chains to produce 6-APA required for preparing semisynthetic penicillins. Most of the enzymes, we discuss are secreted by microbial species outside the cell environment hence extracellular type.

Many intracellular enzymes like restriction endonucleases, Taq polymerases, ligase etc. have application in genetic engineering.

The list of microbial enzymes and their various industrial applications is increasing (See Table : 1.5).

Table 1.5 : Commercial applications of microbial enzymes

Industry	Applications	Enzymes used
(1) Baking and milling (Bakery)	(1) Reduction of dough viscosity (2) Acceleration of fermentation process (3) Increase in loaf volume. (4) Maintenance of freshness and softness.	Amylase, protease
(2) Beer (Brewing)	Mashing	Amylase
	Chillproofing	Protease
	Improvement of fine filteration	β-glucanase
(3) Cereals	(1) Produced baby foods (2) breakfast foods	Amylase
(4) Coffee-cocoa	(1) Coffee bean fermentation.	Amylase
	(2) Preparation of coffee concentrates	Pectinase
(5) Confectionary	Manufacture of soft-centre candies	Invertase
(6) Corn Syrup	(1) Manufacture of high maltose syrup	Amylase
	(2) Production of glucose from corn syrup	Amyloglucosidase
	(3) Conversion of corn syrup to a sweetner fructose	Glucose isomerase
(7) Dairy	(1) Residual H_2O_2 removal from milk	Catalase
	(2) Manufacture of protein hydrolysates	Protease
	(3) Production of whole milk, ice-cream and frozen deserts	Lactase
	(4) Curdling of milk	Protease
	(5) Cheese preparation	Rennets
(8) Poultry	(1) Dried eggs—glucose removal	Glucose oxidase
(9) Flavours	(1) Clarification	Amylase
	(2) Oxygen removal	Glucose oxidase
(10) Fruit Juices	(1) Clarification (2) improving extraction yield	Pectinases
	(3) Debittering of citrus	Naringinase
(11) Laundry and dry cleaning	Detergent/washing	Protease, Amylase Lipase

contd. ...

... contd.

Industry	Applications	Enzymes used
(12) Leather	Dehairing and bating	Protease
(13) Pharma-ceutical and clinical	(1) Digestive aids (2) Cleaning of wounds	Amylase, Protease Streptokinase
(14) Photography	Recovery of silver	Protease
(15) Soft drinks	Stabilization of citrus terpenes and oxygen removal	Glucose oxidase and catalase
(16) Textiles	Desizing of fabrics	Amylase, Protease
(17) Meat	Meat tendering	Protease
(18) Paper	Starch liquification for paper coating	Amylase

1.3.4 Microbial Transformation

Microbial cells may be used to convert a compound into a structurally related compound, having more economical value as exemplified by Steroid transformation. Microbial enzymes catalyze the various reactions in transformation that include dehydrogenation, oxidation, hydroxylation, dehydration and condensation as wall as isomerization, amination, deamination and decarboxylation.

Microbial transformation processes are advantageous over chemical transformation processes: (i) They have high degree of specificity and sensitivity. (ii) It required low temperatures. (iii) It has no need for polluting heavy-metal catalysts.

In one example, the steroid drug cortisone require 37 reaction steps for chemical synthesis. But Scientists at the Upjohn company (1952) discovered a strain of *Rhizopus arrhizus* which could hydroxylate the progesterone. This finishes the cortisone synthesis in eleven steps and reduces the cost of production.

Similar microbial transformation processes have been applied to synthesis of other steroid drugs like aldosterone, Prednisolone and predinsone. Kieslich (1982) cited use of biotransformation techniques for production of other high-value pharmaceuticals like novel antibiotics and prostaglandins. Production of vinegar from ethanol, sorbose from sorbitol in vitamin C synthesis are also examples of microbial transformation processes.

For the microbial transformation reactions to occur, a large biomass culture of microbial species is required. It catalyzes a single reaction. The technology can be modified by use of immobilization techniques.

The immobilized whole cell or enzyme on an inert support can catalyze the biotransformation reaction. Some of microbial transformation products are mentioned in Table : 1.6.

Table 1.6 : Microbial transformation

Microorganism	Biotransformation	Product
1. *Rhizopus arrhizus*	Steroid transformation	Cortisone

contd. ...

... contd.

Microorganism	Biotransformation	Product
2. *Arthrobacter simplex*	Steroid transformation	Prednisolone
3. *Gluconobacter suboxydans*	D-sorbitol to L-sorbose	Vitamin "C"
4. *Acetobacter aceti*	Ethanol to acetic acid	Vinegar
5. *E. coli* for penicillin acylase	Penicillin G to 6APA	6 APA for semi-synthetic penicillin

1.3.5 Recombinant Products

The advent of recombinant DNA technology has extended the range of potential microbial fermentation products. It is possible to introduce genes from higher organisms into microbial cells such that the recipient cells are capable of synthesizing foreign (or heterologous) proteins. Examples of the hosts for such foreign genes include *Escherichia coli, Saccharomyces cerevisiae* and other yeasts as well as filamentous fungi such as *Aspergillus niger* var. *awamori*. Products produced in such genetically manipulated organisms include interferon, insulin, human serum albumin, factor VIII and factor IX, epidermal growth factor, bovine somatostatin and bovine chymosin. Important factors in the design of these processes include the secretion of the product, minimization of the degradation of the product, and the control of the onset of synthesis during the fermentation, as well as maximizing the expression of the foreign gene.

Examples of some therapeutic products produced through recombinant DNA technology are listed in Table : 1.7.

Table 1.7 : Examples of some therapeutic products produced through recombinant DNA technology

Products	Application
Interferon	Cancer and viral infection
Human urokinase	Used in vascular disorder
Insulin	Treatment of diabetes
Human factor IV	Clotting factor for haemophilia
Lymphokines	Immune functioning
Serum albumin	In surgery
Attenuated pseudorabies virus antigen	Vaccine against rabies
Tissue plasminogen activator	In treatment of heart attack
Somatostatin	Treatment of human growth disorder

1.3.6 Food Industry Products—"Fermented Foods"

A very wide range of innumerable products of the food industry, such as sour cream, yoghurt, cheeses, fermented meats, bread and other bakery products, alcoholic beverages, vinegar, fermented vegetables and pickles, etc., are produced through microbial fermentation processes. The efficiency of the strains of the organisms used, and the processes are being continuously improved to market quality products at more reasonable costs.

A list of some selected fermented foods is given in Table 1.8.

Table 1.8 : Some selected fermented foods used in the various countries of the world

Fermented food	Substrate	Microorganism	Country/Region
Ang-Kak (red rice)	Rice	*Monascus purpureus*	China
Gari	Cassava	*Corynebacterium manihot* *Geotrichum candidum*	West Africa
Idli	Rice and urad	*Leuconostoc mesenteroides* *Streptococcus faecalis*	India
Kaffir beer	*Sorghum caffrorum* or *Eleusine coracana*	*Lactobacillus delbrueckii* *Saccharomyces cerevisiae*	South Africa
Kefir	Milk	*Lactobacillus* sp. and yeast	USSR
Tape ketan, tape katella	Rice	*Ankylomyces rouxii* *Endomycopsis burtonii*	Indonesia
Yoghurt	Milk	*Streptococcus thermophilus* *Lactobacillus bulgaricus*	Worldwide
Cheese (processing)	Milk	*Penicillium roquefortii* *P. camembertii*	Worldwide
Sake (rice wine)	Rice	*A. oryzae*	Japan

1.3.7 Elicitors—Induced Products

It is possible to induce production or enhance production of a compound in cultures by using **elicitors**, which may be microorganisms. For example, *Saccharomyces cerevisiae* was an efficient elicitor in the production of glyceollin (*Glycine max*) and berberine (*Thalictrum rugosum*). *Rhizopus arrhizus* trebled diosgenin production by *Dioscorea deltoidea*. The production of morphine and codeine by *Papaver somniferum* was increased 18 times by *Verticillium dahliae*.

1.4 THE FERMENTATION PROCESS OUTLINE

Figure 1.2 illustrates the component parts of a generalized fermentation process. Although the central component of the system is obviously the fermenter itself, in which the organism is grown under conditions optimum for product formation, one must not lose sight of operations upstream and downstream of the fermenter. Before the fermentation is started the medium must be formulated and sterilized, the fermenter sterilized, and a starter culture must be available in sufficient quantity and in the correct physiological state to inoculate the production fermenter. Downstream of the fermenter the product has to be purified and further processed and the effluents produced by the process have to be treated.

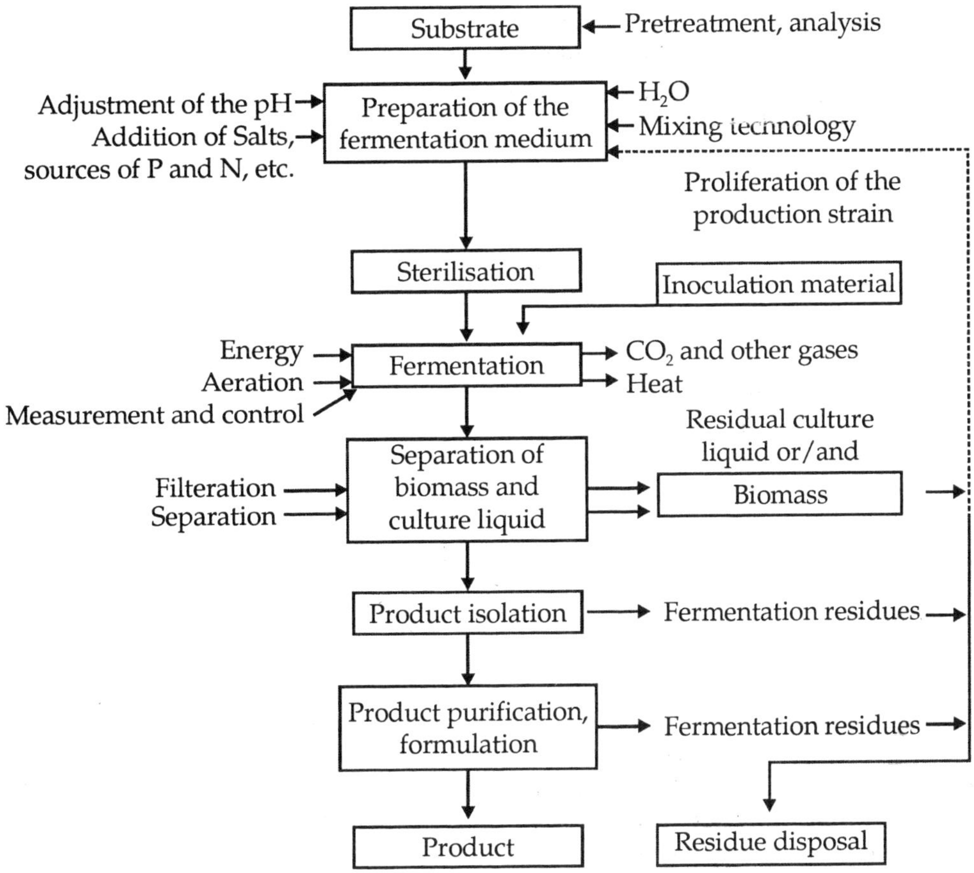

Fig. 1.2 : Flow sheet of generalized fermentation process.

1.5 THE CHRONOLOGICAL DEVELOPMENT OF THE INDUSTRIAL FERMENTATION TECHNOLOGY

The chronological development of the fermentation Industry may be represented as five overlapping stages as discussed follows:

1.5.1 Stage-I (Prior to 1900)—The Ancient Development

Fermentation as an art is one of mankind's earliest achievements in the field of applied microbiology. The products were confined to potable alcohol (wine), bread, vinegar, cheese, yoghurt etc.

Wine and beer making is thought to have been practised at least since 10,000 B.C. in Egypt. About 4000 B.C. Egyptian used Brewer's yeast for carbon dioxide production in the leavening of bread. *Spirulina* algae was harvested for food consumption from alkaline ponds by the ancient Aztecs in Mexico. Some of the oriental fermented food products and sauces based on microbial enzymatic fermentation reactions were known using surface culture method many thousands years ago. About 5000 B.C. earlier fermented milk products cheese by using calf rennet were practiced. In the same way production of vinegar and distilled potable spirits date back to 1000 B.C.

Production of modified foods and alcoholic beverages involving microbial fermentation on large-scale dated from early 1700s, when wooden vats of 1500 barrels capacity were used as fermenter. Early breweries in England had attempted process control in fermentation reaction by the use of thermometer in 1767 and the development of heat exchanger in 1801. By the mid-1800s the role of yeasts in alcoholic fermentation was demonstrated independently by Cagniard-Latour, Schwann and Kutzing. Louis Pasteur studied the role of various types of microbes in specific type of fermentation reaction.

Hansen (1800) at the Carlsberg brewery propagated single isolated yeast cell to obtain pure cultures. These pure yeast cultures were used as strarter cultures in alcoholic fermentation. Though some British breweries still use mixed cultures at the present time.

Vinegar production was originally carried out by leaving wine in shallow bowls or partially filled barrels. Slow oxidation of ethanol in wine by its natural flora produced good quality vinegar. By late 1800s to early 1900s vinegar generators were developed in the vessel packed with a column of inert material like wood shavings, coke, charcoal etc., wine or beer were allowed to trickle. This trickling also provided oxygen required for vinegar fermentation reaction. Use of pasteurised wine as initial fermentation medium and inoculation with 10% of good quality vinegar had given better yield of vinegar product without any contamination problem.

Thus by the beginning of 20th century, the concept of aerobic fermentation process was established. Use of pasteurised fermentation medium and pure cultures of fermentative yeast or use of good quality vinegar as inoculum established the concept of aseptic fermentation technique. Aeration by spray of wine medium for trickling purpose and use of thermometer for temperature showed some developments in process control. Various types of fermentation techniques like surface culture methods, batch fermentation and solid state fermentation (Vinegar production in generator filled with solid inert materials) were also developed before 20th century.

On the pharmaceutical side mouldy meat or bread had been employed in folk medicine for thousands of years to heal wounds and treat infections. In retrospect, the beneficial effects of such treatments was undoubted due to antibiotic production by mould present.

1.5.2 Stage-II (1900-1940)—The Modern Era

Between the years 1900 to 1940s the fermentation technology was developed for industrial production of citric acid, glycerol, lactic acid and acetone-butanol. Most of these products are actually primary metabolites produced by microbes. Yeast biomass production for baker's yeasts and ethanol fermentation processes were developed.

Fed-batch culture technique was introduced to avoid condition of oxygen limitation. Aeration system was improved by supplying air through steam sterilized spargers and mechanical stirring was introduced by stirrers with impeller blades. Government's demands for glycerol and acetone during World War I for use in explosives manufacture had great contribution for development of microbial fermentation technology. Neuberg worked on glycerol fermentation. Weizman's efforts for acetone-butanol fermentation led to the establishment of truly aseptic fermentation and good inoculum in form of pure cultures.

This has reduced the risk of phage infection during fermentation. Vertical cylinders with hemispherical tops and bottoms constructed from mild steel ware used as fermenters which can be steam sterilized. 2000 hl fermenters were used as large seed tanks to develop inoculum in aseptic condition. Thus fermentation technology for large-scale production of organic solvent in steam sterilized fermenters was a major advance in the establishment of modern era of development. This initiated to introduce the aerobic fermentation processes in the 1940s.

1.5.3 Stage-III (1940-1960)—The Accelerated Era of Development

The modern era of development in fermentation technology continued. Again during World War II the Government demand for penicillin had taken special interest in development of fermentation technology. Penicillin antibiotic as a miracle drug became very popular in treatment of wound infections in soldiers. Fungal production of penicillin is an aerobic process. Submerged culture fermentation technology equipped with aeration-agitation devices was established. Use of steam sterilized air at rate of 200 cu. ft./min. reduced risk of contamination. To solve the problem of lower yield of antibiotic by natural isolates; strain improvement programmes were performed. It specifically took interest in mutation and selection programmes to obtain a high yielding strain of *Penicillium*. Solvent extraction techniques for recovery of penicillin was another major advance at that time. Pilot plant facilities also aided for testing of new technology on semi-production scale. Invention of sterilizable oxygen electrodes control loop system for process control of various fermentation parameters like pH, temperature and foam production and later on computerized process control technology was invented in penicillin fermentation processes.

This fermentation technology was applied for production of various metabolites like Streptomycin, Vitamin B_{12}, Gibberlic acid—a plant growth hormone; riboflavin by yeast, amino acid production and enzymes which involve aerobic fermentation processes. For over-production of these metabolites various combinations of mutations and fermentation process control were established. Now new improved techniques used aim to stimulate cell uptakes of new nutrient materials, to prevent side reactions, to induce and activate biosynthetic enzymes, to reduce and inhibit enzyme activity which degrade the product and finally to facilitate the extraction procedures for product recovery.

1.5.4 Stage-IV (1960-1978)—The Golden Era of Development

In early 1960s a number of multinational companies developed the production technology for microbial biomass as the source of SCP or animal feed protein. The largest mechanically stirred fermentation vessels in the range of 80,000 to 1,50,000 dm^3 were used. SCP production by growing yeast biomass on petroleum hydrocarbon resulted in development of jet fermenter and pressure cycle fermenters. These fermenters eliminated need of agitator-mechanically stirring.

Continuous SCP production technology by adding fresh medium to the fermentation vessel and removing culture fluid from vessel at same flow rate was applied. This was found more economic for large scale. Idea of recycling of nutrients from the waste effluent was practiced by ICI. ICI formulated SCP project on methanol. It utilized a continuous 3,000,000 dm^3 pressure-cycle fermenter. The continuous fermentation operation for time periods in excess of 100 days in aseptic condition was succeeded. Continuous sterilization of feed streams, use of computerized system to operate the process control of the sterilization and operation cycle minimized human errors. Pilot plant facilities were utilized for semi-production scale. In India pilot plants for SCP on hydrocarbon were established in collaboration of British Petroleum Company in Baroda and Jorhat in Assam.

For good quality product genetically engineered strains were used in production. Production technology for various fermentations involving primary/secondary metabolites, was tried by use of recombinant cells to get high yield of product.

1.5.5 Stage-V (1978 onwards)—The Present Era

The fifth stage in the development of fermentation technology started by use of *'in vitro'* genetic manipulation techniques. This allowed transfer of gene of interest between unrelated organisms. Human gene for insulin was transferred into *E. coli*. Recombinant *E. coli* for large scale production of interferon; hepatitis subunit vaccine by recombinant yeast cells, tumour-necrosis factor production by recombinant microbial cells were succeeded. Batch/continuous fermentation cultivation was used. Invention of Hybridoma technology has been used for large scale production of Monoclonal antibody. Various diagnostic kits were developed on basis of monoclonal antibodies. A large number of recombinant products are being marketed as health care products.

❑❑❑

2

KINETICS OF MICROBIAL GROWTH AND FERMENTATION

2.1 INTRODUCTION

A clear understanding of microbial growth kinetics is necessary if a large scale fermentation process is to be properly managed. The understanding of microbial fermentation processes requires the study of the kinetics of these processes, i.e. of the rates of production of cells or their metabolites and the effects of various factors on these rates, irrespective of whether the cells are growing or non-growing.

With advancing knowledge of the functions of enzyme systems and the regulation of growth and product formation, the kinetics of microbial processes has become progressively more exact; this in turn facilitates the generalisation and control of microbial processes on an industrial scale and the enhancement of product yields.

This chapter deals with the kinetics of growth and product formation in batch, continuous and fed-batch culture of microorganisms and with the factors affecting these processes.

2.2 KINETIC CHARACTERISTICS OF MICROBIAL GROWTH AND CULTURE PROCESSES

Differences lie in microorganisms with respect to their growth and production of products. Hence, the microorganisms are cultured in different types of vessels in various ways. Therefore, to get the desired product, microorganisms are grown as batch, fed-batch or continuous cultures.

2.2.1 Batch Culture

Batch culture is the simplest method. A desired microbe is grown in a closed culture system on a limited amount of medium of microbial culture. The laboratory grown microorganism in ordinary flask is basically a batch culture.

In batch culture, growth phase of microorganisms passes through many stages. A microbe grows in the medium until the nutrients are exhausted or toxic metabolites secreted by it reach to inhibitory level. From the beginning of inoculation to the end, microbial culture passes through several stages or phases. (Figure 2.1).

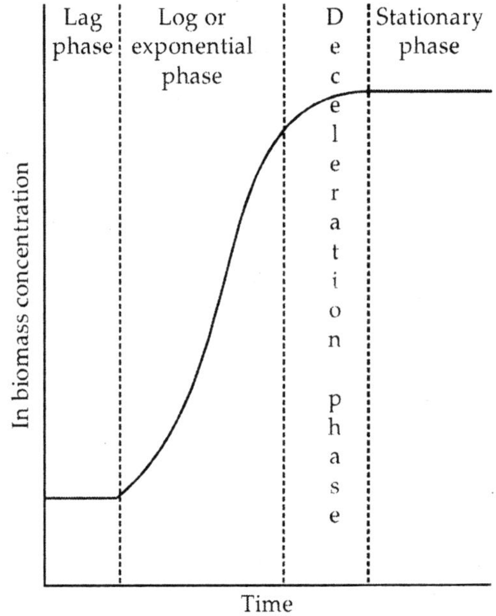

Fig. 2.1 : Growth of a typical microbial culture in batch conditions.

After inoculation, the microbe takes some time to adjust in the new environment according to size of fermenter, and hence does not grow in the medium. Thus, the time taken for adaptation before to come to its active growth is known as **'lag phase'**. The microorganism grows luxuriently till nutrients are present.

Therefore, nutrient-dependent logarithmic or exponential active growth and, thereby, increase in biomass is known as **'log'** or 'exponential phase'. As soon as the levels of nutrients decrease, growth of culture is gradually slown down. The stage of retarding the growth to reach to stationary phase is known as **'deceleration phase'**. However, during stationary phase microorganisms do not grow and thus, fail to increase their biomass. Ultimately the number of microbial cells declines due to accumulation of toxic metabolites. This stage is known as **'death phase'**. At the end, the amount of biomass depends upon the nutritional components and ability of microorganisms to utilize the substrates and convert into biomass.

Bu' Lock *et al.* (1965) have proposed different terminology for these growth stages. They used 'trophophase' for the log phase and 'idiophase' for the stationary phase of batch culture. This typical growth curve is known as 'sigmoidal growth curve'.

During the log phase of culture, growth rate of the microorganisms reaches to its maximum (max). However, after depletion of a substrate, growth rate decreases and finally is ceased. Monod (1942) has demonstrated the relationship of growth rate and concentration of the rate-limiting substrates by the following formula:

$$\mu = \frac{\mu \max [s]}{K_s + [s]}$$

where, μ = Growth rate constant,

$[s]$ = Concentration of limiting substrate,

K_s = Saturation constant value of limiting substrate [s] at which the growth rate is half of the maximum growth rate (μ max).

Microbial Growth Kinetics and Specific Growth Rate—During log phase when cells utilise nutrients and grow to increase biomass, the growth behaves similar to autocatalytic reaction. At this phase growth rate is proportional to cell mass of that period. During this time the rate of 'cell mass increase' (dx/dt) is equal to the specific growth rate (μ) and cell concentration.

$$\frac{dx}{dt} = \mu x \qquad ...(1)$$

where, x = cell concentration (mg/cm^{-3})

t = time of incubation (hours)

μ = specific growth rate (h^{-1})

Equation (1) may be written as

$$\mu = \frac{dx}{dt}.\frac{1}{x} \qquad ...(2)$$

The specific growth rate (μ) acts as an index of rate of cell growth in that very environment. If you plot a graph between dx/dt and x, and determine the slope of straight line, you can calculate the specific growth rate. The value of specific growth rate can be converted into doubling time (i.e. time required by a cell to divide and double its number). It gives a better appreciation of the meaning of these values.

In the beginning of log phase (t = 0) the cell biomass is X_0. Assume that all the cells of batch culture require the same time to double their number (doubling time), then on integration equation (2) you can calculate doubling time as below:

$$\mu.dt = \frac{1}{X}.\mu x$$

or $\log x = ut + K$ (K = constant of integration) ...(3)

When t = 0 from equation (3), we get

$$\log X_0 = K$$

Putting the value of K in equation (3), we get

$$\log X - \log X_0 = \mu.t$$

$$\log \frac{X}{X_0} = \mu.t \qquad ...(4)$$

$$\therefore \qquad td = \frac{1}{\mu} \text{ in } \frac{X}{X_0}$$

When $t = td$ (doubling time)

Then $X = 2X_0$

$$\therefore \qquad t_d = \frac{1}{\mu} \text{ In } 2$$

$$\text{or} \qquad t_d = \frac{\text{In}2}{\mu} \qquad ...(5)$$

If $t = td$ (doubling time), and X = double of X_0, then

$$\text{In}2 = \mu.\text{td}$$

The μ is inversely proportional to t_d. If the microbial culture have high t_d, it will have low μ and *vice versa*. Specific growth rate of $1h^{-1}$ is equivalent to t_d of 0.693. Therefore,

$$t_d = \frac{0.693}{\mu} \qquad ...(6)$$

You can calculate the doublling time of the culture by calculating and substituting in equation (6)

There is another method of calculating the specific growth rate of a microbial culture. Number of microbial cells at different time is counted. If the medium is inoculated with X_0 cells at the time t_0, the number of cells after the time t_t becomes X. The equation (1) may be written as

$$\text{In } \frac{X}{X_0} = \mu.t \text{ or } \text{In } X - \text{In } X_0 = \mu.t$$

After converting the natural logarithms (In) to logarithm to the base 10, you get

$$\frac{\mu}{\frac{\log_{10} X - \log_{10} X_0}{2.303}} = -(t_t - t_0) \qquad ...(7)$$

$$\text{or} \qquad 2.303\left\{\log_{10} X - \log_{10} X_0\right\} = \mu(t - t_0)$$

$$\text{or} \qquad \mu = \frac{(\log_{10} X - \log_{10} X_0)2.303}{t - t_0}$$

This equation is true for biomass concentration (X) as well as cell number (N). If at the time t_0 the microbial culture consists of 10^4 cells/ml and attains 10^{10} cells/ml after 3 hours, the specific growth rate (μ) of the culture can be calculated using equation (3) as below:

$$\mu = \frac{(t - t_0)2.303}{\left(10^{10} - 10^4\right)}(\log 2 = 0.301)$$

$$= \frac{(10 - 4)2.303}{3} = \frac{6 \times 2.303}{3} = 2 \times 2.303 = 4.606\ h^{-1}$$

The specific growth rate of microbial culture is 4.606 h^{-1}. Using the value of μ (i.e. 4.606 h^{-1}) in equation (6) you can derive t_d of the culture

$$t_d = \frac{0.693}{\mu} = \frac{0.693}{4.606} = 0.15 \text{ hour } = 4 \text{ minutes}$$

The t_d of the microbial culture is 4 minutes.

It shows a reciprocal relationship between specific growth rate and doubling time t_d. If μ is high, the t_d will be low and vice versa.

The μ represents the capacity of the microbial culture to grow in a specific environment. The μ of a microbial culture is measured during log phase of growth during which balanced cell growth occurs. In a batch culture the values of μ varies having maximum value at exponential (log) phase of growth. The environmental factors (e.g. pH, temperature, medium composition, aeration, etc.) that affect microbial growth also effect the specific growth rate of the culture. Representative values of μ_{max} of some microorganisms are given in Table 2.1.

Table 2.1 : Representative values of μ_{max} of some microorganisms

Microorganisms	μ_{max} (h^{-1})
Benechea natriegens	4.24
Methylomonas methanolytica	0.53
Aspergillus nidulans	0.36
Penicillium chrysogenum	0.12

2.2.2 Continuous Culture

A continuous culture is that culture where a steady exponential phase for growth of culture retards due to depletion of nutrients, rather than by accumulation of toxic products; it is prevented by addition of fresh medium to the fermenter and removal of spent medium and microbial biomass from it as a result of which the exponential phase of culture is prolonged.

Continuous culture is an open process in which microbial cultures also grow continuously in log phase. One of the nutrients of culture medium is kept limited. Hence, at log phase the cell growth stops as the nutrients of limited quantity is exhausted. In continuous culture, fermented medium is continuously removed from the fermenter.

Therefore, to keep the culture always in log phase, fresh medium is added continuously to the fermenter (before diminishing the nutrients) at the time of removal of medium. Here the rate of supply of nutrients in the form of raw material and removal of products/cells should be volumetrically the same i.e. volume added is equal to volume removed. It means that volume of the medium always remains constant. This should be optimised with different microbial cultures and different growth media.

(a) Principle

In a batch culture, the typical bacterial growth curve includes three transitional periods between growth phases. This means that not all the cells are in identical physiological state. Time is required for some cells to catch up with others. In terms of physiological conditions, the growth curve includes young, actively metabolizing cells on one hand, and cells dying on the other, with the cells in between these extremes. The effect of chemical substances and physical conditions on the organism is not identical in all phases of growth. In a commercial process it is obviously desirable to maintain a microbial population in the logarithmic phase of growth in a constant environment. The growth rate of a culture, and so its physiological conditions may be maintained absolutely constant by the process of continuous culture, usually in a chemostat. The fresh nutrient medium flows into the culture vessel from a reservoir of a sterile medium at a defined and constant rate. The volume in the culture vessel is kept constant (Fig. 2.2) by a device that allows the culture medium, together with the accumulated waste products and, older and dead cells to leave the culture vessel at the same rate. The level of the growth is controlled by maintaining a fixed limiting concentration of a particular nutrient (i.e., growth limiting chemical component) in the medium. The remaining constituents essential for growth of the producer organism are added in the medium in excess of requirement. The growth limiting nutrient is added into the medium at a concentration below that required for maximum growth.

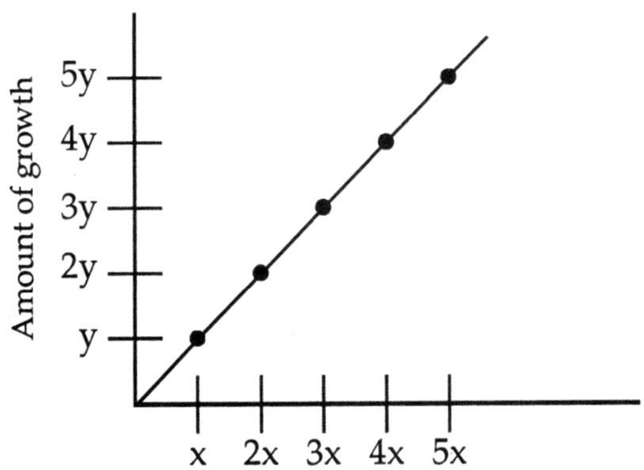

Fig. 2.2 : Cell crop of bacterial culture.

(b) Devices for continuous fermentation

Among the diverse kinds of continuous fermentation, two basic types can be distinguished:

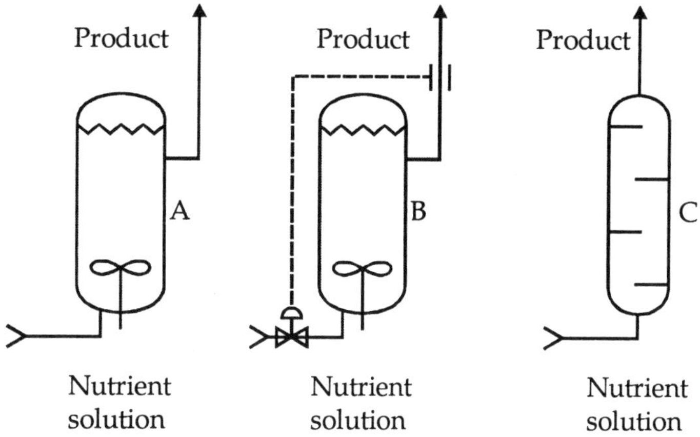

Fig. 2.3: Continuous fermentation in the chemostat (A), turbidostat (B), and plug-flow reactor (C).

1. Homogenously mixed bioreactor—This is run as either a chemostat or a turbidostat. In the chemostat in the steady state, cell growth is controlled by adjusting the concentration of one substrate (Figure 2.3A). Any required substrate (carbohydrate, nitrogen compound, salts, O_2) can be used as a limiting factor. In the turbidostat, cell growth is kept constant by using turbidity to monitor the biomass concentration and the rate of feed of nutrient solution is appropriately adjusted (Fig. 2.3B).

2. Plug Flow Reactor—In this type of continuous fermentation, the culture solution flows through a tubular reactor without back mixing. The composition of the nutrient solution, the number of cells, mass transfer (O_2-supply) and productivity vary at different locations within the system (Figure 2.3C). At the entrance to the reactor, cells must continuously be added along with nutrient solution (usually as a back flow in a bypass from the fermenter outlet or from a second continuous fermentation).

In a continuous process under steady state conditions, cell loss as a result of outflow must be balanced by growth of the organism.

(c) The Kinetics of Continuous Culture

Chemostat is usually operated at different dilution rates. The dilution rate is the ratio of inflowing amount of nutrient medium per hour to the volume of the culture.

Therefore, the dilution rate (D) can be put into

$$D = \frac{F}{V}$$

where D = Dilution rate (h^{-1})

F = Flow rate (dm^3/h)

V = Volume (dm^3)

The change in the cell concentration in the fermenter, in a given period becomes

$$\frac{dx}{dt} = \text{growth - output}$$

$$\frac{dx}{dt} = \mu_x - Dx \qquad (1)$$

Medium is added continuously at a suitable rate and displacement of culture is balanced by the production of new biomass (i.e., output of biomass is balanced by the growth of the cells), the steady state can be expressed as

$$\frac{dx}{dt} = O$$

$$\mu x = Dx$$

$$\mu = D \qquad (2)$$

The steady state condition in a system is controlled by the flow of medium through the system (i.e., Dilution rate), and the mechanism underlying this balance, is essentially, the relationship between μ and residual limiting substrate concentration (Monod, 1942)

$$\mu = \frac{\mu max\, S}{K_S + s}$$

From the equation (2) $\mu = D$

Therefore, $$D = \frac{\mu\, max\, s}{K_s + s} \qquad (3)$$

where s is the steady state residual concentration in the medium, and this can be represented as

$$\bar{s} = \frac{K_s D}{\mu\, max - D} \qquad (4)$$

The substrate concentration in the chemostat is determined by the dilution rate. Virtually, growth depletes the substrate to a concentration that supports the growth rate equal to dilution rate.

The concentration of cells in the system at steady state can be described as

$$\bar{x} = Y(S_R - \bar{s}) \qquad (5)$$

where x represents the steady state cell concentration

From the equations (4) and (5) above,

$$\bar{x} = Y\left(S_R - \frac{KsD}{\mu max - D}\right) \qquad (6)$$

Thus it becomes apparent that, the operational variables (i.e., S_R and D) controls the steady state of biomass concentration.

In the chemostat, maximum stability is attained within the range of dilution rates over which cell density changes only slightly with changes in dilution rates. As D approaches μ max the concentration of residual limiting substrate required to support the increased growth rate increases significantly, and therefore, the residual substrate rises and biomass depletes. Ultimately, a dilution rate is reached at which $\bar{x}$ equals to zero. This dilution rate is referred to as Dcrit. and can be expressed as

$$\text{Dcrit} = \frac{\mu\text{max}.S_R}{K_S + S_R}$$

In a chemostat system, as the substrate limiting factor controls the cell growth concentration, Dcrit., will be slightly lower than the μ max. Continuous fermentation processes have been developed for the production of single-cell protein, antibiotics, organic solvents, starter cultures and for cellulose decomposition (in laboratory studies). Pilot plants or production plants have been installed for continuous production of single-cell protein based on n-alkanes, C_t-compounds and starches.

Industrial processes using continuous fermentation include waste-water treatment as well as the production of beer, glucose isomerase, and ethanol.

Although many fermentations for metabolite production work well as continuous processes at a laboratory scale, only a few processes have proved useful for practical application for several reasons:

- Many laboratory methods operate continuously for only 20 to 200 hours; for industrial use the system must be stable for at least 500 to 1000 hours.
- Maintaining sterile conditions on an industrial scale over a long period of time is difficult.
- The composition of substrates must be constant in order to obtain maximal production. The composition of industrial nutrient solutions varies (e.g. corn steep liquor, peptone, starch).
- When high-yielding strains are used, reverse mutants arise, which can overgrow the production strains in continuous culture.

Continuous cultures also find considerable use in research on ecological processes. Growth at low-substrate levels in a continuous culture is analogous to growth at low-substrate levels in nature, so that the continuous cultures provide excellent models for analysis of ecological processes.

2.2.3 Fed-batch Culture

Basically it is the batch culture which is fed continuously with fresh medium without removal of the original culture medium from the fermenter. It results in continuous increase in volume of medium in the fermenter.

In fed-batch culture the nutrients should be added at the same rate as they are consumed by the growing cells. Therefore, excess of nutrient addition should be avoided. In batch culture when high concentration of substrate inhibits microbial growth, the fed-batch culture is preferred over the former. Hence, in a fed-batch culture substrate is fed at such a concentration that remains below the toxic level. This activity accelerates the cell growth. A high cell density is achieved in fed-batch culture as compared to batch culture.

Fed-batch culture is an ideal process for production of intracellular metabolites in maximum amount. For example, alkaline protease used in biological detergents is produced by the species of *Bacillus*. Batch feeding of nitrogen sources (e.g. ammonia, ammonium ions and amino acids) keeps these substrates at low concentration and induces protease synthesis.

(a) Kinetics of Fed-batch culture

The biomass concentration at stationary phase (X max) in a batch culture, under growth limiting conditions can be described as

$$X \max \underline{\Omega}\, Y.\, S_R$$

The initial inoculum is supposed to be insignificant in comparison with the final biomass.

If the fresh medium were allowed to enter the system at a dilution rate less than X-max, all the substrate gets exhausted.

$$F.S_R \,\underline{\Omega}\, \mu.\frac{X}{Y}$$

Where, X is the total biomass present in the vessel.

The concentration of the cells 'x' remains constant, even if the total biomass 'X' increases with the time.

$$\frac{dx}{dt}\underline{\Omega}\, 0$$

and, therefore $\mu = D$

Such a system then represents a quasi-steady state. As time advances, the volume of culture increases with decrease in the dilution rate. Therefore, D can be represented as

$$D = \frac{F}{V_0 - F.t}$$

where $F =$ flow rate

$V_0 =$ Initial volume of culture

$t =$ time that fed-batch conditions have been operating

The dilution rate can be kept constant by increasing exponentially the flow rate. Most industrial processes operate 'Fed-batch' culture systems with significant extension of productive period of a traditional fed batch process without inherent disadvantages of a continuous process.

(b) Advantages and examples of fed-batch culture

A low but constantly replenished medium has the following advantages:

a) maintaining conditions in the culture within the aeration capacity of the fermenter;

b) removing the repressive effects of medium components such as rapidly used carbon and nitrogen sources and phosphate;

c) avoiding the toxic effects of a medium component; and

d) providing limiting level of a required nutrient for an **auxotrophic** strain.

Production of baker's yeast is mostly by fed-batch culture, where biomass is the desired product. Diluting the culture with a batch of fresh medium prevents the production of ethanol, at the expense of biomass; the moment traces of ethanol were detected in the exhaust gas.

The production of penicillin, a secondary metabolite, is also by fed-batch method. Penicillin process has two stages: an initial growth phase followed by the production phase called the **'idiophase'**. The culture is maintained at low levels of biomass and phenyl acetic acid, the precursor of penicillin, is fed into the fermenter continuously, but at a low rate, as the precursor is toxic to the organism at higher concentrations.

2.3 KINETIC CHARACTERISTICS OF MICROBIAL FERMENTATION PROCESSES

Before discussing various types of classification of microbial fermentation process, we shall describe some important terms used in the charecterization of these processes.

2.3.1 Productivity, Rates, Yield Constant and Economic Coefficient

Productivity is defined as the concentration of product (cell mass or metabolite concentration) divided by process time. It reflects the mean rate at which the product is synthesized during the process. For instance, the productivity of penicillin biosynthesis in a process which yields 16000 units per ml in 120 h is 133 units ml^{-1} h^{-1}. Fermenter productivity can be expressed in a similar way, e.g. 250 kg dry weight per 200 m^3 per hour.

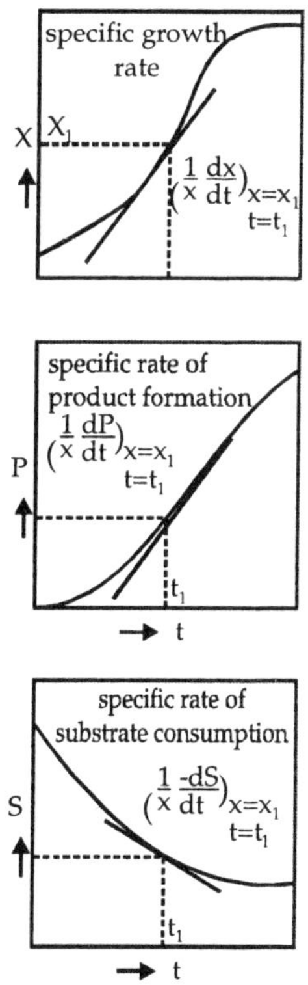

Fig. 2.4 : Specific rates of growth, product formation and substrate consumption. (Aiba *et al.*, 1973)

The **rate of a microbial process** is defined as the momentary rate of concentration change of a certain component. The most widely used parameter is the change in product concentration; in other cases respiration rate, growth rate or substrate consumption rate can be used. All these rates are important and should be included in the characterisation of microbial processes.

The **rates of microbial processes** can be defined by **volumetric rates** or **specific rates.** Typical units of **volumetric rates** are amounts (in mg or g) of a product synthesized in 1 ml or in 1 l in one hour, or amounts of consumed substrate expressed in a similar way. The volume rate at a given moment of batch culture can be determined as the slope of the tangent of an appropriate curve of concentration against time.

Another way of expressing rates of microbial processes is as **specific rates** defined as rates of concentration changes per unit cell mass. **Specific rates** are thus determined by dividing volume rates by the concentration of microorganisms. Volume units are then cancelled and the following typical units of specific rates are obtained: g product per 1 hour per g cells, g substrate consumed in 1 hour per 1 g cells, or g cells per 1 hour per 1 g cells. The last expression represents the specific growth rate. Fig. 2.4 defines the **specific growth rate, specific rate of product formation** and **specific rate of substrate consumption** at $t = t_1$. Ordinates indicate respectively cell concentration X, product concentration P and substrate concentration S, abscissa shows process time t. During batch culture the specific growth rate is seen to be a function of time t, its maximum value corresponding to the growth rate in the exponential phase.

In kinetic and other fundamental studies specific rates are preferred to volumetric rates since they reflect cellular activity; on the other hand, volumetric rates are more suitable for calculations of dimensions of culture devices.

Economically important quantities are the **yield constant** and the **economic coefficient.** The former quantity represents the amount of cell mass or product per unit of consumed substrate, the latter relates the cell mass or product to the total amount of introduced substrate.

2.3.2 Classification of Microbial Processes

Microbial processes can be divided into several groups according to different criteria, especially according to the relationship between substrate consumption and product formation and according to the type of reactions involved.

Relationship between substrate consumption and product formation:

a) Product synthesis proceeds stoichiometrically relative to the consumption of carbon sources which provide energy for growth; changes in rates of product formation are parallel with changes in rates of carbon consumption.

b) Product synthesis is only indirectly connected with the consumption of the carbon source which provides energy for growth.

c) Product synthesis is apparently unconnected with the consumption of carbon sources.

An example of a process of type (a) is the production of ethanol by yeasts and the production of lactic acid. In these cases growth depends completely on the ability of the microorganisms to obtain energy from glycolysis. Hence microbial growth and product formation proceed in parallel only during anaerobic culture. In the presence of oxygen the substrate is metabolised in a different way and yields different products. The main product in aerobically growing yeast is cell mass and the rate of its synthesis parallels substrate consumption.

In processes of type (b) the microbial growth is removed in time from product formation although the carbon source provides both energy for growth and carbon for product formation. Typical examples of these processes are the biosynthesis of citric or itaconic acid and some amino acids.

Processes of type (c) exhibit no apparent relationship between the rate of consumption of energy-yielding substrate and the rate of product formation. Examples of these processes are the biosynthesis of antibiotics, vitamins and enzymes. These substances are often excreted by cells after growth termination. If the carbon source is added in excess, no change in the level of product is observed, in contrast to the preceding group. Since the products of this group are often complex molecules, the yields are usually enhanced by adding to the medium, at a certain time, substances such as precursors.

2.3.3 Types of Reaction

(a) Simple reaction—Substrates are converted into products in a fixed stoichiometry without accumulation of intermediates. Examples of these processes are the growth of yeasts and bacteria or the enzyme-catalysed conversion of substrate by a suspension of precultured microorganisms.

(b) Simultaneous reaction—Substrates are converted to products in variable stoichiometric proportion without accumulation of intermediates. The relative rates of formation of these products vary with substrate concentration. An example of this process is the accumulation of polysaccharides or fats dependent on nitrogen source concentration.

(c) Consecutive reaction—Substrates are converted to products with accumulation of an intermediate; this is found, e.g., in the conversion of glucose to gluconic acid by microorganisms lacking gluconolactonase, or in the biosynthesis of antibiotics.

(d) Step-wise reaction—Substrate is completely converted to an intermediate before this is converted to the product, or several substrates are selectively converted to product in preferential order. Thus, in a medium containing two substrates, e.g. a hexose and a pentose; microorganisms may consume first the hexose and then the pentose, a phenomenon known as diauxy. An example of the first type is the oxidation of glucose to 5-ketogluconic acid. Bacteria first convert all the glucose to gluconic acid which is then transformed to 5-ketogluconic acid.

It should be noted that the above schemes may not describe all microbial processes since these processes may be highly complex. Thus microbial processes characterised by product formation apparently unconnected with growth or the consumption of a carbon source may, under different processing conditions, change so that product synthesis is connected with growth or carbon source consumption. For example the metabolic curves determined during penicillin biosynthesis (Fig. 2.5) indicate an obvious presence of a two-phase process. In the first phase lactose, i.e. carbon source, is utilised only for the synthesis of cell mass and only after growth has ceased does significant antibiotic synthesis occur.

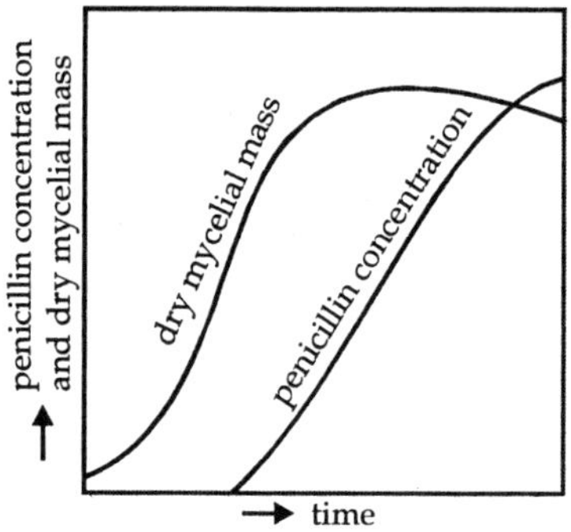

Fig. 2.5 : Penicillin production unconnected with growth.

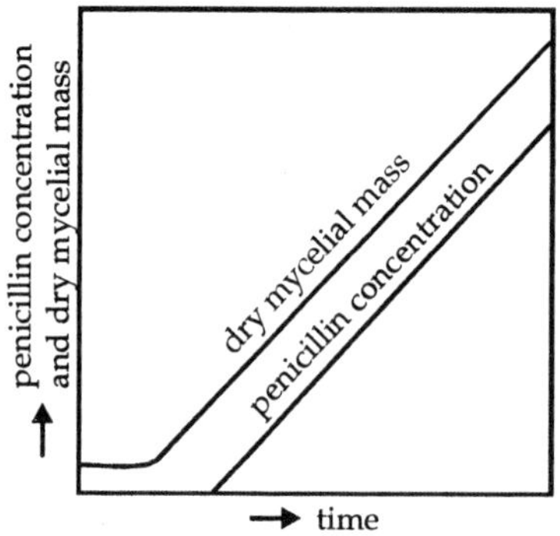

Fig. 2.6 : Penicillin production connected with growth.

When, however, a continuous feed of glucose or sucrose is used instead of lactose, cell mass growth may be observed in parallel with penicillin formation (Fig. 2.6); product formation is then connected with growth.

❑❑❑

3

MICROBIAL SCREENING

3.1 INTRODUCTION

Since prehistoric times, humans have exploited microorganisms for their own use. By means of trial and error, people developed strains of microbes that were used in production of beverages, food, and antibiotics without knowing that microbes were the responsible agent. With the discovery that microorganisms existed, and the subsequent development of culture methods, came the birth of fermentation technology, industrial microbiology and modern biotechnology.

Expanded development of microbial screening and cultural techniques has brought us to a point where microbially produced products are a major part of our life. While production of beverages and food remains a large part of fermentation industry, the discovery of penicillin in first half of 20th century revolutionized microbial screening for other useful products such as antibiotics, vitamins, enzymes, organic acids, solvents and other specialty chemicals.

During the past fifty years, screening of industrially important microorganisms has evolved steadily, although somewhat haphazardly, while still relying on the original underlying techniques of enrichment culture, selective media, and sheer labour. Classical methods are still used extensively with modification resulting from the use of chemical analogs, coupled colourimetric reactions, and advances in instrumentation.

This chapter focuses on the concept of microbial screening, development of primary screening techniques and significance of secondary screening studies.

3.2 CONCEPT OF MICROBIAL SCREENING

Screening may be defined as the use of highly selective procedures to allow the detection and isolation of only those microorganisms of interest from among a large microbial population (Casida, 1968).

Screening techniques test large number of microorganisms in one or few steps and discards most of the valueless organisms, while at the same time detects the small percentage of useful microorganisms present in the population. The screening procedure saves a lot of time and labour without the necessity of extensive studies to be carried out on each individual organism.

The type of screening technique used depends on

(a) the type of microorganism that is described,

(b) particular product of interest, and

(c) source from which microorganisms are to be obtained.

Microbial screening is performed at two levels:

(I) Primary screening, followed by (II) Secondary screening.

- **Primary screening** isolates few organisms from a large population which produces the product of interest
- **Secondary screening** procedures further test these organisms and access the potential of these organisms for commercial fermentations which of course very few microorganisms have it. The secondary screening discards those organisms which do not have real value for industrial processes.

3.3 SELECTION STRATEGY AND TECHNIQUES

Microorganisms are used to produce hundreds of commercial products valued in the hundreds of billions of dollars worldwide. Products produced by naturally occurring microbes have the advantage of being considered "natural" themselves and therefore are more easily approved for marketing than genetically manipulated organisms. For this reason, most fermentation industries favour the selection of naturally occurring microorganisms for production purposes, making a viable and efficient screening program a necessity.

The diversity of microorganisms holds the promise of many new technologies, including remedies for diminishing environmental pollution, that could substantially impact our standard of living. The belief that the microbe holds the solution to many problems is exemplified by **Perlman's "Laws of Applied Microbiology"**, in which he states that **microorganisms are capable of any task,** and that, **"If you take care of your microbial friends, they will take care of your future."**

The selection of industrial microorganisms is systematically approved using a strategy similar to the one shown in figure 3.1.

The key elements of a selection strategy are:

(a) Defining the activity of interest;

(b) Surveying known microbes with that activity;

(c) Developing enrichment and screening protocols;

(d) Identifying sources of new microorganisms; and

(e) Developing screening methodology.

Primary screening is predominantly a qualitative endeavour in which a large population of organisms is screened directly or indirectly for a specific type of activity.

Secondary screening is both qualitative and quantitative in that its objective is to determine the precise activity of the organisms, to verify production of compounds and to evaluate the production potential of screened organism.

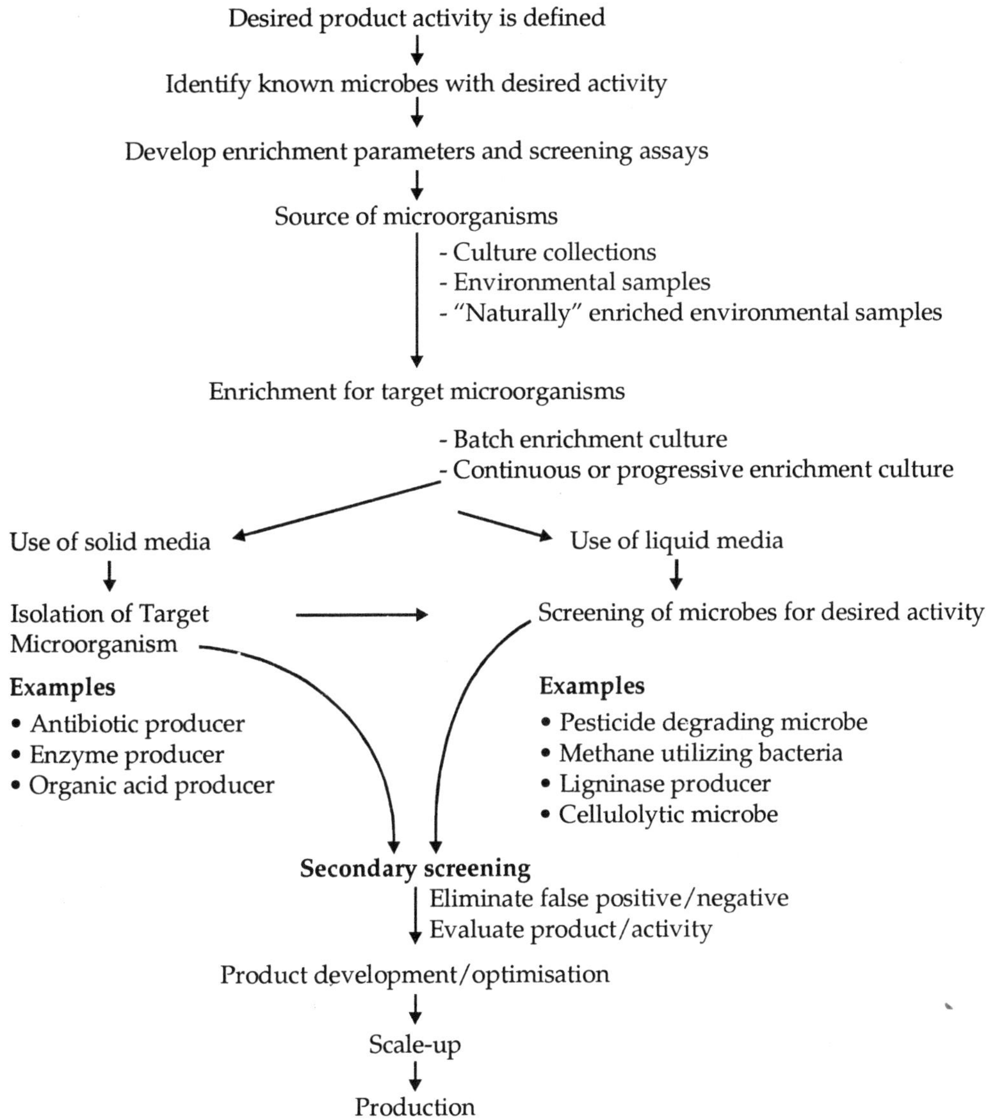

Fig. 3.1 : General schematic of an industrial screening programme.

3.4 ISOLATION OF INDUSTRIALLY IMPORTANT MICROORGANISMS

Primary screening programs involve highly selective procedures for detecting and isolating individual species from a mixed population.

Isolation is the first stage in the primary screening for industrially important microorganism. It involves the obtaining of (i) pure culture followed by their (ii) assessment to determine which carryout the desired reaction or produce desired product.

3.4.1 Criteria (being) Important in the Choice of Organism—(Bull *et al.* 1979).

(1) The nutritional requirement/characteristics of the organism—Process of fermentation be carried out using a very cheap medium, e.g. the use of methanol as an energy source. This requirement may be met by suitably designing the isolation medium.

(2) The Optimum temperature of the organism—The use of an organism having an optimum temperature above 40°C reduces the cooling costs of a large-scale fermentation. During isolation procedure, use of such temperature is beneficial.

(3) The reaction of the organism with the equipment to be employed and the suitability of the organism to the type of process to be used.

(4) The **stability** of the organism and its amenability to genetic manipulation make possible to get high yield strains by strain-improvement programmes.

(5) The **productivity** of the organism, measured in its ability to convert substrate into product and to give a high yield of product per unit time.

(6) The ease of **Product recovery** from the culture. (Downstream processing)

The organism most well suited to an economic process chosen on the basis of results based on tests subsequent to the isolation (Point 3, 4, 5, & 6). Before the commercial operation of process, the toxicity of the product and the organism has to be assessed.

3.4.2 Sources of Industrial Microorganisms

The organism may be isolated in some way from natural environment. Industrial microbiologists may also choose to buy the culture having desired characteristics from Culture Collection Centres. E.g.

(1) NCTC — National Collection of Type Cultures, London, U.K.
(2) ATCC — American Type Culture Collection, Maryland , U.S.A.
(3) NCIMB — National Collection of Industrial and Marine Bacteria. – ABERDEEN, U.K.
(4) CMI — Commonwealth Mycological Institute. Surrey, U.K.
(5) NCYC — National Collection of Yeast Cultures, Norfolk, U.K.
(6) IP — Institute Pasteur, Paris, France.
(7) IMTECH — Institute of Microbial Technology, Chandigarh, India.
(8) NCIM — National Collection of Industrial Microorganisms, NCL, Pune, India.

It is worthwhile to purchase cultures demonstrating the desired characteristics and then to develop it. The ideal isolation procedure commences with an environmental source—soil, sewage, marine mud, sea-water, fermenting or spoiled vegetables, grass, insect droppings etc; which are highly probable to be rich in the desired types. Soil is so designed as to favour the growth of those organisms possessing the industrially important characteristics. A special ecological niche may help to isolate an industrially important microorganisms, e.g. Hot springs for thermophiles. Alkaline spring to isolate alkalophile *B. subtilis* for alkaline protease used in detergent industry. Isolation of organism may be carried out by growing them on particular substrate, in the presence of certain compounds or desired procedure may be used to select the culture which show the characteristic of high frequency production.

Primary screening or first screening involves use of highly selective procedures for detection and isolation of organism having industrial potential. Ideal selection techniques are rapid, sensitive, predictive, inexpensive, easy to handle many samples and effective for broad range.

3.4.3 Principles Involved in Isolation Methods

(A) Isolation methods using Selection of the desired character—like methane utilization, pesticide degradation and enzyme production—

(1) Enrichment Liquid Culture—Biomass for SCP.

(2) Use of Solidified Media—Enzyme producer.

(B) Isolation procedure using methods selectively not utilizing the desired characteristics—

(a) Screening for the production of antibiotic

(b) Screening for the pharamacellogically active compounds.

(c) Screening for the production of growth factor.

(d) Screening for the production of polysaccharide.

3.4.3.1 Isolation Methods Based on Selection of Desired Characters

(a) Enrichment Liquid Culture technique for batch cultures—Enrichment culture is a technique resulting in an increase in the number of a given organism relative to the number of other types in the original inoculum. The process involves taking a mixed population (Soil dilution) and providing conditions either suitable for the growth of the desired type; as per example-methanol utilizing bacteria and Cellulose-decomposing microbe by making provision of particular substrate respectively methanol and cellulose or inclusion of certain inhibitors.

The growth of the desired type results in the modification of the medium. The sub-culturing is repeated by inoculating the enriched culture into identical fresh medium. Desired organism is isolated by spreading a small inoculum of enriched culture onto solid medium.

Organism having industrial potential isolated by batch-enrichment and purification on solid media perform poorly in continuous industrial fermentation processes. Use of continuous enrichment culture provides an organism in pure or mixed stable culture adopted to continuous culture in industry.

(b) Continuous Enrichment Culture Technique—Technique has been used for the isolation of organisms to be used **for biomass** production.

(i) The culture that can grow using the desired carbon source

E.g. a mixed culture using methane as the carbon source in a continuous enrichment was isolated and shown to contain mixture of one methylotroph and number of non-methylotrophic symbionts which result in mixed culture fermentation—*Methylophilus methylotrophus* for "SCP" —"Pruteen".

(ii) Technique has been used to isolate an *Arthrobacter* sp. producing a yeast-lysing enzyme complex.

(iii) To acclimatize soil isolates to the detoxification of high concentration of Parathion pesticide, continuous enrichment process has been used.

Advantages

(1) Isolation of an organism capable of growth on a simpler medium make basis of **cheaper commercial process** and the process should be more resistant to contamination.

(2) Use of **high temperature** during **isolation** results in the isolation of a thermophilic strain minimizing cooling problems in the subsequent process.

Limitation

Use of continuous enrichment process may **lead to washout of the inoculum before an adapted culture is established.**

To solve this problem, (1) Isolation process can be started in batch culture using a **20% inoculum** and as soon as growth is observed, the culture should be transferred and subsequent purification and stabilization of the enrichment performed in continuous culture or (2) Use of Turbidostate or Two stage chemostat during continuous isolation process can solve the problem of early wash out. Two Stage chemostat is more advantageous. The first stage of the system is used as a continuous inoculum for the second stage and consisted of a large bottle containing a basic medium inoculated with a soil infusion. Continuous inoculation is employed until an increasing absorbance is observed in the second stage.

(c) Use of Solidified media

(i) Isolation of enzyme producers—Solidified media have been used for the isolation of certain enzyme producers—

—Amylase producing bacteria and fungi

—Species of *Bacillus* producing alkaline protease.

Technique involves the use of a selective medium incorporating the substrate or the enzyme which encourages the growth of producing types.

(a) Use of starch agar medium for amylase producing bacteria

(b) Isolation of bacteria producing alkaline protease (Enzyme used in detergents).

Aunstrup, *et al.* isolated species of *Bacillus* producing alkaline proteases. Soil samples were pasteurised to eliminate non-sporulating organisms and then spread onto surface of agar (at pH 9-10) containing a dispersion of an isoluble protein. Colonies which produced a clear zone to the digestion of the insoluble protein were taken to be alkaline protease producers.

Test systems for screening of metabolites (antibiotics and some enzymes) are briefly listed in Table : 3.1.

Table 3.1 : Test systems for screening of metabolites

Product sought	Test system
Antibiotics	Agar plates with strains of test microorganisms,. Inhibition zones as indicator of activity
β-Lactamase-resistant antibiotics	Agar plates with test microorganisms to which β-lactamase has been added
Proteases	Agar plates with casein, selection of colonies which produce clear zones on the turbid plates
Amylases	Agar plates with starch, selection of colonies after staining with iodine
Lipases	Agar plates with oil emulsion, selection of colonies after precipitation of free fatty acids with Ca^{2+}
Phosphatases	Agar plates with phenolphthalein-diphosphate and pH indicator, selection based on colour change
NAD	Agar with microorganism auxotrophic for NAD

(ii) Isolation of Organic acid producers—Microorganisms producing organic acids or amines from various carbon substrates often can be detected by the incorporation of a pH- indicating dye, such as neutral red or bromo thymol blue into a poorly buffered agar nutrient medium. The production of acid /amines, can be indicated by a change in the colour of dye in the vicinity of the colony.

An alternate procedure for detecting organic acid production involves the incorporation of calcium carbonate in the medium so that organic acid production is indicated by a cleared zone of dissolved calcium carbonate around the colony.

This procedure need further testing by paper chromatogr762arphy to confirm the acid or base product of actual interest.

(iii) Screening for Enzyme Inhibitors—Enzyme inhibitors can be screened by the zone inhibition reaction using enzyme containing agar, e.g. Trypsin inhibitor and Proteinase inhibitor.

(iv) Screening for other Metabolites—

1. Lipid (Secretion).—Detection by auxanographic zone using lipid auxotroph of *Saccharomyces cerevisiae* [Kenyon and Hegemann, 1970]

2. NAD—Detection by auxanographic zone using *Haemophilus influenzae* (Witholt, 1972).

3. Estrogen (from steroid)—red colour formation by Spraying p-nitrobenzene diazonium fluoroborate (1% in 50% acetic acid) [Vezina *et al.*, 1969].

4. Chlorostatin—Chlorosis zone of *Euglena gracilis* [Mayer *et al.*, 1969]

3.4.3.2 *Isolation Methods Based on Selectively not Utilizing Desired Character*

Isolation of organism involved for production of antibiotics, growth promoters and large concentration of amino acids and nucleotides uses media that allow maximum expression of their genetic material.

(a) Screening for the production of antibiotics—Antibiotics are antimicrobial substances derived from fungi and bacteria exemplified by penicillin (*Penicillium chrysogenum*). Detection of antimicrobial action of antibiotic may be achieved by growing the potential producer on agar plate in the presence of test organism, or organism against which antimicrobial action is required. Production of inhibitor is indicated by the inhibition of the test organism.

Microbial isolate may be grown in liquid culture and the cell-free broth tested for antimicrobial activity. The use of liquid medium in initial screening tests has the advantage of avoiding any anomalies in the relation of result of agar tests to submerged cultures studies, though it takes up considerably more space and facility.

(i) Antibiotics with confined spectra may be isolated by using a combination of test organism—*Bacillus subtilis* and *Streptomyces viridochromogenes* or *Clostridium pasteurianum* (ii) Kirromycin group of antibiotics have been detected by this method. This technique identifies antibiotics with a low activity against *B. subtilis* and a high activity against other test organisms.

Specific Screening technique for detection of novel antimicrobial:

(i) Beechum Pharmaceuticals discovered—Isolation of β-lactamase inhibitor, i.e. clavulanic acid using a screen which detected synergy between ampicillin and the test sample in their inhibition of β-lactamase producing *Klebsiella.*

(ii) Glaxo Laboratones have utilized an *in vitro* enzyme screen designed to detect inhibitors of Carboxypeptidase which is involved in the cross-linking of peptidoglycan of bacterial cellwalls. This work has led to detection of many novel antibiotics.

(b) Screening for Pharmaceutically Active compounds—Umezawa (1972) pioneered the concept of screening for microbial products capable of inhibiting key enzymes of human metabolism. Logic that if a compound inhibits a key enzyme *in vitro,* it may have pharmacology action *in vivo*. Active compounds detected in an "*in vitro*" screen may then be used in animal-test- systems. Hamill (1982) has reviewed this area and the success of techniques is indicated in Table : 3.2.

Table 3.2 : Pharmacological Screens based on the Enzyme-target approach

Target Area	Target Enzymes
Catecholamine Synthesis	Tyrosine hydroxylase
	Dopamine β-hydroxylase
	Monoamineoxidase
	Catechal-O-methyltransfarase
Antihistamine	Histidine-decarboxylase
	Histamine- N-methyl transferase
Antilipidaemic	3-HMG-CoA reductase condensing enzyme.
5-Hydroxytryptamine synthesis	Tryptophan hydroxylase
Proteolysis	Trypsin

contd. ...

... contd.

Target Area	Target Enzymes
	Plasmin.
	Papain
	Chymotrypsin
	Cathepsins A, B and D.
	Pepsin.
Antihypertension	Angiotensin-converting enzyme.
Prostaglandin synthesis	Prostaglandin synthetases
Glycohydrolases (antidiabetic)	α-amylase, Sucrase
	β-galactosidase
	Neuraminidase

(c) Screening for the Production-of-Growth factors.—The production of growth-factors such as amino acids and nucleotides is a difficult characteristics to employ as a selective force in an isolation procedure. Organisms are isolated on random basis. Detection of growth factor producer depends on the stimulation of the growth of auxotrophic bacteria by the isolate.

Screening procedure for *Amino acid producer* (Bacterium)— The majority of amino acid producers belong to the genera—*Arthrobacter, Microbacterium, Brevibacterium, Micrococcus* and *Corynebacterium.* Fungi can be excluded by incorporation of antifungal compound such as cycloheximide in the isolation media. Initial screen for production of amino acids is a qualitative one normally carried out on solid media. Isolation plates containing 30-50 colonies are replica plated onto solid medium capable of supporting amino acid production and grown for 2 to 3 days.

Developed colonies were killed with UV light overlaid with agar containing suspension organism auxotrophic for the desired product(Assay bacterium). After 16 hours incubation at 37°C, the assay organism grow as a halo around any producing colony on the plate and can be isolated from master plate. Qualitative assays can be performed further where the organism is grown in liquid culture and cell-free broth estimated for the amino acid concentrate.

3.5 SECONDARY SCREENING

Secondary screening qualitatively and quantitatively expands information gained from the primary screening program. Secondary screening represents a major commitment of time and resources to determine whether a given microorganism possesses the potential for use in a new or improved industrial fermentation. As such, secondary screening must serve to eliminate worthless microorganisms and fermentation procedures from further consideration as quickly and as inexpensively as possible.

Secondary screening involves many of the same techniques used in primary screening. Usually small flasks are utilized during early stages, small laboratory fermenters during intermediate stages, and pilot-plant fermenters for scale-up studies during later stages. Much more elaborate experimentation is required for secondary than for primary screening. So, primary screening does not provide much information needed in setting up a new fermentation process.

The significance of secondary screening can be realized by a thoughtful understanding of the following realities associated with secondary screening—

1. It is very useful in sorting out microorganisms that have real commercial value from many isolates obtained during primary screening. At the same time, microorganisms that have poor applicability in a fermentation process are discarded. It is advisable to discard poor cultures as soon as possible, since such studies involve much labour and high expenses.
2. It provides information whether the product produced by a microorganism is a new one or not. This may be accomplished by paper, thin layer, or other chromatographic techniques. The compound under consideration is compared with previously known compounds.
3. It gives an idea about the economic position of the fermentation process involving the use of a newly discovered culture. Thus, one may have a comparative study of this process with processes that are already known, so far as the economic status picture is concerned.
4. It helps in providing information regarding the product yield potentials of different isolates. Thus, this is useful in selecting efficient cultures for the fermentation processes.
5. It determines the a optimum conditions for growth or accumulation of a product associated with a particular culture.
6. It provides information pertaining to the effect of different components of a medium. This is valuable in designing the medium that may be attractive so far as economic consideration is concerned.
7. It detects gross genetic instability in microbial cultures. This type of information is very important, since microorganisms tending to undergo mutation or alteration in some way may lose their capability for maximum accumulation of the fermentation products.
8. It gives information about the number of products produced in a single fermentation. Additional major or minor products are of distinct value, since their recovery and sale as by-products can markedly improve the economic status of the prime fermentation.
9. Information about the solubility of the product in various organic solvents is made available. This knowledge is useful in the recovery and the subsequent purification of the product.
10. Chemical, physical and biological properties of a product are also determined during secondary screening. Moreover, it reveals whether a product produced in the culture broth occurs in more than one chemical form.
11. It reveals whether the culture is homofermentative or heterofermentative.
12. Determination of the structure of the product is done. The product may have a simple, complex, or even a macromolecular structure.

13. With certain types of products (e.g. antibiotics) determinations of the toxicity for animal, plant or man are made if they are to be used for therapeutic purposes.
14. It reveals whether microorganisms are capable of a chemical change or of even destroying their own fermentation products. For instance, microorganisms that produce the adaptive enzyme, decarboxylase, can remove carbon dioxide from the amino acid, leaving behind an organic amine.
15. It tells us something about the chemical stability of the fermentation product.

3.6 FUTURE POTENTIAL AND NEEDS OF MICROBIAL SCREENING

The exciting progress being made in the areas of protein and genetic engineering will hopefully substantially impact screening programs in the future. For now, genetic engineering and protein engineering have a greater impact on industrial microbiology at the level of product optimisation. Microbial proteins such as the alkaline protease subtilisin have been engineered for increased stability and changes in substrate specifity. Takagi *et al.* successfully used site-directed mutagenesis to develop enhanced activity in a subtilisin protease. Many industries and research institutes are working extensively to develop enhanced genetic promoters for increased gene expression and subsequent increased product formation.

An example of the type of techniques that may prove useful for screening is the utilization of colony hybridization to detect DNA sequences specific for hydrocarbon-catabolic pathways. A similar screen based on hybridization of DNA extracted from environmental samples (dot blot) can be used to look for microbes with particular DNA sequences. New developments with polymerase chain reaction (PCR) should increase the sensitivity of such techniques in the future.

The potential of microbial screening for industrial products lies in the tremendous microbial diversity of the earth. The organisms are there: we must develop the techniques to select for and isolate them. The largest impediment to this is the lack of research dedicated to the development of screening techniques. This is a multi-disciplinary problem that requires the expertise of chemists and engineers as well as microbiologists.

3.7 MAINTENANCE AND PRESERVATION OF INDUSTRIAL CULTURES

Associated with industrial cultures is the problem of maintaining them for various periods in a viable condition. This is necessary so as to eliminate genetic instability, protect against contaminants and retain its characters. Most bacteriological laboratories usually maintain a large collection of pure cultures, frequently referred to as the stock-culture collection. Microbial species isolated and characterised by microbiologists are deposited in these centres. The stock cultures generally retain all the characteristics. Stock culture collection centres have been established throughout the world to help microbiologists in obtaining pure cultures for various studies. Some of the centres also help in classifying a newly isolated organism.

3.7.1 Maintenance Methods for Stock Cultures

Once a pure culture is obtained it must be kept pure. One of the most frequent ways in which erroneous results and conclusions are obtained in microbiology is by the use of faulty or contaminated cultures. Cultures of organisms of interest that are maintained in the laboratory for study and reference are called stock cultures. The **stock culture** must be maintained so that it is free from contamination, retains viability, and remains true to type.

The most common way of maintaining stock cultures is either on agar slants or in agar stabs.

(A) Agar slants

The slants are inoculated and incubated until reasonable growth is obtained, and then they are usually placed in a refrigerator, care being taken that the culture tubes are well sealed (with melted wax) so that they will not dry out. With hardy organisms such as spore formers, Gram-positive cocci, yeasts, and fungi, agar slant cultures maintained in this way may retain viability for weeks or months. It is the usual practice to transfer cultures to fresh slants at regular intervals (Sub-culturing).

(B) Agar stabs

For some organisms agar stabs are better for preserving viability than agar slants. The tube for an agar stab, called an **agar deep**, is allowed to stand vertically while the agar is solidifying rather than at an angle as for slants. Inoculation is done with a straight needle, after being touched to the culture, is stabbed into the agar all the way to the bottom. During incubation, the culture grows in a thin zone along the line of the stab.

The advantage of agar stab procedure is that waste products such as acids that are produced by the organism spread out into the agar from the zone of growth and are diluted. Accumulation of waste products is one of the main reasons for the dying off of many stock cultures. Although less total growth is obtained in a stab than on slant, enough growth occurs so that new cultures can be initiated when needed.

3.7.2 Methods of Preservation

A considerable amount of research has been performed to develop improved conditions for the preservation of bacteria. Not all species respond in a similar manner to a specific condition or process. A method useful for one species may not be applicable to another. It is also essential that the method of preservation and maintenance conserve all the characteristics of the species as they were at the time of preservation.

There are basically two major methods for preservation and maintenance of cultures—

(a) Methods in which organisms are in a state of continuous metabolism and,

(b) Methods in which organisms are in a state of suspended metabolism.

3.7.2.1 Organisms in Continuous Metabolism State

It depends on repeated sub-culturing of organisms onto fresh nutrient medium when the formerly used medium gets dried or is used up. This includes preservation of organisms by following methods—

(A) Periodic Transfer to Fresh Media—All microbiology laboratories preserve microorganism on agar slants. The slants are incubated for 24 hours or more, and then stored in a refrigerator. These cultures are periodically transferred to fresh media. The time interval at which transfers are made varies with the organism and the conditions of growth.

(B) Overlaying Cultures with Mineral Oil—The agar slants are inoculated and incubated until good growth appears. They are then covered with sterile mineral oil to a depth of 1 cm. above the tip of the slanted surface. This method has the unique advantage that one can remove some of the growth under the oil with a transfer needle, inoculate a fresh medium and still preserve the initial culture. The simplicity of this method makes it extremely attractive. Some species have been preserved satisfactorily for 15-20 years by this method.

(C) Storage in Sterile Soil—This method is widely used for preserving spore forming bacteria and fungi. Spore suspensions are added to sterile soil (sterilized for 2-3 hours at intervals of 1-2 days) and the mixture is dried at room temperature and stored in refrigerator. Bacterial cultures maintained by this procedure have been found viable for 70-80 years.

(D) Saline Suspension—Sodium chloride in high concentrations is frequently used as an inhibitor of bacterial growth. Bacteria are suspended in one per cent salt solution (Sublethal concentration) in screw cap tubes to prevent evaporation. The tubes are stored at room temperature. Whenever needed, the transfers are made on agar slants and incubated.

3.7.2.2 *Organisms in Suspended Metabolism State*

This depends on some sort of drying or preservation at low temperature to a level at which the metabolism is slowed down or stopped, but revival of culture is possible. It includes—

(A) Drying in Vacuum—The organisms are dried over calcium chloride in a vacuum and are then stored in the refrigerator. The organisms survive longer than when air dried.

(B) Lyophilization (Freeze – Drying)—Lyophilization is the most effective procedure for the preservation of cultures. Many species of bacteria preserved by this technique have remained viable and unchanged for more than 20 years. In this process (Figure 3.2) the cells are dried rapidly, while they are frozen. The bacterial suspension is placed in small vials, which are then immersed in a mixture of dry ice and alcohol, at a temperature of –78°C. The vials are immediately connected to a high-vaccum line. This dries the organisms while still frozen. Finally the vials are sealed off in a vacuum with a small flame. These cultures can then be stored for several years at 4°C. This method is also employed for the preservation of sera, toxins, enzymes and other biological materials. To revive microbial cultures, it is merely necessary to break open the vial aseptically, add a suitable sterile medium, and after incubation make further transfers. This method greatly reduces the danger of contamination. An additional advantage is the requirement of a small storage area.

(C) Use of Liquid Nitrogen—The ready availability of liquid nitrogen (temperature of –196°C) has provided the microbiologists another means for maintaining the cultures. In this procedure the cultures are frozen with a protective agent (glycerol or dimethyl sulfoxide) in sealed vials. The frozen cultures are then kept in liquid nitrogen refrigerator. This process has been successful with many specimens which cannot be maintained by lyophilization.

(D) Storage in Silica Gel—Both bacteria and yeast can be stored in silica gel powder at low temperature for a period of 1-2 years. In this method, finely powdered, heat sterilized and cooled silica powder is mixed with a thick suspension (Paste) of cells, mixed and stored at low temperature. The basic principle in this technique is quick dessication at low temperature which allows the cells to remain viable for longer period.

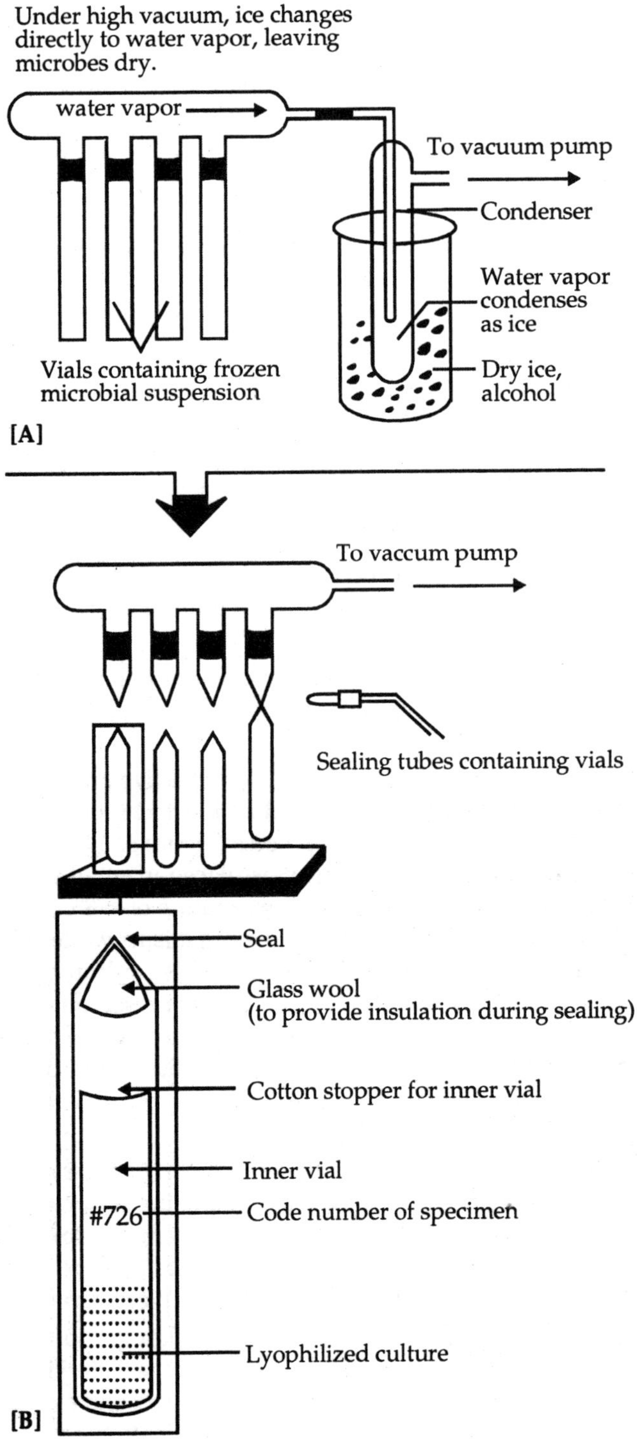

Fig. 3.2 : Lyophilization process of preservation of microorganisms. (A) Small cotton-plugged vials containing frozen suspensions of microorganisms are attached to a condenser and a high-vaccum pump. This system dehydrates the specimen while it is in the frozen state. (B) After the specimen is dehydrated, the tubes containing the vials are sealed while still under vaccum. Details of an individual lyophilized specimen are shown enlarged. This lyophilized culture of microorganisms will remain viable for years.

3.8 ROLE OF CULTURE COLLECTION CENTRES

A culture collection either can be a specialized collection of a restricted group of organisms or it can be a generalized collection that attempts to maintain a wide range of organisms of interest. The purpose of a culture collection is to maintain cultures in a viable contaminant - free state and to have them available for immediate shipment at nominal cost to research workers, teachers, students, or industry. The individual worker thus need not to maintain cultures that he uses rarely.

The maintenance of a large and diverse culture collection is not simple, since different organisms must be stored in different ways. A high degree of skill and knowledge is necessary. The price charged for a stock culture may often seem high, but it should be noted that the purchaser is paying for the immediate availability of a contaminant- free culture. It must be emphasized, however, that the investigator should not accept blindly the reliability and purity of a stock culture received from a culture collection centre. Immediately upon receipt, the culture should be carefully checked, and if it does not meet specifications, it should be returned. Since culture collections often maintain thousands of different cultures, it is only natural that errors may occur, leading to the distribution of faulty cultures.

Culture collection centres also render the services of identification and confirmation of any new isolate obtained by researcher.

Organisms are listed in culture collections by species name, but they are also often given numerical designations. Not all pure cultures of the same species are identical in all ways. Each isolate of the same species is usually designated as a strain, and one strain is distinguished from another by use of arbitrary numbers or letters. The culture number used by the culture collection is unique number referring to a single strain of the species. Since strains of the same species may differ in important ways, it is necessary to specify the exact strain number when referring to work carried out with a specific stock culture.

3.8.1 World-wide Status of Culture Collection Centres

The increasing use of microorganisms for various purposes has made it obligatory that the culture of great value should be maintained to provide viable and stable cultures. Hence all microbiologists need to preserve cultures for a variety of reasons. The role of maintenance of cultures is performed by the culture collections. The basic functions of culture collections, therefore, are collection, maintenance, and supply of reliable, authentic cultures. In addition to the above basic activities, culture collections provide a number of other services also.

There are 566 culture collections listed in their directory by the World Data Centre on Microorganisms (WDCM) and in India there are more than 50 culture collections and National Collection of Industrial Microorganisms (NCIM), Pune is the largest amongst these collections and American Type Culture Collection (ATCC), USA is the largest in the world.

The diversity of culture collections and wide interest of their users has led to need to coordinate activities and World Federation for Culture Collections (WFCC) provides this international coordination.

The collection, maintenance, characterization and distribution of cultures generates a great deal of information that must be effectively documented. For this purpose Microbial Strain Data Network (MSDN) has established link through computers between the culture collections all over the world and specific information and retrieval of data can be obtained. Characterisation, identification, preservation and patent services make an important contribution to the development of biotechnology and the biotechnological importance of culture collections can be exploited only if information on the properties of holdings is known, properly documented and easily accessible.

3.8.2 List of some Important Culture Collection Centres

(I) At Abroad—

(A) **For all types of microorganisms** (Algae, Moulds, Yeast, Bacteria, Actionomycetes, Protozoa, Animal Viruses, Bacteriophages, Cell lines)—

(1) American Type Culture Collection [ATCC], 12301 Parklawn Drive, Rockville, Maryland, 20852, USA

(2) Carolina Biological Supply Company [Commercial Service], 2700 York Road. Burlington, North Carolina 27215, USA

(B) **For Protozoa and Algae—**

(3) Culture Collection of Protozoa and Algae. 36 storey's way, Cambridge University, Cambridge CB3 ODT, England

(4) General Biological Supply House [Commercial Service], 8200 S. Hoyne Avenue, Chicago, U.S.A.

(C) For Algae—

(5) The Culture collection of Algae and Microorganisms. Institute of Applied Microbiology (IAM), Uni. of Tokyo, 1- Chome, Yoyoi, Bunkyo- Ku, Tokyo, Japan.

(6) University of Texas culture collection of Algae, Uni. of Texas, Austin Texas, U.S.A.

(D) For Fungi—

(7) International Mycological Institute (Formerly CMI), Bakeham Lane, Egham, Surry TW 209 TY, U.K.

(8) Centralbureau voor Schimmelcultures (CBS), Osterstaat 1, Baarn, Netherlands.

(9) IHEM Culture collection
Place Croiok due sud 3-B-1348, Louva in-la-Neuve-Belgium.

(E) For Yeast—

(10) National collection of Yeast cultures. Brewing Industry Research Foundation, Lyttel Hall, Nutfiel, Near Redhill, Surrey, U.K.

(F) For Actinomyceters—

(11) Institute for Research of New Antibiotics, USSR Academy of Medical Sciences, Bolshaia Pirogovskaia 11, Moscow, Russia.

(12) Botanical Department, Uni. of North Carolina, Coker Hall, Chapel Hill, North Carolim 27514, U.S.A.

(G) For Bacteria—

(13) The Wellcome Bacterial Foundation
Wellcome Foundation Ltd., Longley court,
Beckenham, Kent, U.K.

(14) National Collection of Industrial Bacteria
Torry Research Station, P.O. Box 31, 135 Abhey Road, Aberdeen AB 9 8 DG. Scotland.

(15) National collection of Plant Pathogenic Bacteria
Plant Pathology Laboratory. Milton Road, Harpenden., Hert fordshire, Scotland.

(16) National collection of Dairy organisms
National Institute for Research in Dairying, Shinfield, Reading Berkshirel, U.K.

(17) The Upjohn Stock Culture Collection
The Upjohn Company, 301 Henrietta street, Kalamazoo, Michigan 49001, U.S.A.

(18) LMG Collection, Laboratorium voor Microbiologie
Rijks Universiteit Gent K.L. ledeganck straat 35, B-900 Gent, Belgium

(19) Culture collection Laboratory of Microbiology
Technical University, Julianalaan 67 A, Delft, Netherlands.

(20) Institute Pasteur, 25 Rue du Docteur Roux, 75724 PARIS Cedex 15, France

(21) Centre de Collections de Types Microbeins, , 19, Rue Cesar-Roux, Lausanne, Switzerland.

(22) Microbiological Type - Culture collection, 4- 54,
Juso-Nishinocho, Higashiyodogawa-ku, Osaka, Japan.

(23) Culture collection Unit, Northen Utilization Research Branch,
U.S. Deptt of Agriculture, Peoria 5, Illinois, U.S.A.

(24) National collection of Type- cultures, Central Public Health Laboratory,
Colindale Avenue, London, U K.

Address your inquiry to **"The curator"** of above culture collection centres for taking services from them.

(II) In India

(A) For all types of microorganisms

(1) National Collection of Industrial Microorganisms [NCIM], C.S.I.R. National Chemical Laboratory, Pune - 411008

(2) Microbial Type Culture Collection and Gene Bank [MTCC], Institute of Microbial Technology [IMTECH], Post Box No. 1304, Sector 39-A, Chandigarh-160 014

(B) For Fungal Cultures

(3) Herbarium Cryptogame Indie Orientalis, Division of Mycology and Plant Pathology, I.A.R.I., New Delhi - 110 012

(C) For Yeast

(4) Fermentation Technology Laboratory, Indian Institute of Science, C.V. Raman Road, Bangalore - 560 012

(D) For Actinomycetes—(Helping only in identification services)

(5) Department of Microbiology, Bose Institute, 93/1, Acharya Prafulla Chandra Road, Kolkata - 700 009

(E) For Cyanobacteria—

(6) National Facility for Marine Cyanobacteria, Bharathidasan University, Tiruchirapali - 620 024

❑❑❑

4

Gene Manipulation Techniques

4.1 INTRODUCTION

Molecular genetic techniques can be used to manipulate genes in order to alter the expression and production of microbial products, including the expression of novel recombinant products. "Classical" (*in vivo*) genetic techniques are essentially limited by two factors. Firstly, they can only be applied to the existing genetic complement of an organism, i.e. they are restricted to naturally occurring genes or relatively minor modifications of these genes. It is not possible to get an organism to make a product totally foreign to that organism using these techniques. Secondly, with classical techniques, one can only work on the basis of phenotype and mutants are selected by their effect on the observable characteristics of the organism. This not only limits the changes that can be selected, but also means that it can be difficult to determine the nature of the mutation that has caused the alternation in the phenotype.

This discovery of **recombinant DNA technology** (also known as **gene cloning** or **in vitro genetic manipulation**) has dramatically broadened the spectrum of microbial genetic manipulations. The basis of this technology is in the use of restriction endonucleases and DNA ligases as a means to specifically cut and paste fragments of DNA. In this way, foreign DNA fragments can be introduced into a vector molecule (a plasmid or a bacteriophage), which enables the DNA to be replicated after being introduced into a bacterial cell.

4.2. CONSTRUCTION OF RECOMBINANT DNA

Today many areas of biological research require the use of recombinant technology in one form or the other. The ability to modify and clone genes has become an essential tool in research and has accelerated the rate of discovery and the development of bioindustries.

The basic steps in DNA cloning involve the following steps (Fig. 4. 1).

1. A fragment of DNA is inserted into a carrier DNA molecule, called a *vector* to produce a recombinant DNA.
2. The recombinant DNA is then introduced into a host cell, where it can multiply and produce numerous copies of itself within the host. The most commonly used host is the bacteria, although other hosts can also be used to propagate the recombinant DNA.
3. Further amplification of the recombinant DNA is achieved when the host cell divides, carrying the recombinant DNA in their progenies, where further vector replication can occur.
4. After a large number of divisions and replications, a colony or clone of identical host cells is produced, carrying one or more copies of the recombinant DNA. The recombinant DNA is now said to be cloned.
5. The colony carrying the recombinant DNA of interest is then identified, isolated and analysed.

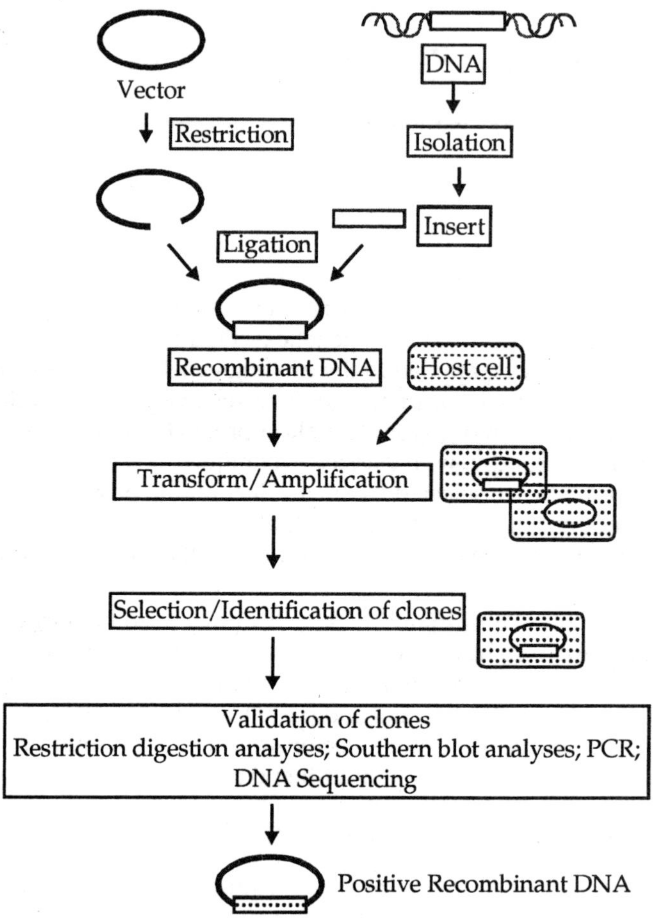

Fig. 4.1 : Basic steps in gene cloning.

From the schematic diagram, one can appreciate that DNA cloning is a relatively straightforward procedure. Essentially the basic principles in DNA cloning involve cutting a piece of DNA (restriction digestion) and pasting the restricted DNA into a vector (ligation). In order to produce more of this recombinant DNA for easy manipulations, the DNA is then introduced and propagated in host cells (transformation and amplification). Once the clone containing the desired DNA is selected, the recombinant plasmid can finally be isolated and further explored (analysis).

4.2.1 Enzymes used in Gene Manipulations

Recombinant DNA technology will not be possible had it not been for the discovery and exploitation of the functions of restriction and modifying enzymes. Within a cell, these enzymes participate in essential processes such as DNA replication and degradation of undesirable DNA. A number of bacterial strains have been identified that can produce these enzymes in high quantities. After purification from the bacterial cell extracts, these enzymes are still able to carry out their natural reactions when the appropriate artificial condition is provided. Currently, a number of these enzymes are produced commercially by recombinant DNA technologies, and are supplied as highly purified and well-characterized enzyme preparations.

The enzymes that are used in gene manipulations can generally be classified into 5 classes based on their functions:

1. **Nucleases,** which cut or degrade DNA molecules.
2. **Polymerases** can copy and make new strands of DNA.
3. **Ligases,** which join pieces of DNA molecules together.
4. **Modifying enzymes,** which modify the DNA by adding or removing chemical groups.
5. **Topoisomerase,** which remove or introduce supercoils from covalently closed-circular DNA.

The functions of these enzymes and how they are used in DNA cloning will be described briefly. Three classes of enzymes that are more widely used in DNA cloning will however be described in detail. A point to note is that although most enzymes can be classified into a particular class of function, a few can display multiple functions. For example, polymerases have both the functions of making new strands of a DNA as well as degrading it.

4.2.1.1 Nuclease

The function of this class of enzymes is to degrade or digest DNA by breaking the phosphodiester bonds that link one nucleotide to the other in a DNA strand. There are two types of nucleases:

(a) **Exonuclease,** which digests nucleic acids at either ends of the DNA.

(b) **Endonuclease,** which digest nucleic acids from within the DNA.

The main difference within different endonucleases or exonucleases is their ability to digest either single-stranded or double-stranded DNA. For example, while the enzyme *Exonuclease* III (*Exo* III) is able to cleave double-stranded DNA, the enzyme *Exonuclease* VII (*Exo* VII) is only able to cleave single-stranded DNA. Similar distinctions can be applied to endonucleases. Whereas the endonuclease DNAase I, isolated from bovine pancreas, is able to digest double-stranded DNA, the mung bean nuclease (from the sprouts of mung bean) can only digest single-stranded DNA. Both DNAse I and mung bean nuclease are examples of endonucleases that cleave DNA at non-specific sites. There is yet another group of endonucleases that can only cleave DNA at restricted sites defined by specific DNA sequences. This group of enzyme is known as restriction endonucleases and is the most commonly utilized enzymes in recombinant DNA technology.

Restriction Endonuclease

This group of enzymes was first discovered when it was observed that some strains of bacteria were immune to bacteriophage infection. It is now known that these strains of bacteria had produced endonucleases to restrict the propagation of new phage particles by digesting the foreign DNA before it can replicate. During this process, the bacteria protect their genome by modifying their DNA through methylation to prevent the action of these degradative enzymes.

These restriction endonucleases have been given names consisting of 3 italicized letters (representing the bacteria from which the restriction enzymes isolated from), followed by a letter to identify the strain (if applicable), and end with a Roman numerical to distinguish the different enzymes isolated from the same bacteria. Some examples are given below to illustrate this nomenclature:

***Eco*R I** is the first restriction/modification enzyme isolated from *Escherichia coli strain* R.

***Hind* III** refers to the third system isolated from *Haemophilus influenzae strain* R_d.

***Bam*H I** was the first restriction/modification system isolated from *Bacillus amyloliquefaciens* H.

To date, more than 2900 restriction enzymes are known, and they have been generally classified into Type I, Type II or Type III specific classes. Types I and III enzymes are complex, and are not commonly used in genetic engineering. These enzymes cleave the DNA at a substantial distance from their recognition sites. Type II restrictions enzymes, on the other hand, play a central role in recombinant technology.

The central feature of the Type II restriction enzymes is that each enzyme has a specific recognition sequence. The DNA is cleaved either *within* or *very near* to the recognition sequence. It is important to note that these enzymes are very specific and will cut the DNA only when their recognition sequences are present. The recognition sites of Type II enzymes are usually palindromic (two fold symmetry) and are 4, 5, 6 or more base pairs long. Some examples of the most commonly used restriction endonucleases are shown in Table : 4.1.

Table 4.1 : Some examples of the frequently used restriction endonucleases. The sequences shown are that of the positive strand of DNA from the 5′ to 3′ direction

Enzyme	Recognition site	Blunt or cohesive end
Hind III	A/AGCTT	cohesive
*Bam*H I	G/GATCC	cohesive
Bgl II	A/GATCT	cohesive
Pst I	CTGCA/G	cohesive
Sma I	CCC/GGG	blunt
Pvu II	CAG/CTG	blunt
Alu I	AG/CT	blunt
Not I	GC/GGCCGC	cohesive

Note that some restriction endonucleases, for example *Bam*H I, cut the DNA asymmetrically (as shown by the slash). This results in each fragment having four unpaired bases (in this case GATC) at both ends, referred to as "sticky" or cohesive ends. These single-stranded regions are complementary and will therefore tend to pair with DNA fragments with compatible sequences. For example, in the case of *Bam*H I-restricted DNA, the DNA can form complement pairing with either a *Bam*H I or *Bgl* II restricted DNA.

Some enzymes, such as *Pst* I, leave a four-base sticky end at the 3′ end of the fragment, while others, such as *Sma* I and *Pvu* II, cleave in the centre of the recognition sequence, generating blunt-ended fragments.

4.2.1.2 *Polymerases*

Polymerases are enzymes that can synthesize new strands of nucleic acids that are complementary to an existing piece of DNA or RNA. These enzymes can only do so when the template has a double-stranded region that can act as a primer to initiate nucleic acid polymerization.

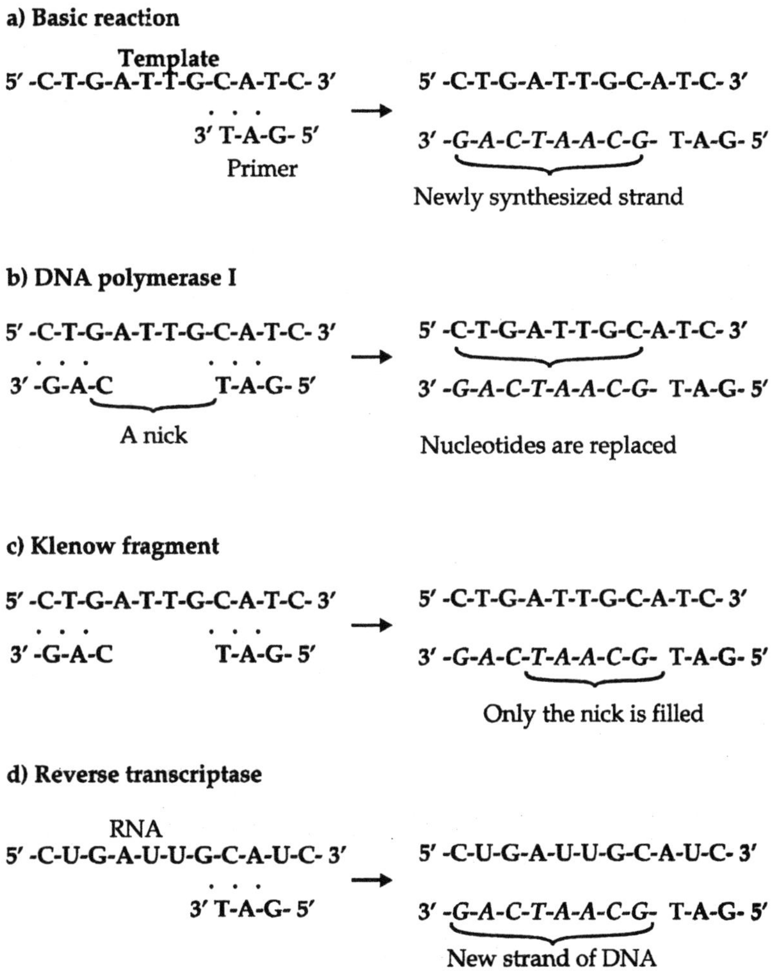

Fig. 4.2 : The major polymerases used in DNA cloning.

There are 3 main types of polymerases that are frequently used in recombinant DNA technology (Fig. 4.2). The first, DNA polymerase I, attaches itself to the single-stranded portion of a largely double-stranded DNA and initiates synthesis from the 5' end of the DNA, replacing the old strand as it proceeds. One can therefore see that this enzyme has dual functions: DNA synthesis and degradative activities.

The second most commonly used polymerase enzyme is the Klenow fragment, which is essentially derived from DNA polymerase I. The DNA polymerase I enzyme has been modified to remove the DNA degradative function, retaining only the DNA polymerization function. This modified enzyme, called the Klenow fragment, is thus able to synthesize new DNA strands complementary to the single-stranded portion. It cannot, however, replace the preceding double-stranded DNA as it does not have the nuclease activity. The major application of the Klenow fragment is in DNA sequencing.

The third type of polymerase enzyme important in recombinant DNA technology is reverse transcriptase. Unlike the two other polymerases, the reverse transcriptase uses RNA as the starting template. The ability of this enzyme to synthesize a DNA strand complementary to an RNA template is crucial to a technique called cDNA cloning.

4.2.1.3 *Ligases*

DNA ligases serve to link DNA strands together by forming phosphodiester bonds between the 5′-phosphate and the 3′-OH termini of the discontinuous strands. In the cell, these enzymes serve to repair single-stranded breaks (nicks) that arise during DNA replication. The enzymes can also join together 2 individual pieces of DNA together. The most commonly used DNA ligase is the T4 DNA ligase that is purified from *E. coli* infected with the T4 bacteriophage. The use of these enzymes in DNA cloning will be described in the later part of the chapter.

4.2.1.4 *Modifying Enzymes*

These enzymes modify the DNA by either adding or removing a chemical group. There are numerous modifying enzymes, but the 3 most commonly used ones are:

(a) **Alkaline phosphatase** removes a phosphate group from the 5′ end of the DNA. This is usually done to prevent vector religation.

(b) **Polynucleotide kinase** acts in reverse of the alkaline phosphatase by adding a phosphate group to the 5-terminus of a DNA.

(c) **Terminal transferase** adds on one or more nucleotides on the 3′ end of a DNA.

4.2.1.5 *Topoisomerases*

This set of enzymes is able to change the conformation of a closed circular DNA by adding or removing supercoils. These enzymes are used mainly in study of DNA replication, and are not too widely used in DNA cloning.

4.2.2 Use of Restriction Endonucleases in Recombinant DNA Technology

Restriction digestion of DNA is one of the most fundamental techniques in recombinant DNA technology. A typical DNA restriction digestion reaction is illustrated in Fig. 4.3. The amount of DNA to be digested will depend on the nature of the experiment. For example, if a routine cloning experiment is to be performed, typically 1-2 μg of DNA is sufficient. However if one is constructing a genetic library, more starting DNA (10-20 μg) is needed. The second point to consider is the choice of the restriction endonuclease to be used. There are hundreds of commercially available restriction enzymes, supplied as purified enzymes with known concentrations. By convention, I unit of enzyme is defined as the amount required to digest 1 μg of DNA in 1 hour at 37°C. Before adding the enzyme, the DNA solution must be adjusted to provide a suitable reaction condition to ensure optimal enzyme activity. Most enzymes function in pH 7.4, but require different Mg^{2+} concentration and ionic strength (usually provided by NaCl). All commercial enzymes are supplied with their respective buffer in a 10X working concentration. This concentrated buffer will have to be diluted to a 1 X concentration in a digestion reaction. The optimal temperature for most restriction endonucleases to function is at 37°C, although there are some enzymes that require a lower temperature to function optimally (e.g. *Sma* I is used at 25°C).

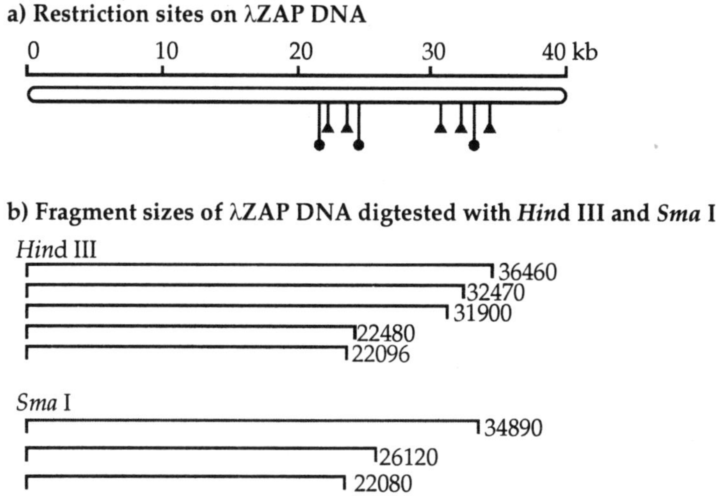

Fig. 4.3 : Restriction of the λZAP DNA using *Hind* III and *Sma* I. The sizes of each fragment are indicated in base pairs.

To illustrate this technique, let us consider digesting a sample of λZAP DNA (of concentration 0.5 μg/μL) with *Hind* III (Fig. 4.4).

We would start with 2 μg of DNA for restriction digestion. Hence 4 μL of the DNA sample is needed for the reaction, which is performed in a typical 1.5 mL microfuge tube. Next, we would need to provide the proper environment for the enzyme to work in. A typical restriction digestion reaction is carried out in a 20 μL volume. Therefore, we would need 2 μL of a 10X *Hind* III buffer solution. The restriction endonuclease can now be added. A commercial *Hind* III enzyme is typically supplied at a concentration of 10 U/μL. Since we are using 2 μg of DNA, a 0.5 μL of the enzyme (5 U total) is more than sufficient to completely digest the DNA. Since the total reaction volume required is 20 μL, 9.5 μL of water is finally added to the reaction mix. The digestion reaction is then carried out at 37°C for 1 hour to allow for complete digestion of the DNA. At the end of the incubation period, the enzyme must be destroyed to prevent accidental digestion of other DNA molecules that may be added at a later stage. This can be achieved by simply denaturing the enzyme by heating the reaction mix at 37°C for 10 min, or adding EDTA to chelate the Mg^{2+}, thus rendering the enzyme inactive.

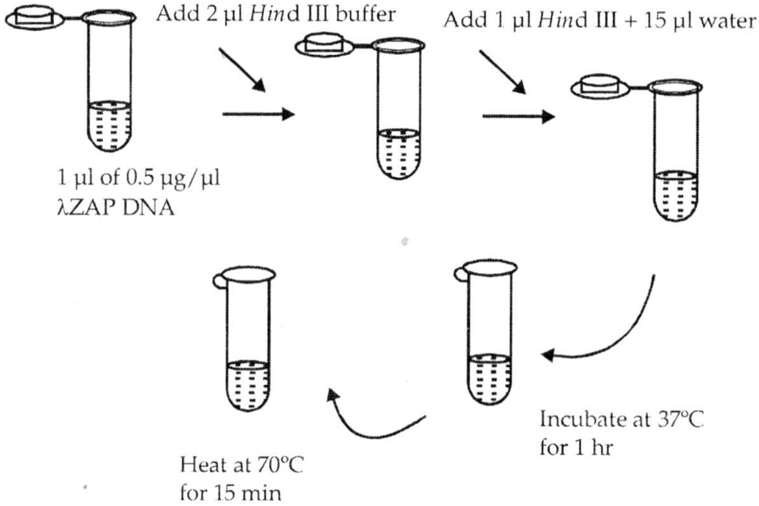

Fig. 4.4 : Restriction digest of λZAP DNA with *Hind* III.

4.2.3 Analysis of the Result of Restriction Endonuclease Digestion

As can be predicted from Fig. 4.3, restriction digestion of the DNA will yield multiple DNA fragments, the sizes of which will depend on the distance between the specific restriction sites for *Hind* III. Clearly, a method is needed to separate the DNA fragments in order to determine the number and sizes of the different fragments. One of the easiest methods to achieve this is by using the technique of gel electrophoresis. Eventually we would also like to be able to acquire more information about the DNA in terms of the relative positions of different restriction sites within the DNA. This is called a restriction map.

4.2.3.1 Separation of DNA Molecules by Gel Electrophoresis

Nucleic acids have net negatively charged phosphates with pKa of ~2. Hence in neutral conditions (pH ~7), the molecules will have an overall negative charge. Thus when the DNA molecules are subjected to an electric field in a buffer, the molecule will migrate to the positive end. The DNA molecules, to a great extent, possess a constant charge to size ratio, irrespective of their sizes. Hence nucleic acids of all sizes will migrate at a similar rate in liquids and it is therefore not possible to separate them in liquids. One way to separate the DNA fragments according to their sizes is by carrying out electrophoresis in gels that act as molecular sieves. Gels are semi-solid matrices of polymers entrapping a buffer. Since the DNA molecules have to weave through this matrix mesh to get to the positive end, the sizes of' the molecules become a factor; the smaller the DNA fragment, the faster it can migrate through the gel. Two of the most commonly used gel matrices are agarose and polyacrylamide, or a mixture of both. In practice, the composition of the gel matrix will determine the sizes of the DNA to be separated. For example, a 1% agarose gel is used to separate DNA fragments of sizes 500-10,000 bp (Table : 4.2).

Table 4.2 : Guide to resolution of linear DNA versus gel concentrations

Agarose (%)	Resolution for linear DNA (kb)
0.5	1.0–40
0.8	0.5–30
1.0	0.3–20
1.5	0.2–5
2.0	0.1–2

If separation of smaller pieces of DNA fragments is required, differing by just bases, then a polyacrylamide gel can be used. The resolution of double-stranded DNA fragments using nondenaturing polyacrylamide gel is shown in Table 4.3:

Table 4.3 : Resolution of DNA in polyacrylamide gels

Polyacrylamide (%)	Double-stranded DNA (bp)
3.5	100–1000
5.0	100–500
8.0	60–400
12	50–200
20	5–100

4.2.3.2 *Visualization of DNA Molecules in a Gel*

The easiest way to visualize DNA fragments in a gel is by using fluorescent dyes. A number of fluorescent dyes can be used for this purpose. One of the most commonly used is ethidium bromide, EtBr. EtBr molecule intercalates (inserts) between the stacked bases of double-stranded nucleic acid or nucleic acid stacked into some secondary structures, as in the case of RNA. The intercalated EtBr will emit a fluorescent light when excited at the UV range (200-400 nm), indicating the position of the DNA fragment. It is important to note that prolonged exposure of the DNA to UV will induce strand breakage within the DNA. Other fluorescent dyes, for example SYBR Gold and SYBR Green, are also available for DNA visualization.

4.2.3.3 *Estimation of the Sizes of DNA Molecules*

In gel electrophoresis, the DNA fragments will be separated according to their sizes. The most common way to estimate the sizes of DNA is by including a DNA marker in the electrophoretic run. DNA markers are standard restriction digests, usually of λ DNA, comprising fragments of known sizes. As the sizes of the DNA markers are known, the fragment sizes in the experimental digest can be estimated by comparing the positions of the DNA bands in the two tracks.

There are a number of DNA markers available commercially catering to various ranges of DNA sizes. The choice of markers used in electrophoresis is dependent on the size range of the experimental DNA fragments. For example, if the size of a PCR product, less than a kilobase long, is to be estimated, then it would be more appropriate to use a 100 bp (size range of 100-500 bp) DNA marker rather than a λ DNA/ *Hind* III marker (size range of 125-3,000 bp).

4.2.3.4 Mapping the Positions of Different Restriction Sites

The next step in a restriction analysis is to determine the relative position of different restriction sites within the DNA, i.e. to build a restriction map of the DNA. This is necessary if one wants to locate and cut a particular region of the DNA.

To construct a restriction map, a series of restriction digests need to be performed. As a guide in choosing the types of restriction enzymes to use, one should first identify one or more unique restriction sites in the vector (i.e. 5′ or 3′ to the DNA insert) to anchor a relative map position. A set of digestion experiments is then performed, using one restriction enzyme per tube. The number and sizes of the fragments produced by each restriction endonuclease must then be determined by gel electrophoresis. Next, a series of double digestions, where the DNA is restricted with two restriction endonucleases, needs to be done. If the enzymes share similar conditions required for optimal enzymatic activity, then the double digestion can be done in a single-step reaction (i.e. the enzymes are added at the same time). However, if the enzymes do not share similar enzymatic buffer conditions, then the digestion has to be done one after the other. Comparing the results of the two types of digestion will allow many of the restriction sites to be mapped within a piece of DNA (Fig. 4-5).

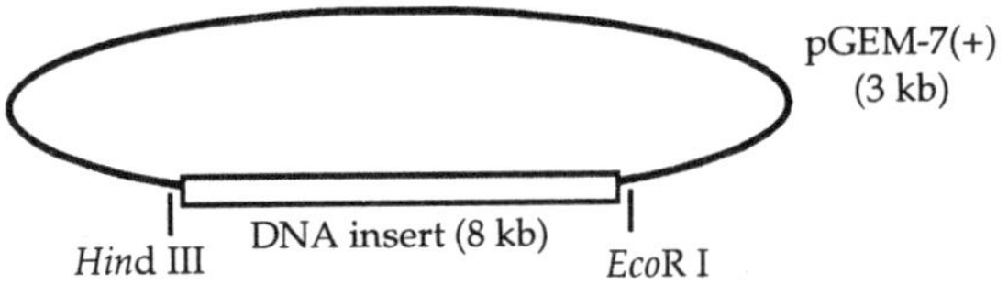

Enzyme	**Number of fragments**	**Sizes (kb)**
*Eco*R I	3	2.0, 4.0, 5.0
Hind III	2	3.0, 8.0
*Eco*R I + *Hind* III	5	1.0, 1.0, 2.0, 3.0, 4.0

Conclusion:

1. The number of restriction sites for each of enzymes is : *Eco*R I 3
Hind III 2
2. The *Hind* III and *Eco*R I sites can be mapped as follow:

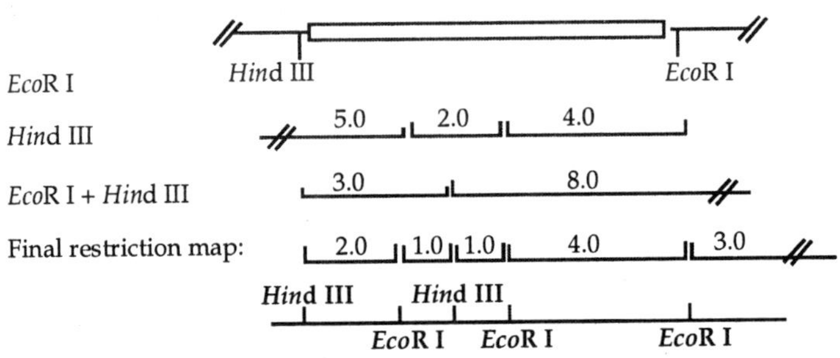

Fig. 4.5 : Construction of a restriction map.

4.2.4 DNA ligase and its use in Recombinant DNA Technology

The final step in the construction of a recombinant DNA molecule is the joining of the DNA fragment and the vector in a process known as ligation. As mentioned in Sec. 4.2.1, this reaction is mediated by DNA ligases.

Figure 4.6 shows two examples of a ligation reaction, one between blunt-ended DNA fragments and the other between DNA with cohesive ends. In general, ligation reactions between cohesive ends are more efficient compared to blunt end ligations. This is generally governed by the kinetics of the two ends of the DNA molecules meeting each other before a ligation reaction can proceed. In the case of cohesive end ligations, the single-stranded sequences from both strands of the DNA form a complement with each other, thus forming a stable albeit transient structure for the ligase to work on.

In contrast, without the aid of complementary sequence formation, blunt-ended DNA molecules rely on chance association to bring the two ends together. In order to increase the efficiency of ligation, blunt-end ligations usually have to be done in the presence of higher amounts of DNA molecules. Alternatively, we can modify the blunt ends of DNA fragments to form cohesive ends by adding short nucleic acids in the form of linkers or adaptors.

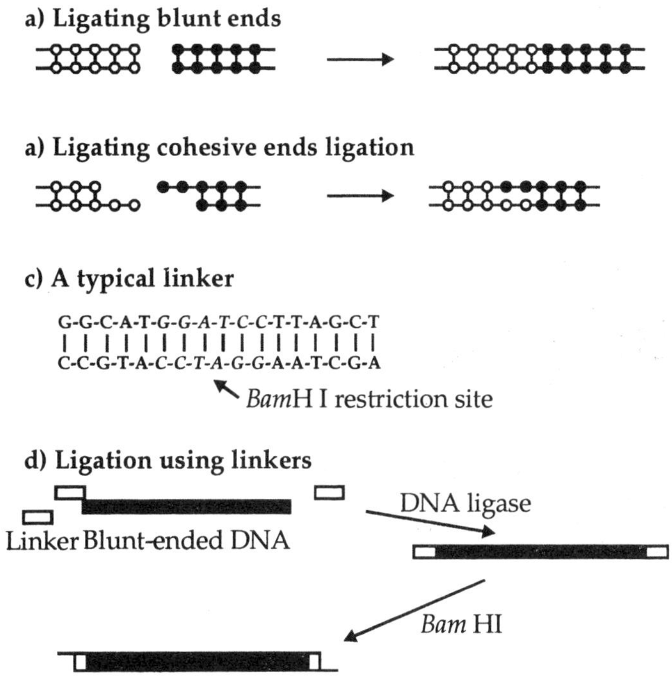

Fig. 4.6 : DNA ligation.

4.2.4.1 *Linkers*

Linkers are short, blunt-ended pieces of double-stranded DNA of known sequences, with one or more restriction sequences inserted within it (Fig. 4.6). The linkers, which can be synthesized in large quantities, are attached to blunt-ended DNA molecules by DNA ligase. Although this is a blunt end ligation, this reaction can be performed with high efficiency as the linkers can be added in high concentrations. It is important to realize that the use of high linker concentrations would inevitably cause multiple linkers to be attached to either side of the DNA molecule. However, restriction digestion with the appropriate enzymes will create cohesive ends within the linkers. This modified DNA-linker molecule will now be ready to be ligated to a cloning vector restricted with the appropriate restriction enzymes.

4.1.4.2 *Adaptors*

Adaptors are similar to linkers in that they are also double-stranded small pieces of nucleic acids that are synthesized *in vitro*. They are however different from linkers in that they do not have blunt ends, but are made with cohesive ends instead (Fig. 4.7). To prevent the formation of adaptor concatamers during ligation, the 5′ cohesive ends of the adaptors could be modified to incorporate a 5′-OH group instead. Hence, the enzyme DNA ligase will not be able to make phosphodiester bonds between the adaptors (recall that DNA ligase only makes a phosphodiester bond between a 5′-phosphate terminus and a 3'-OH terminus; see Sec. 4.2.1). Hence, the adaptors can be ligated to the DNA molecules but not to themselves. After the adaptors have been attached, the 5′-OH cohesive ends are then modified to 5′-phosphate by adding a phosphate group using the enzyme polynucleotide kinase (see Sec. 4.2.1). The modified DNA-adaptor molecules are now ready to be inserted into an appropriate cloning vector.

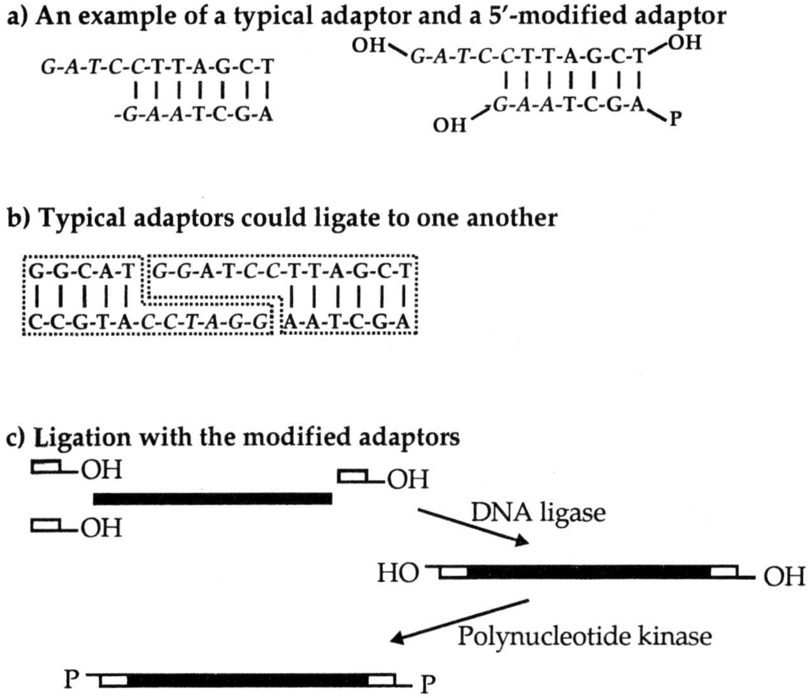

Fig. 4.7 : Use of adaptors in the cloning of DNA.

4.3 HOST ORGANISMS

The use of living cells as hosts to propagate and amplify recombinant DNA is one of the cornerstones in molecular biology. Among the many hosts used in biotechnology, bacteria have been and still are extensively used to produce genetic vectors as well as in the production of recombinant proteins. Other hosts including yeast, plant, insect and mammalian cells, are also exploited for the production of recombinant proteins for specialized purposes. Every host has its own unique properties that can be harnessed for different uses.

One of the objectives of molecular biotechnology is to produce therapeutic proteins. Today, many of the commonly used bacterial strains have been selected or engineered to make them suitable for specific purposes. As an example, the host restriction or modification systems of many laboratory strains have been rendered deficient in these systems and hence allow the propagation of unmodified DNA (e.g. DNA fragments from ligation or PCR products).

E. coli is the abbreviated name for the bacterium in the Family Enterobacteriaceae called *Escherichia* (Genus) *coli* (Species), *E. coli* represents approximately 0.1% of the total bacteria within an adult intestine. This bacterium is a small rod-shaped organism with a circular genome of about 4500 kb in length. The whole genome has now been sequenced. Its genetic property has been the subject of intensive research over the last 30 years and a wide variety of mutants have been isolated and characterized. With the recent availability of their genomic sequences, it is now possible to carry out targeted mutations to produce novel engineered strains for recombinant DNA studies and protein production. The rapid growth of *E. coli* both in minimal and rich media makes it an ideal host for the propagation of recombinant DNA and proteins *in vivo*. Almost all laboratory strains are derivatives of the wild type isolates K-12 or B strains and generally do not carry *Eco*R I or other type II restriction systems.

4.3.1 Growth

There are two common ways to grow *E. coli* in the laboratory: on solid or in liquid media. These media are designed to attain both high biomass and yield of recombinant DNA. This is especially useful for low yield vectors (e.g. BACs) where the number of copies in one cell can be very low in commonly used media, e.g. Luria broth (LB). The development of rich media has been successful for both fed-batch fermentations and small-scale applications. Some of these media go by the labels of "Superbroth", "Terrific Broth" and H15. By obtaining a high yield of vectors, there is therefore no need to use large centrifuges for subsequent manipulations of the host cells. It is also possible to affect the solubility of recombinant proteins by manipulating the growth parameters and media contents.

4.3.2 Handling, Culturing and Storage

It is very important, when working with bacteria, that the cultures are kept pure and properly stored. Note that air-borne bacteria can often confound proper storage. Hence, it is advisable to work in the flow hood. All glassware and media should be sterilized either by autoclaving or filtering through 0.2 or 0.4 μm filters.

Pure cultures can be obtained by propagating bacterial cultures from single, isolated colonies on solid agar plates. This procedure is a powerful and simple method to obtain single, pure species. If the stock is contaminated with another strain and there is a mean to verify the correct strain, then it is possible to plate out and select for the correct strain. For example, XL-l blue cells (Stratagene) harbour a transposon that confers tetracycline resistance. If this cell culture is contaminated with another strain without this genetic marker, then by simply plating out into individual clones on a selection media (e.g. tetracycline), the contaminant can be eliminated.

The bacterial cultures can be stored for long periods in stab vials, as frozen glycerol cultures or as lyophilized cultures. Stabs should be transferred every 2 years whereas frozen glycerol cultures at –80°C or lower can be stored indefinitely. Storage at -20°C in a common-use freezer will result in lose of viability due to fluctuations in temperature with frequent opening and closing of the freezer.

For frequent laboratory use, bacteria can be streaked out on solid agar plates and stored at 4°C. It is advisable to replate these cultures every 1-2 month to maintain viability.

4.3.3 Introduction of Genes into Bacterial Host Cells

The ability to transfer recombinant DNA into a living host is a necessity in molecular biology. In nature, genetic transfers are common occurrences. There are 3 mechanisms of genetic transfer in prokaryotes:

- **Conjugation**—transfer via "conjugative plasmids" requiring cell-cell contact.
- **Transduction**—transfer by bacteriophage. Plasmids (usually smaller than the phage genome) can be encapsidated into phage proheads to produce plasmid-transducing agents. This form of gene transfer is highly efficient.
- **Transformation**—introduction of "naked" DNA into bacteria cells. This is one of the most commonly used methods to introduce recombinant DNA into host cells.

4.3.3.1 *Transformation*

This method of transferring genetic material (as free DNA) to recipient cells does not require the contribution from intact donor cell (e.g. need of pili, i.e. the "sex organs" of bacteria). This method is not restricted to plasmids alone but can also be used to transform large DNA molecules. Figure 4.8 illustrates the two general methods for transforming bacteria:

(a) A chemical method utilizing salt and heat shock to promote DNA uptake.

(b) Electroporation utilizing electric shock to facilitate DNA entry.

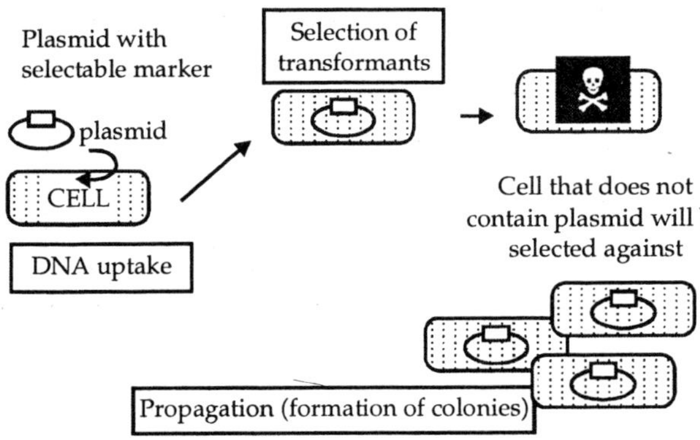

Fig. 4.8 : Transformation of bacterial cells using plasmid DNA.

To assess the success of a transformation, there are three related steps that need to be considered: the entry of the DNA (uptake), production of a selectable phenotype (e.g. antibiotic resistance) and the propagation of these transformed cells to form visible colonies.

4.3.3.2 *Competence State*

In most host cells, including *E. coli*, competent cells do not occur naturally. For these cells, a competent state has to be induced artificially. Competence is usually a consequence of a nutritional shift-down (decrease of the availability of one or several nutrients, or of available energy) and usually result in decrease in the rate or a blockage of DNA synthesis. The exact mechanism is not well understood.

Several growth media and conditions have been developed empirically to induce competence. The kinetics of competence varies greatly from one strain to another (i.e. some strains may show competence for several hours and drops off with time). A number of metabolites can stimulate the establishment of competence in some strains. Although the biochemical mechanisms are not clearly understood, important modifications of the cell envelopes take place during development of competence.

(a) Chemical induced competence

This method of inducing competence is used extensively in many laboratories. There are a number of available procedures using various chemicals including calcium chloride, cobalt chloride, rubidium chloride and others. The transformation efficiency ranges from 10^6 colony forming units (cfu)/μg to greater than 10^9 cfu/μg of DNA, rivaling electroporation. As with other means of transformation, it is highly dependent on the strains used.

The strategy is as follows: cells are grown overnight to act as the starting inoculum. A small portion of this will be used to seed a fresh batch of media, and are then grown with high aeration. When the cell growth is in the logarithmic phase, the cells are collected and processed. Here, the cells are chemically treated to induce competence. All along the process, the cells are treated gently and in the cold to prevent lysis. It is thought that these chemical treatments facilitate the recruitment of channels where DNA molecules can enter. In the process of DNA uptake, the molecules are incubated with the cells at 4°C for a period of half to one hour. During this time, DNA molecules, being predominantly negatively charged, complex with cations (calcium, rubidium or cobalt) and precipitate onto the cell surface. A brief heat-shock treatment enhances the uptake of these complexes through the channels. There is still no concrete evidence that this actually happens although all the observations are consistent with this model of DNA uptake.

The cells can be prepared and kept frozen in glycerol and stored at –70°C for at least 6 months with little loss in efficiency. Batches of the cells prepared in exactly the same manner will vary. Hence it is essential to determine the transformation efficiency of every batch made. Several factors will affect the transformation efficiency, including the density of the cells at harvest, the absolute requirement for maintaining the cells at 0-4°C throughout the processing and the period of time at 4°C before freezing down for storage.

Competent cells of high efficiency for transformation can be prepared in the laboratory. However, this method is rather involved and tedious, requiring experience in treating the cells. It is no longer necessary to prepare competent cells if the usage is low. There are now commercially available competent cells that come in the form of supercompetent or ultracompetent cell preparations.

Since some portions of the host cells will not have taken up the DNA, it is necessary to be able to select for the recombinant colonies. This selection marker, usually in the form of an antibiotic resistance gene, is carried by the transforming DNA. Plating the transformed bacteria onto selective media would thus enable the transformants to be distinguished from the non-transformants. The next challenge is to distinguish colonies containing the recombinant DNA from the ones that contain self-ligated vectors. With most cloning vectors, insertion of the DNA fragment would destroy the integrity of a marker gene present in the vector. Recombinant colonies can then be identified by the characteristics that are not displayed by the host cells due to the inactivated gene.

(b) Electroporation

In general, this method produces better transformation efficiency and is more successful with many different cell types including plants, yeast, insects and mammalian cells that are resistant to chemical methods. Large molecules are also electroporated with high efficiency. Thus, this method can also be used to transfer drugs or even whole genome into cells of all types. The maximal efficiency occurs when the pulse is strong enough to yield 40-80% cell death. Therefore for electroporation, one needs a higher number of cells to start off in order to obtain a reasonable number of transformants. Like chemical methods, each cell type requires an optimal set of parameters to produce a maximum number of transformants. These are usually determined empirically and many protocols have been published.

Electroporation (electrically induced pore formation) is currently viewed as a bilayer membrane phenomenon, where short (μs to ms) electrical field pulses cause the transmembrane voltage to rise to about 0.5-1.0 V, thus causing pore formation. The membrane is then reversibly permeabilized, thereby enhancing the transport of molecules across the membrane.

The electroporators generally have two components: a generator to provide voltage pulses of electrical fields and a chamber in which cells are subjected to the electrical fields (Fig. 4.9). The transient power required by the generators to deliver the electrical energy to the chambers can easily far exceed the electrical power available in the laboratory outlets. To circumvent this limitation, electrical energy is first stored in capacitors. The accumulated voltage is then discharged, at high power, into the chamber.

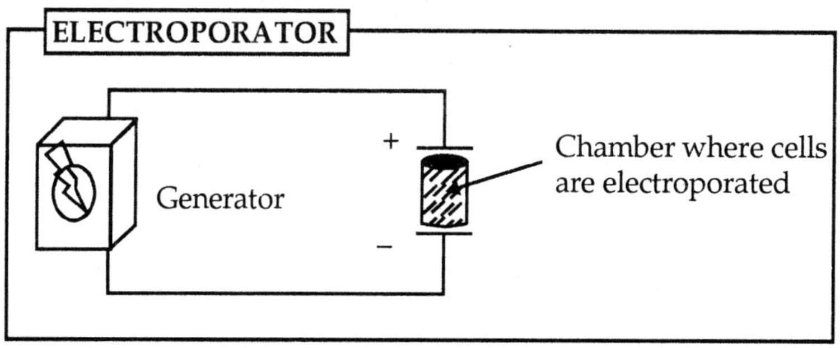

Fig. 4.9 : Delivery of DNA into cells by electroporation.

Bacteria require much higher field strengths (6-25 kv) than most eukaryotic cells. For electroporation, the density of cells to be used should be approximately 1-2 × 10^6 cells/mL. However, this may give rise to a significant proportion of fused cells. This is particularly true with cells that do not possess cell walls (e.g. mammalian cells). This process is called *"electrofusion"*.

The electrocompetent cells are prepared in the same way as the cells prepared by chemical competence, except that only water or a 10% glycerol solution is used throughout the preparation. As with chemical methods, the efficiency will vary between different batches of cells, and between different strains. In the preparation of electrocompetent cells, all centrifugations must be gentle and all manipulations must be carried in the cold (0-4°C).

4.3.3.3 Transfection of Bacterial Cells

There are two ways by which recombinant DNA in a phage vector can be introduced into bacteria: transfection and *in vitro* packaging. Transfection is a process equivalent to transformation, except that the DNA to be transferred is a phage DNA rather than a plasmid. As in the transformation process, the phage DNA is incubated with the competent bacterial cells and the DNA uptake induced by heat-shock. Alternatively the phage recombinant DNA can also be packaged *in vitro,* using the necessary purified phage proteins, to form mature phage particles, which can now infect host cells. This method of genetic transfer is more efficient than transfection with the recombinant phage DNA. When the infected cells are spread onto a solid agarose media, cell lysis caused by the propagated phage particles, can be visualized as plaques on a lawn of bacteria.

4.3.4 Other hosts

Beside *E. coli,* the other commonly used hosts include yeast, plant, insect and mammalian cells in culture. These hosts are eukaryotes and are generally used for the expression of specific proteins rather than for propagating recombinant DNA. Protein expression and the controls involved will be discussed in the latter part of the chapter.

Unlike prokaryotes, eukaryotes can carry out a number of post-translational modifications of expressed proteins. These include:

1. Proteolytic processing of precursor proteins.
2. Specific glycosylations.
3. Post-translational modifications of amino acids.
4. Correct disulfide bond linkages.

Precisely how an expressed protein is modified post-translationally is highly dictated by the type of eukaryotic host used. Hence, with certain proteins, the correct choice of a host is essential.

4.3.4.1 Yeast as Expression Systems

The common baker's yeast (*Saccharomyces cerevisiae*) has been used extensively as a host to express cloned eukaryotic genes. There are a number of reasons for this. This is a single cell and is extremely well known genetically (the whole genome has been completely sequenced) and physiologically. It is easier to manipulate and can be grown in both small culture vessels and large-scale bioreactors. There are several strong promoters and naturally occurring plasmids that have been isolated and well characterized. This organism is capable of carrying out some post-translational modifications. The most important factor is that this strain of yeast has been in use in baking and brewing industries, and has been listed by the U.S. Food and Drug Administration as a "generally recognized as safe" (GRAS) organism. Thus, the use of this organism for the production of human therapeutics (e.g. drugs or pharmaceuticals) does not require as much verification as compared to unapproved host cells. A number of proteins have been produced in this host including, Hepatitis B and HIV vaccines, insulin, blood coagulation factor XIIIa and others.

Other strains of yeast are also used for the purposes of recombinant protein production. They include *Kluyveromyces lactis* (used commercially to produce lactase), *Schizosaccharomyces pombe, Pichia pastoris* and *Hansenula polymorpha.*

4.3.4.2 *Cultured Insect Cells as Expression Systems*

Fruit fly (*Drosophila melanogaster*) and baculoviruses are currently used to produce high amounts of recombinant proteins. In the former case, a number of cell-lines (e.g. S2) have been developed and have attained immortalized non-tumorgenic growth. These cells in culture can then be transfected with vectors with the appropriate promoters.

Baculoviruses infect invertebrates, including many insect cells. This virus does not infect humans and hence, is explored as a potential host to produce therapeutic proteins. As this virus harbour a large genome, a number of steps is involved in generating recombinant baculovirus which will then be used to infect insect cells. The infected cells in culture will then produce the protein of interest.

4.3.4.3 *Plants as Expression Systems*

There are good reasons to develop plants that express foreign substances. The addition of certain gene(s) often improves the agricultural, horticultural or ornamental value of a crop plant. These "transgenic" plants can act as living bioreactors for the inexpensive production of many economically important metabolites or proteins. There are a number of effective DNA delivery systems and expression vectors that can be used on a range of plant cells.

4.3.4.4 *Mammalian Cells as Expression Systems*

Immortalized mammalian cells grown *in vitro*, has been invaluable in the studies of gene expressions and mechanisms underlying many diseases (e.g. cancers). Many mammalian cells from a variety of body tissues have been successfully grown in plastic wares bathed in specialized liquid media in controlled environment (37°C; 5% CO_2/95% O_2), often supplemented with bovine (cow) serum. After a defined number of generations, these cells die in a programmed fashion. Occasionally, some cells escape death and they thrive in the environment and begin to develop immortality. This does not necessary mean that the cells are "cancerous". A cell-line can stay immortal and not be a tumour. Tumorigenic cells can stay immortal in culture and when transplanted into immune deficient mice, develop into full-blown cancers. There are a large number of mammalian cells used for molecular genetic applications. Some of these are also used to produce recombinant proteins.

4.4 VECTORS

Plasmids are widely distributed throughout prokaryotes. These are self replicating entities (replicons), independent of the chromosome. They can be stably inherited in an extrachromosomal state, but are generally dispensable. However, not all extrachromosomal nucleic acids are necessarily plasmids. Some of the heterogeneous circular molecules found in *Bacillus megaterium* and the replicative forms of the filamentous phages (M 13, Fd, f1) are not plasmids. However, parts of these replicative forms have been incorporated into the backbones of many vectors.

Most plasmids exist as *double-stranded* circular DNA. The replication of all plasmids is, to an extent, dependent on their host. Plasmids also exhibit selective replicative capabilities in specific hosts. This is referred to as the host-range of the plasmids. Therefore not all hosts can harbour the same plasmids.

The development of recombinant DNA technology has enabled the modifications of existing plasmids to improve their use as cloning vectors. There are many types of cloning vectors including plasmids, fosmids, bacteriophages and others. They all have common features that make them suitable for use as vehicles for genetic engineering. These vectors can also be easily modified to enable them to express heterologous proteins efficiently.

4.4.1 Some General Requirements for a Cloning Vector

1. All vectors must be able to replicate within a host cell if these are to be used as **"shuttle vectors"**. Some **episomal vectors** (vectors that are not integrated into the chromosomes) do not seem to replicate in mammalian cells (inappropriate origin of replication), but can be effectively used to express recombinant genes. Some mammalian expression vectors, however, do replicate under intense selection pressures.
2. There must be efficient ways of introducing vector DNA into a host cell.
3. There must be ways to isolate the vector DNA from the host DNA.
4. The presence of specific restriction sites in the vector that can be used for cloning an insert DNA.
5. The vector should have at least a selectable marker in the vector to indicate the presence of the vector in host cell. This can be antibiotic or other xenobiotic markers, or merely reporter groups, for e.g. luciferase, green fluorescent protein (GFP) and β-galactosidase.
6. It is desirable to have unique restriction enzyme sites within a selectable or screenable marker so that the presence of a DNA insert in the cloning vector can be selected or screened.

A number of these vectors are available commercially. They can be used for cloning in prokaryotes, yeast, baculovirus, mammalian and other cells (e.g. plants). These vectors come in many different names. They are designed to have specific functions (host specific, expression, recombination) and can encode inserts of different sizes (Table 4.4).

Table 4.4 : Some examples of vectors and their maximum insert sizes

Vectors	Size of inserts (kb)
Plasmids (pBR322)/Phagemids (pGEM, pBSK)	0-10
Bacteriophage λ (insertion vectors; e.g. λ gt 10, λ gt 11, Lambda Zap, Unizap XR, Zap-express)	0-10
Bacteriophage λ (replacement vectors; Lambda Fix II, Lambda DASH II, EMBL, Charon)	9-23

contd. ...

... contd.

Vectors	Size of inserts (kb)
Cosmid vectors (SuperCos 1, pWE15, COS)	25-45
Bacteriophage P 1	60-100
PAC (PI artificial chromosomes)	100-150
BAC (bacterial artificial chromosomes).	100-300
YAC (yeast artificial chromosomes)	200-2000

4.4.2 Vector Systems

Vectors can be classified as "shuttle vectors" and "expression vectors". Shuttle vectors are used generally for the propagation of DNA. The prokaryotic vectors for carrying DNA inserts, which are of high copy numbers, are regarded as shuttle vectors. By simple manipulations, these vectors can be made to express proteins as well. The distinctions between these two types are gradually eroding, as new vectors that can be used for both propagation and expression have been designed. Some of these vectors are packaged into viruses for high efficiency gene transfer (infection as opposed to transduction or transformation).

There are basic features and principles that all vectors rest on. The vectors available today are rather versatile, with many variations of a common theme. The designs of eukaryotic vectors have evolved from those found in prokaryotic systems, often borrowing and wedding specific functions. The earlier plasmids are derivatives of naturally occurring sequences that are host compatible, e.g. pBR322. Nowadays, many commercial companies, including Promega, Invitrogen, Clontech, Stratagene and others, have spliced together sequences from phage DNA and plasmids, giving rise to a modified type of plasmid called *phagemid*. Phagemids are therefore plasmids that have been synthetically manipulated to contain a small segment of sequences from a filamentous ***phage*** (M13, fd or f 1).

4.4.3 Prokaryotic Cloning Vectors

A generic prokaryotic cloning vector has some of the following common features:

1. An origin of replication : *ori*.
2. An antibiotic resistance gene: e.g. ampicillin, kanamycin, tetracycline, etc.
3. A selectable marker: e.g. colour selection with β-galactosidase (*lacZ*).
4. An available multiple cloning sites (MCS).
5. Other accessories:
   (i) Promoters for *in vitro* syntheses of RNA: SP6/T7/T3.
   (ii) f1 origin to allow production of single-stranded DNA.
   (iii) MCS sequences that allow unidirectional deletion of inserts.

Let us examine the structures of two commonly used vectors in the laboratory that are commercially available, Bluescript (pBSK; Stratagene) and pGEM (Promega) (Fig. 4.10). They are known as phagemids, and are designated as **pBSK**(+/-) and **pGEM-XZf**(+/-), where

X denotes vector number, f the presence of phage origin of replication, and (+/-) the orientation of the single-stranded DNA produced.

(a) **pGEM series.** This series comprises the vectors pGEM-3Z, pGEM-4Z, pGEM-3Zf(+/-), pGEM-5Zf(+/-), pGEM-7Zf(+/-), pGEM-9Zf(-), pGEM-1 Zf(+/-) and pGEM-13Zf(+). The parental plasmid for this series is the pUC vector. These vectors contain the pMB1 origin of replication. The early series of pGEM vectors (e.g. pGEM-3Z and -4Z) do not contain an f1 origin for the production of single-stranded DNA. This feature, plus the addition of new MCS sequences, is incorporated into the later generation of pGEM vectors to allow for increased versatility and convenience in gene manipulations.

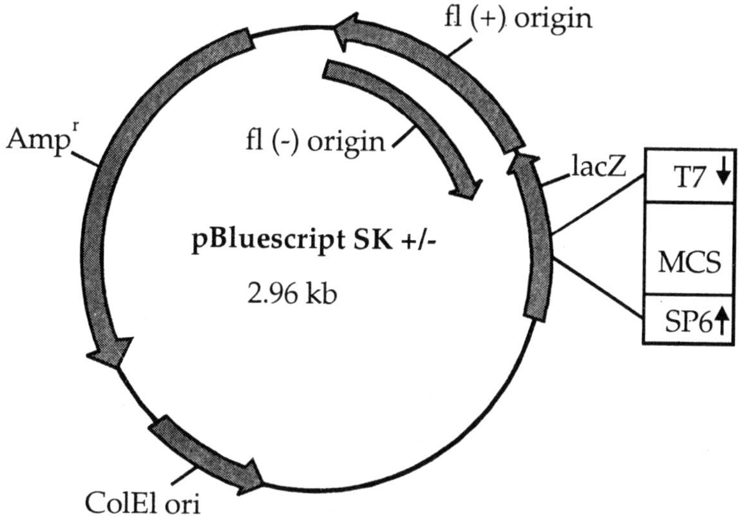

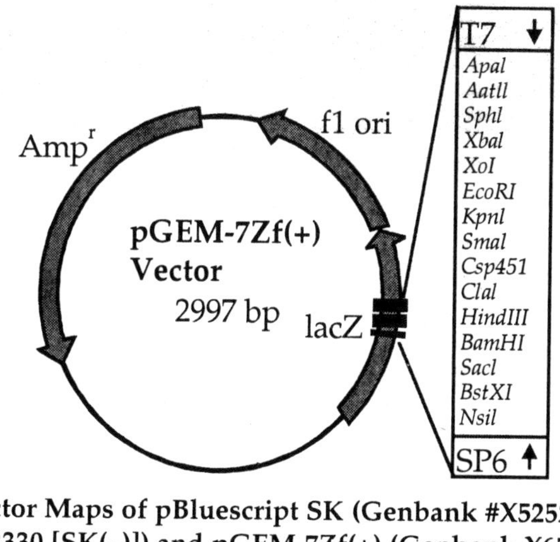

Fig. 4.10 : Vector Maps of pBluescript SK (Genbank #X52528 [SK(+)] and #X52330 [SK(–)]) and pGEM-7Zf(+) (Genbank X65310). Adapted from Stratagene and Promega, 2001

(b) **pBIISK(+/-)/pBIIKS(+/-).** Both are the same vectors, except that the orientation of the MCS site has been reversed between the T7 and T3 promoters. These phagemids contain the ColEl of origin of replication. The pBlISK vectors are derivatives of pUC19 (which is a derivative of pBR322) and, like the pGEM vectors, replicate in high numbers because of the lack of the pBR322 *rom* gene.

(c) Both pGEM and pBIISK/KS vectors use ampicillin resistance as a selection marker. Therefore transformants will acquire resistance to ampillicin.

(d) Both vectors have phage promoters; T7/SP6 in the pGEM series, and T7/T3 in pBIIKS/SK. The promoter sequences flank the insert and contain all the information required for the initiation of RNA transcription by T7, T3 or SP6 RNA polymerase. These polymerases do not cross-initiate and hence, can be used to produce highly specific RNA transcripts. The RNA transcripts produced can be used for a number of studies. For example, the transcripts can be used as riboprobes for nucleic acid hybridizations, as a template for *in vitro* translation of proteins, or in gene expression studies.

(e) Both vectors have two origins of DNA replication, one which is specific for *E. coli* (ColEI for BSK and pMB I for pGem), and the other from the single-stranded DNA phage f 1. The ColEl or pMB1 origins are used for plasmid replication. However, following phage f1 infection, the other origin is used and single-stranded DNA is produced. This is useful for single-stranded DNA sequencing and some mutagenesis strategies.

Early cloning studies were rather cumbersome. The most common strategy was to subclone the DNA fragment from a plasmid or lambda vector, into filamentous bacteriophage M13 vector for the production of a single-stranded DNA template that can then be used for sequencing. Once sequencing has been completed, the template had to be recloned into another specialized vector, such as a vector designed for efficient *in vitro* transcription, or an expression vector for protein expression. If site-directed mutagenesis experiments were to be carried out, the mutant forms of interest had to be then subcloned at each step. These rather time-consuming steps have now been eliminated though the development of vectors that possess, in a single construct many features required for specialized DNA manipulations.

(f) **Genetic markers.** Both vectors have convenient ways to select for the presence of DNA insert colorimetrically. This selection method is based on the biochemical properties of the *lac* operon (see Sec. 4.4.4).

(g) **Protein expression.** Foreign DNA encoding protein sequences can be expressed in these two vectors, as long as the sequences are in-frame to the expression of β-galactosidase gene, which is under the control of the *lac* promoter. An adaptation of this strategy is the use of pGEM-T vector where random sequences of a protein can be incorporated and expressed as a fusion protein of β-galactosidase. It is advisable to use a host with the $lacI^q$ present in order to increase the yield of proteins produced.

The T7 promoter can also be exploited for protein expression. By inserting the foreign DNA encoding protein sequences under the control of the T7 promoter and adding a ribosome-binding site (RBS) plus an ATG start site, it is possible to keep the expression tightly turned off until a T7 RNA polymerase is expressed in certain host cells [e.g. BL21(DE3)].

(h) **Gene mapping.** The unique *BssH* II restriction sites flanking the T3/T7 promoters of pBlISK offer an easy and rapid method to map large and small DNA inserts. To do this, complete *BssH* II digestion of the clone is first carried out, followed by partial restriction digestion with the restriction enzymes of choice. The partial fragments from the restriction digestion will then be separated by agarose gel electrophoresis, transferred to a nylon membrane, probed with labeled T3/T7 to determine the lengths of the partial fragments. The different lengths of the digested fragments identify the location of the restriction enzyme sites relative to the T3/T7 positions.

(i) **MCS unidirectional deletion of insert.** An alternative to the above partial restriction map analysis is simply to exploit the use of the MCS. If the insert DNA has restriction sites similar to that in the MCS, complete restriction will result in the excision of some vector and insert sequences. If sequencing is carried out from one direction, it will then be possible to gain internal sequence information.

(j) **Unidirectional deletions with *Exonuclease* III.** The restrictions sites have been arranged to allow unidirectional deletions of the cloned insert with *Exo* III. This enzyme is a 3-5′ single-stranded exonuclease and require double-stranded DNA template. Hence, *Exo* III does not digest 3′ end single-stranded overhangs, but exclusively digest the 3′ ends of blunt ends or 5′ overhangs. Therefore, double digestion with a 3′, and a 5′ or blunt end producing enzyme, followed by *Exo* III digestion, and the subsequent removal of the remaining single-stranded DNA by mung bean nuclease will result in unidirectional deletions of the insert DNA. This method allows the serial deletion of large inserts for sequencing from one direction.

4.4.4 *Lac operon*: A Model for the Control of Gene Expression

A number of cloning and expression vectors utitize the *lac*Z system as either a selection marker or as a reporter gene. It is therefore important to have a basic understanding of this system.

The *lac* operon found in *E. coli* is one of the best-known examples of a bacterial operon (Fig. 4.11). Structurally it consists of 3 genes: *lac*Z (coding for β-galactosidase), *lac*Y (coding for a permease which is needed for uptake of lactose) and *lac*A (coding for an enzyme called thiogalactosidase transacetylase). A regulatory region containing the promoter site is found at the 5' end, overlapping with a sequence known as the operator.

The *lac*I^q repressor protein binds to the operator region that physically overlaps with the promoter (which drives the *lac* operon) and tile first 20 or so bases to be transcribed, thus preventing the initiation of transcription. The *lac*I^q gene is not part of the *lac* operon. The bound *lac* repressor protein has an affinity for a molecule called allolactose, which is a derivative of lactose. This binding to allolactose is allosteric and will prevent the repressor from binding to the operator region. Therefore, in the absence of lactose, the repressor is active and binds to operator, preventing transcription. However in the presence of exogenous lactose, the repressor is inactivated and the operon is then expressed. The cell then produces the enzymes that are necessary for the metabolism of lactose. As lactose will be metabolized as soon as induction begins, thus diminishing the inductive effect, it is not a good inducer of the *lac* operon. In practice, a synthetic analogue of lactose that will not be broken down easily by β-galactosidase is used instead. An analogue that is commonly used in the laboratory is isopropyl-l-thio-β-galactosidase (IPTG). Not all substances that are substrates for breakdown by β-galactosidase, can act as inducers as they are not recognized by the repressor. One example of such a chemical is 5-bromo-4-chloro-3-indolyl-β-D-galactopyranoside, commonly known as X-gal.

Some *E. coli* have a modified *lac*Z gene, lacking a segment referred to as *lac*Z′ which codes for the α-peptide portion of β-galactosidase. These bacteria can only synthesize the full active enzyme when they harbour a plasmid that carries the missing *lac*Z′ segment. This phenomenon is called α-complementation and is exploited in a technique used to distinguish a recombinant colony within a background of non-recombinants. In this technique, the transformed bacterial cells are plated onto solid media containing X-gal and IPTG. The active β-galactosidase enzyme breaks X-gal down to form a product that is deep blue in colour. Since the MCS site resides within the *lac*Z′ gene in most plasmids, insertion of a DNA fragment into the vector will inactivate the β-galactosidase enzyme, resulting in the formation of white colonies instead. This selection technique is also applicable to distinguish bacterial cells infected with recombinant bacteriophage DNA that has the *lac*Z′ gene in its genome (e.g. λZAPII vector).

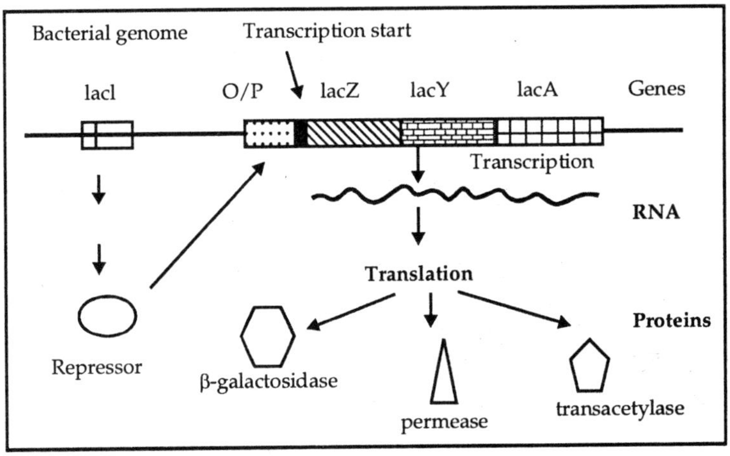

O/P = operator and promoter

Fig. 4.11 : The *lac* operon.

4.4.5 λ Bacteriophage Vectors

Two problems hinder the use of plasmids for all genetic manipulations: the severe size limitation of the insert DNA, and the efficiency of gene transfer. Although the efficiency of transformation has dramatically increased by the use of electroporation and specific chemical treatments, it remains a hindrance. As an alternative, infection as a way of introducing DNA into cells show efficiency close to unity (100%). Attention was therefore directed at bacteriophage λ as it possesses a number of desirable qualities required for cloning, a highly efficiency method of gene transfer as it infects the bacterial cells, and the ability to insert large DNA fragments (~20 kb) by manipulating the genome.

To design vectors based on this phage, it is necessary to insert the DNA of interest into the λ replicating sequences (replicon) *in vitro*. The resulting recombinant DNA must be able to infect *E. coli* at high efficiency. The latter requirement is met by developing an *in vitro* packaging system, which mimics the way in which wild-type λ DNA is packaged in a phage coat protein, resulting in high infection efficiency.

Two major strategies are used in generating λ vectors: replacement vectors and insertion vectors. As the name implies, replacement vectors are those which have some segments of its sequences deleted to accommodate the sequences of interest. It is observed that DNA molecules between the sizes of 37 to 52 kb can be stably packaged into the phage head. Exploiting this finding, removing the region of λ genome that is nonessential for growth (central portion where the genes are necessary for integration and recombination) and inserting the DNA of interest to make up the required size, foreign DNA of up to 23 kb in length can be easily incorporated. Hence, these vectors are commonly used for genomic cloning (Fig. 4.12).

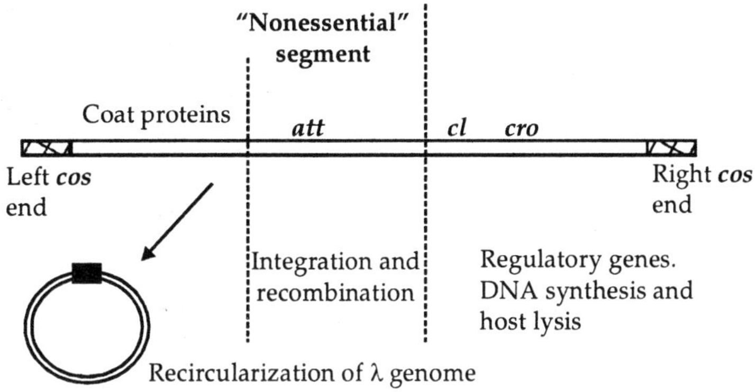

Fig. 4.12 : The bacteriophage λ as a cloning vector.

For cDNA (complement DNA generated from mRNA) cloning, where the sizes of DNA fragments are generally not as large (<10 kb), these sequences can be cloned into modified λ vectors (insertion vectors).

With an insertion vector, the nonessential segment has been deleted, and the two arms ligated together. Therefore this vector has at least one restriction site into which a DNA fragment can be inserted. The newer versions of λ insertion vectors (e.g. Lambda ZAP II vector) allow easy *in vivo* excision of insert DNA in a phagemid form. This is enormously convenient as it is easier to isolate and purify plasmid DNA as opposed to the purification of phage DNA.

The *in vivo* excision depends on the DNA sequences of the phagemid (BSK) and the presence of a variety of proteins, including filamentous phage (M13)-derived proteins. The M13 phage proteins recognize the f1 bacteriophage "origin of replication" found within the pBluescript sequences. This f1 origin of replication can be divided into 2 overlying parts: (a) the site of initiation (I), and (b) the site of termination (T) for DNA synthesis. These two regions have been subcloned separately into the lambda ZAP II vector. The lambda phage vector is made accessible to the M13-derived proteins by simultaneously infecting a particular strain of *E. coli* (the SOLR strain) with the lambda phage vector and the bacteriophage. The "helper" proteins (from M13) then recognize the initiator DNA that is within the lambda vector. One of these proteins nicks one of the two strands of DNA and a new DNA synthesis begins, duplicating the DNA sequences downstream of the nick site. This synthesis continues through the cloned insert until the termination signal is encountered. The single-stranded DNA molecule is then circularized by the gene II product of the M 13 phage. This results in a circular DNA molecule containing the DNA sequences between the initiator and terminator sites. In this case, this includes all sequences of the pBluescript and the insert DNA (Fig. 4.13).

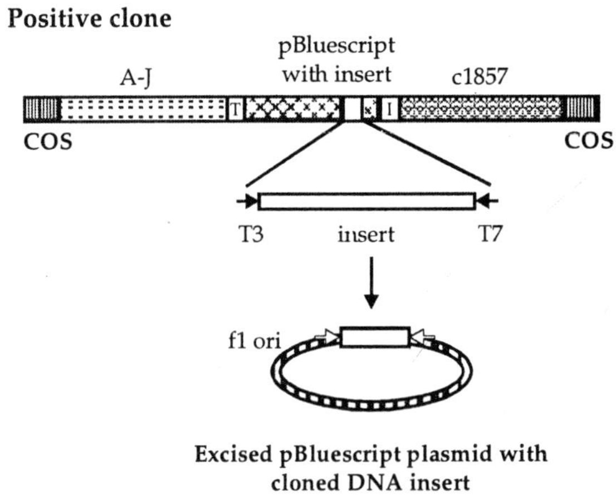

Fig. 4.13 : The newer versions of λ insertion vectors (e.g. Lambda ZAP II vector) allow easy *in vivo* excision of insert DNA in a phagemid form.

4.4.6 Other Vectors

4.4.6.1 Cosmid Vectors

Effective packaging of the λDNA molecules can only occur if the *cos* sites are present. It is now known that only a small region around the *cos* site is required for recognition by the packaging system. Plasmids have been constructed containing a fragment of the λDNA including the *cos* site. These are called ***cosmids*** and are used in conjunction with the *in vitro* packaging system. The packaging into the phage head requires a large amount of DNA. If the vector is only 5 kb in length, then to be incorporated into the phage head, an insertion of between 32 to 47 kb of DNA (gene of interest) is required. This large amount of DNA cannot be inserted into any of the known λ vectors. Note that after packaging *in vitro*, the phage can infect the appropriate host. The recombinant cosmid DNA is injected and circularizes like a phage DNA, but replicates as a normal plasmid without expressing phage proteins. Selection of transformants is done on the basis of a vector-drug-resistance marker carried on the plasmid portion.

4.4.6.2 Bacterial Artificial Chromosome (BAC) Vectors

Most of the commonly used vectors for propagating small inserts have high copy numbers (discussed in other sections). One disadvantage of possessing high copy numbers is that these vectors often show structural instability of inserts (especially for large inserts and those with repeated sequences common in higher eukaryotes). Thus, this limits the use of these high copy number bacterial vectors.

To overcome these limitations, attention has recently been focussed on vectors that exist in low copy number replicons, e.g. *E. coli* fertility plasmid, the F-factor. This plasmid occurs naturally in many strains and is maintained at a copy number of 1-2 per cell. Vectors based on the F-factor system can accept large foreign DNA fragments (>300 kb). Using electroporation, it is possible to introduce these recombinants into *E. coli* efficiently. However, it must be noted that the copy number will be low within the cells. The consequence of this is that, to prepare these vectors, efficient isolation procedures must be used.

An example of a BAC vector derived from a mini-F plasmid is shown in Fig. 4.14.

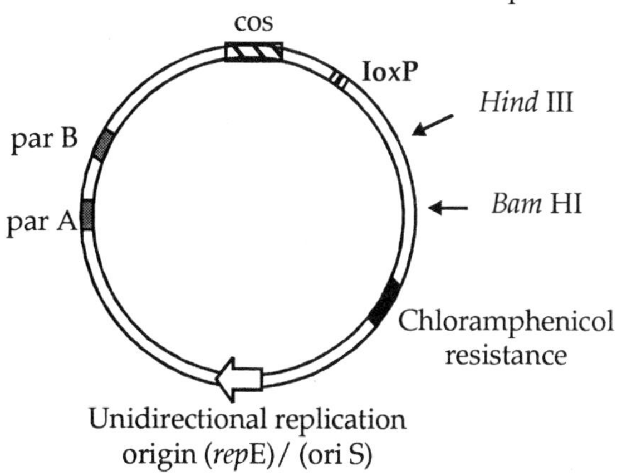

Fig. 4.14 : A BAC vector.

4.4.6.3 Yeast Artificial Chromosomes (YACS)

As the upper size limit of cosmid cloning is about 35,045 kb, to cover the whole human genome in a theoretical contiguous chain would require about 70, 000 clones. What is required is the development of vectors that can accommodate insert sizes even bigger than in the cosmids. This will reduce the number of clones to be handled. The minimum structural element for a linear chromosome is the presence of an origin of replication (autonomous replicating sequence, *ars*), telomeres and a centromere.

A linear molecule that behaves like a chromosome in yeast has been constructed by combining an *ars* and a centromere (both from yeast) with telomeres from Tetrahymena (Fig. 4.15). These YACS allow the insertion of large pieces of foreign DNA (- 1 Mb) into this linear artificial chromosome.

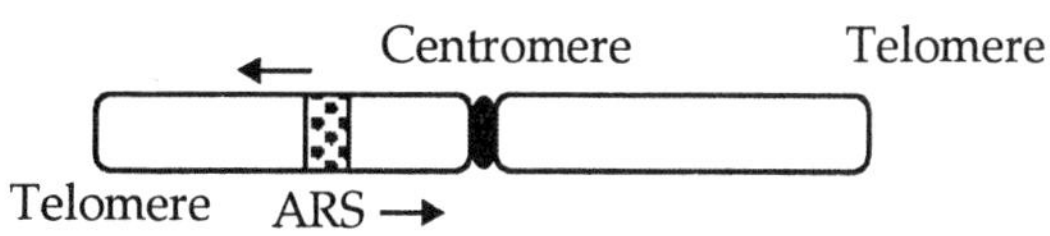

Fig. 4.15 : Yeast artificial chromosome as a cloning vector.

There are a number of operational problems with YACS. The existence of chimeric DNA segments from different regions of the genome into a single clone is one of such problems. Another problern is that of instability, which tends to delete internal regions in the inserts. Other problems include the difficulty of removing the yeast host chromosome background from the YACS and low yield of DNA from YACS. A number of these concerns are currently being addressed.

4.4.6.4 *Shuttle Vectors*

Gene cloning and associated techniques are also extremely valuable for investigating all aspects of the behaviour of a wide range of bacteria. *E coli* genes are often expressed satisfactorily in other Gram-negative organisms, but the reverse is usually not true. Genes from *Pseudomonads,* for example, are only poorly expressed in *E. coli* and have to be analysed in their original hosts. In many cases, the techniques available for these organisms are limited. For example, transformation (or electroporation) frequencies may be relatively low, and the methods for screening gene libraries may be difficult. It is therefore advantageous to carry out the initial cloning of a gene using *E. coli* as the host organism. After the desired DNA construct has been obtained, it can be recovered and then inserted into the target host. This procedure requires a vector, which can replicate in either organism. Certain plasmids are capable of existing in a variety of hosts and can function as shuttle vectors. A shuttle vector carries replication origins for two different plasmids on the same vector (Fig. 4.16). This will facilitate preliminary manipulations in a well-characterized system, such as *E. coli,* and then transfer the clone to the organism of interest. If the clone is to be expressed in both hosts, then it may also be necessary to provide two promoters as well.

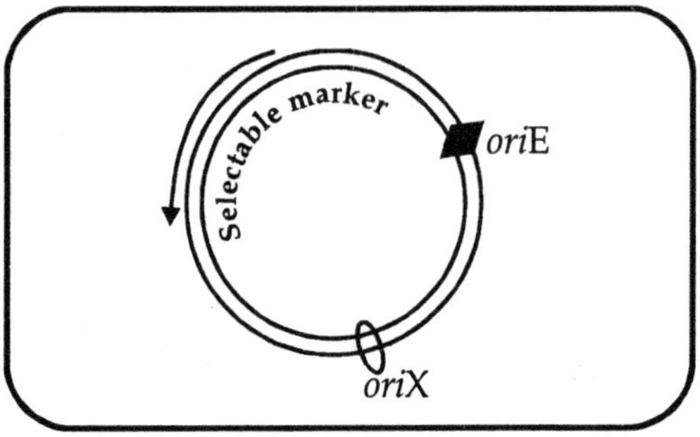

In *E. coli,* plasmid replication is initiated using *ori*E.

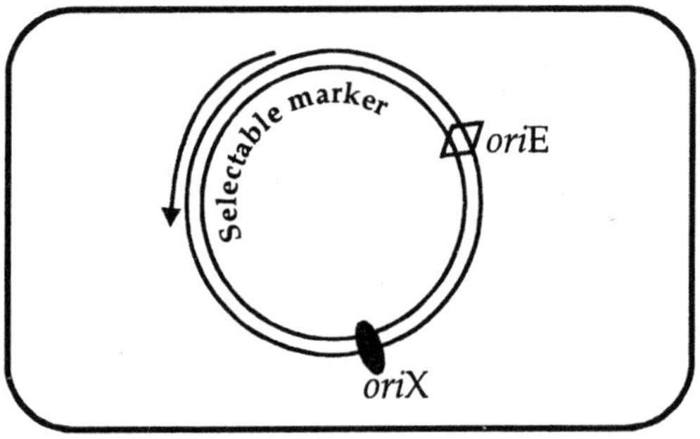

In the alternative host, the plasmid replication is initiated using the second origin of replication, *ori*X.

Fig. 4.16 : Shuttle vectors with dual origins of replication.

4.4.6.5 Broad Host Range Vectors

Broad host range plasmids can be used as an alternative to dual-origin shuttle vectors. These plasmids belong to two incompatibility groups, P and Q/P4. P group plasmids are usually derived from *Pseudomonads* and the prototype of this group is RN (also known as RPI or RK2). RN is a broad host range plasmid and is able to replicate in some Gram-positive bacteria and in most Gram-negative bacteria. The P group of plasmids contain replication and transfer functions which are widely dispersed on the bacterial chromosome, thus making it extremely difficult to construct lower molecular weight derivatives which could be used as cloning vehicles. One example is pRK2501, which is 11 kb in length, but lacks all transfer and mobilization functions. Since this vector is transfer-deficient and non-mobilizable, it satisfies important requirements for the biological safety of vectors with broad host-range. A system, which employs two plasmids, has been developed for cloning in *Rhizobia* (Fig. 4.17). The first plasmid, pRK290, is the cloning vehicle which is transfer-deficient but still possess all mobilization functions (Tra^- Mob^+) and can replicate equally well in *E. coli* and *Rhizobium* because it contains the RK2 replicon. This vector can be mobilized by a second plasmid residing in the donor cells, and hence can be transferred to

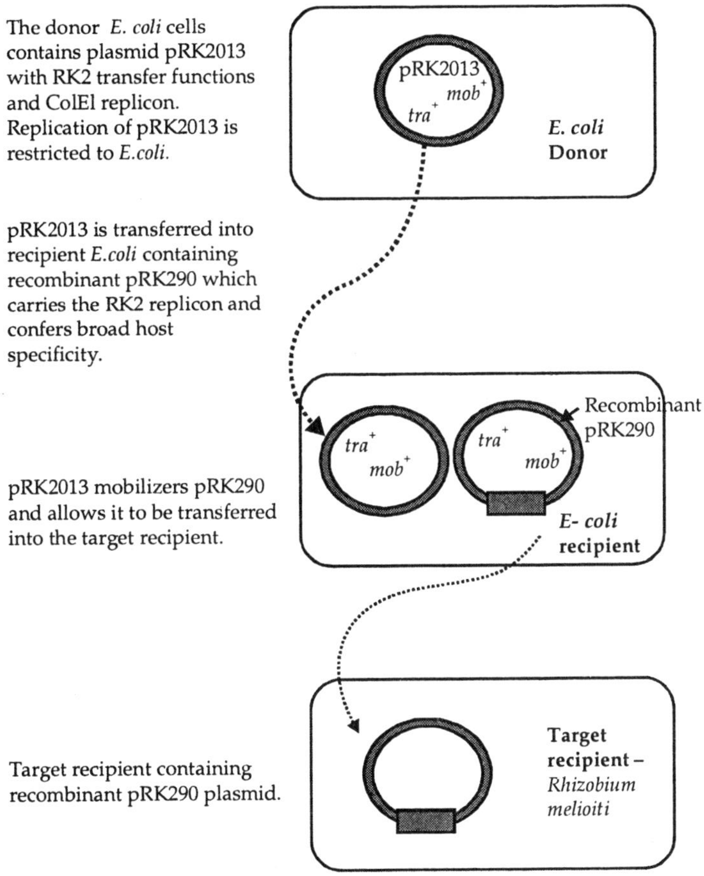

Fig. 4.17 : Use of broad host specificity plasmids for gene transfer into other host organisms.

suitable recipient cells, e.g. *Rhizobium*. The mobilizing plasmid in the donor is pRK2013, which possesses all transfer functions and also contains the neomycin resistance region of RK2 and the ColEl replicon, which restricts its replication, in *E. coli* but not in *Rhizobium*. When this plasmid enters *E. coli* cells, the transfer functions of pRK2013 are expressed and mobilize plasmid pRK290, which can replicate in *Rhizobium* cells.

4.4.7 Prokaryotic Expression System

In general, expression vectors should have the following properties:

1. **A promoter,** which marks the start of gene transcription. This sequence is recognized by the RNA polymerase.
2. **Inducible elements,** which provides the ability to control the level of protein expression.
3. **Accessory elements** for stability allows for the maintenance of a stable transcript.
4. **Transcriptional termination,** which marks the end of a gene where transcription should stop.
5. **Affinity tag** that allows the protein to be easily isolated and purified (optional).

The promoter is the most important factor of an expression vector. In all expression vectors, the cloning site is located just downstream from a promoter and ribosome binding site (RBS), allowing heterologous regulation of gene expression. The important sequences for *E. coli* RNA polymerase are the TTGACA and TATAAT (Pribnow box) sequences found at position -35 and -10, respectively. The most commonly used promoters are those from the *lac* or *trp* operons, a combination of the two called the *tac* promoter (*trp* and *lac*), and the λ_{PL} promoter from phage λ. These promoters are inducible and, in the case of the *lac* promoter, transcription or derepression of the gene is brought about by the addition of IPTG. The ability to control levels of expression is important: continuous high levels of expression may affect plasmid stability or the protein product may be toxic to the host organism.

Efficient translation of mRNA in prokaryotic cells cannot occur without a ribosomal binding site (RBS). This includes an initiation codon, AUG, together with a consensus sequence, AGGAGGU, which typically lie 3-12 bases upstream of the AUG. This is the Shine-Dalgarno sequence (SD). The sequence of the RBS, together with the secondary structure of the mRNA, has an effect on translation efficiency of mRNA. To obtain expression of foreign genes in *E. coli*, it is necessary to incorporate a RBS into the recombinant DNA construct. Furthermore, the SD sequence must be located at the optimal distance from the translation start codon.

More sophisticated expression vectors, with more efficient expression signals, are used when it is necessary to obtain very high levels of formation of the product of the cloned gene, for example in commercial processes. Extremely high levels of specific protein, representing perhaps 50% of total cell protein, can be produced using such a vector. However, diverting so much of the cell's resources into a product that is useless as far as the cell is concerned will slow down the growth rate (which will also exacerbate plasmid instability). The usual strategy for overcoming this problem is to use a promoter that is subject to repression, so that the bacteria can be grown to high cell density before product formation is induced by altering the growth conditions.

The choice of a promoter will depend on the requirements of the expression. For example, if it is desired that the cloned DNA be regulated to prevent inhibition of host cell growth, the *trp*, *lac* or λ promoters might do well. The appropriate repressor molecule is supplied as a regulator. The *tac* promoter, a combination of the -35 region of the *trp* promoter and the -10 region of the *lac* promoter, also has a high efficiency of expression. Many phage promoters are also very efficient for in protein expression. For example, the phage T5 promoter out-competes all *E. coli* promoters, thus allowing the expression of large amounts of the recombinant protein in *E. coli*.

An example of a prokaryotic expression vector that is designed for high protein expression and easy purification is the pQE-9 which belongs to the pQE series of vector (Fig. 4.18).

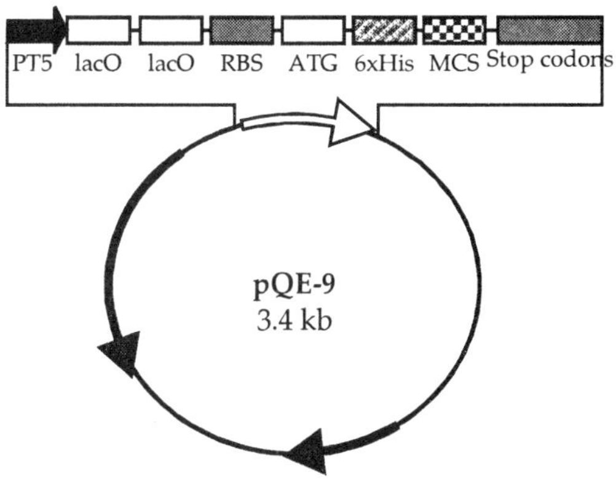

Fig. 4.18 : An example of a prokaryotic expression vector that is designed for high protein expression and easy purification. Adapted from Qiagen 2001.

This vector uses the Phage T5 promoter, which can be recognized by the *E. coli* RNA polymerase. The incorporation of a double *lac* operator repressor system provides a tightly regulated, high-level expression of recombinant proteins in *E. coli*. In addition, the vector contains a 6xHis tag segment that will allow easy purification of the expressed recombinant products *in vitro*.

Phages such as T3, T7 and the *Salmonella* phage SP6, code for their own RNA polymerases. Each of these polymerases specifically recognizes its own particular promoter and ignores those of other phages or the host bacterium. Thus, by using such a phage promoter to drive the expression of a foreign gene, the expression of recombinant protein will be absolutely dependent on the availability of the phage RNA polymerase. By regulating the availability of the phage RNA polymerase using an inducible promoter, such as the λ_{PL} promoter, which is repressed by providing the cI repressor *in trans*, allows the expression of the recombinant protein to be tightly regulated.

Sometimes, a particular mRNA transcript made from cloned DNA is poorly translated in the host cell. This lack of efficiency may be due to an increased susceptibility to degradative enzymes or to the use of rare codons in the mRNA. The only solution for the latter problem is to change the host cells to one that has more suitable pattern of codon usage or else to use site-specific mutagenesis to change the codons themselves. In the case of the former problem, changing the structure of the mRNA molecule itself can often prevent excessive degradation. In particular, mRNA stability can be enhanced by adding sequences such as REP onto the end of the cloned DNA so that a suitable hairpin loop is formed. The loop prevents access of the exonuclease to the end of the mRNA.

4.4.8 Isolation and Purification of Plasmids

An obvious prerequisite for cloning in plasmids is the ability to purify the plasmid DNA. The strategy is to release the plasmid DNA from the cell and then it purify away from the genomic DNA and cell debris. The step that demands the most attention is the initial steps of cell lysis and neutralization of the reaction. The incomplete lysis of the cells will reduce the recovery and purity of the plasmid DNA. It is important to note that cell lysis must be carried out with minimal physical force. After neutralization, the chromosomal DNA of high molecular weight can be removed together with the cell debris, by high-speed centrifugation, to yield a clear lysate where the plasmid resides. There are a number of ways of isolating plasmids and many of these rely on the alkaline lysis procedure that is both simple and effective.

4.4.8.1 Alkaline Lysis

This is largely based on thc observation that there is a narrow range of pH (12-12.5) within which linear DNA, but not supercoiled DNA, denatures (Fig. 4.19). However, prolonged exposure to this pH will eventually denature even the supercoiled DNA. The method is rather simple but highly effective, even with large vectors like BACs. Plasmid

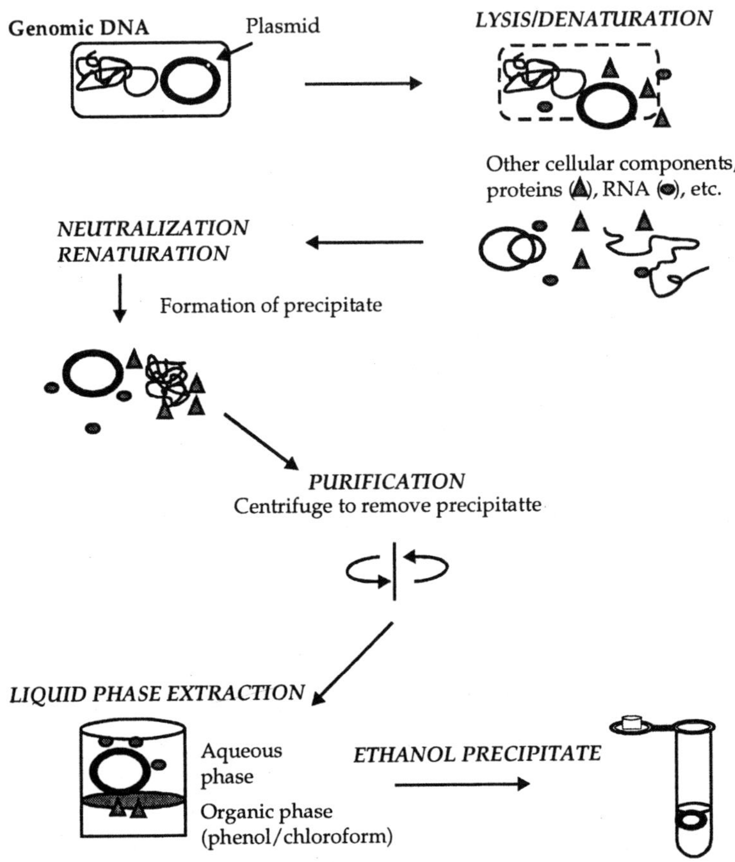

Fig. 4.19 : Isolation of plasmid DNA using alkali lysis.

containing cells are first lysed in high pH and SDS. The lysate is then neutralized. The chromosomal DNA renatures and aggregates to form an insoluble network. The high concentration of acetate also causes the protein-SDS complexes (and excess dodecylsuphate) to precipitate. Provided the pH of the alkaline lysis step is carefully controlled, supercoiled plasmids will remain in the native state. The insoluble precipitates are then removed by centrifugation. The plasmid can be purified by extracting any remaining bacterial proteins using a mixture of phenol and chloroform (organic phase). The purified plasmid, which is retained in the aqueous phase, is then recovered by precipitation using salt and ethanol. This process is known as ethanol precipitation.

4.4.8.2 Boiling Method

This is based on the principle that the supercoiled DNA is not denatured as much as the chromosomal DNA. When the extract cools down, the chromosomal DNA will reanneal but due to its complexity will form insoluble "tangles" and hence precipitate. As in alkaline lysis, this can be removed by centrifugation. Although this method is similar to the above method, it is not highly reliable and is dependent on host strain.

4.5 IDENTIFICATION OF CLONES

The success or failure of a cloning experiment relies mainly on the ability to identify and select for the particular clone of interest. This is especially true if a large collection of recombinant DNA representing the entire genome of an organism (a genetic library) has been cloned. Once this hurdle is overcome and a clone of interest is obtained, the scientist is then able to use a wide number of techniques to explore and extract information from the gene.

There are many ways of identifying and analyzing DNA often with the ultimate objective of gaining an insight into its sequence and function. This section describes some of the methods commonly used to isolate, identify and confirm the clone carrying the gene of interest. One of the simplest but powerful methods to identify the correct recombinant DNA directly is by a technique known as **hybridization probing.** This technique forms the basis for methods such as colony and plaque lifts, as well as Southern blot hybridization. The piece of unknown DNA can also be identified by PCR using specific primers. The DNA can eventually be confirmed by a technique called DNA sequencing.

4.5.1 Hybridization Probing: Its use in Identifying Specific Clones

Hybridization of nucleic acids immobilized on filters occupies a central role in molecular biology. This method has been used extensively in the detection of gene expression and identification of genes. This is based on the properties of strand separations (denaturation), the ability to immobilize nucleic acids on solid matrices and the ability to **probe** for the gene of interest by the formation of a stable duplex between complementary nucleic acid sequences.

In filter hybridization, nucleic acids are bound to a special filter (membrane) in such a way that intramolecular reannealing is prevented, but yet allows the formation of a hybrid with a single-stranded probe added in solution. The probes are incubated with the filter-bound nucleic acids (targets) under conditions favouring the formation of duplexes. Unbound probes are then washed away and the probe-target hybrids are visualized. We shall illustrate the use of this technique in the examples below.

4.5.1.1 *Colony Blotting and Plaque Lift Assays*

Nucleic acid hybridization can be used to identify the DNA of interest contained in either a bacterial colony or a phage plaque. This method has been extensively used to screen cDNA and genomic libraries, constructed in either bacteria (colony blot) or bacteriophages (plaque lifts), for clones carrying the desired genes of interest.

The first step in trying to identify the clones of interest within a library is to transfer the colonies or plaques onto either nitrocellulose or nylon filter membranes (Fig. 4.20). The filter membranes are placed on the bacterial colonies or plaques and marked with Indian ink on the plates for orientation purposes. After a few minutes, colonies/plaques would have been blotted onto the membranes. The membranes are then lifted off the plates, denatured and treated so that the DNA released from the colonies or plaques are immobilized onto the membranes. The single-stranded DNA can then be tightly bound to the membrane by baking it at 80°C if a nitrocellulose membrane is used, or irradiating the membrane with UV if a nylon membrane is used instead.

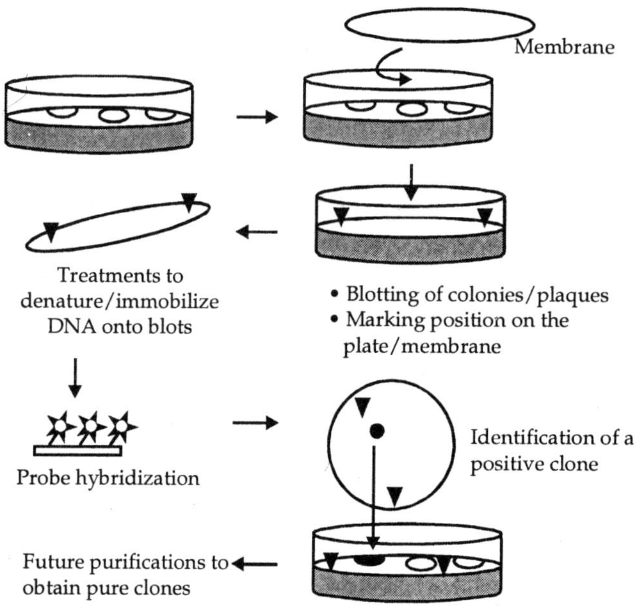

Fig. 4.20 : Identification of positive clones by colony transfer onto a membrane and hybridization probing.

Next, to locate the DNA of interest within a library of DNA, the membranes are then probed using a labeled piece of complementary DNA, in a solution of chemicals that will promote nucleic acid hybridization. Positive colonies or plaques containing related sequences to the probe used are then identified and subsequently rescued from the agar plates.

In this process, unrelated colonies or plaques in close proximity to the positive clone will inevitably be rescued simultaneously. Further serial dilutions and screenings of the colonies or plaques are therefore necessary to isolate pure clones. These clones are then amplified for the isolation and purification of the insert DNA. Finally the DNA fragments can then subcloned into other vectors of choice for further characterizations, for example restriction mapping and sequencing.

4.5.1.2 *In vitro labeling of Probes*

The probes used to identify target sequences can be in the form of double stranded DNA, single-stranded DNA (synthetic oligonucleotides) or as RNA. These probes can be labeled either internally or at the terminus end. Two competing methods can be used to generate these probes: radioactive and non-radioactive. The principles underlying both methods are the same except for the differences in the tags and detection methods used.

In general, long probes are usually labeled internally to achieve high specific activities. These probes generally provide the strongest hybridization signals. A major disadvantage of internally radiolabeled probes is that it suffers from radioautolysis. Non-radioactive labeling detection methods are becoming more acceptable and in some cases, can replace radioactive methods. However, this is not without limitations. More steps are generally required for non-radioactive methods and this may introduce more errors. In addition, more optimization of the experimental conditions is required. A major advantage of using non-radioactive methods is that it is not hazardous and the probes generated can be stored indefinitely.

Below are two examples of the commonly used technique for specific probe labeling. The principles are essentially the same for both radioactive and non-radioactive formats. All methods listed here are commercially available as kits.

(a) Nick translation—The term "translation" in this method must not be confused with the translation of mRNA into proteins. Translation here refers to the progressive extension of the newly made strand of DNA by the polymerase.

Single-stranded nicks (internal cleavage of one strand of the duplex DNA) are first generated at random in the double-stranded DNA using low levels of DNAase I. DNA polymerase I is then added. At the nicked sites, the 5′ to 3′ *exonuclease* activity of DNA polymerase I remove nucleotides from the 5′ end of the nick, while the 5′ to 3′ polymerase activity adds dNTP and labeled dNTP to the 3′ end (see Sec. 4.2). Hence, the nick is translated downstream, with the labeled nucleotides replacing the nucleotides located between the original and the new position of the nick. This results in a nick-translated probe that is labeled on both strands with interspersed sequences of the newly synthesized labels and the original (unlabeled) DNA. This method works best with probes of 500 base pair or more. The incorporated dNTP can either be radiolabeled nucleotides or chemically modified non-radioactive nucleotides (Fig. 4.21).

Exonuclease [5′→3′]/ *Filling in* [5′→3′]
DNA polymerase I

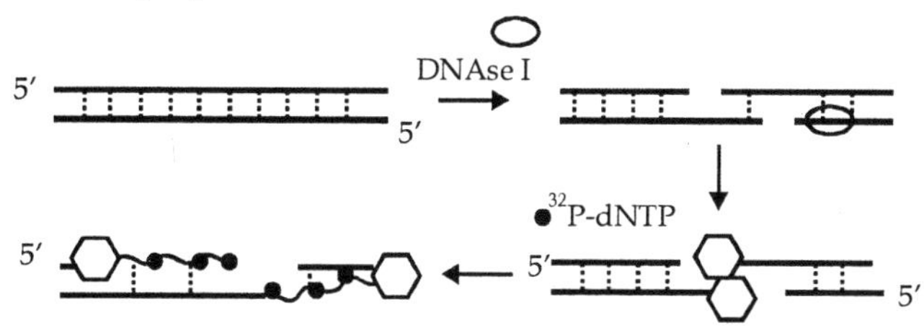

Fig. 4.21 : Labeling of probe DNA by nick translation.

(b) Random priming—This uses a large excess of random hexamers (or longer oligomers) as nonspecific primers. Both single-stranded or double-stranded DNA is first denatured. Random hexamers (NNNNNN, where N is any nucleotides) are then incubated with the template where they will hybridize to specific sequences by chance. Next, the Klenow fragment of T7 DNA polymerase will extend the primer-template. The template strand remains unlabeled, whereas the newly synthesized strand(s) are fully labeled by Klenow. As little as 1-5 ng of template DNA is sufficient for synthesis of high-specific probes. It appears that probes less than 500 bp are more efficiently labeled than longer probes. There are commercially available Klenow fragments that do not have the 3′-5′ proofreading exonuclease domain. Thus, these enzymes have higher processitivity (Fig. 4.22).

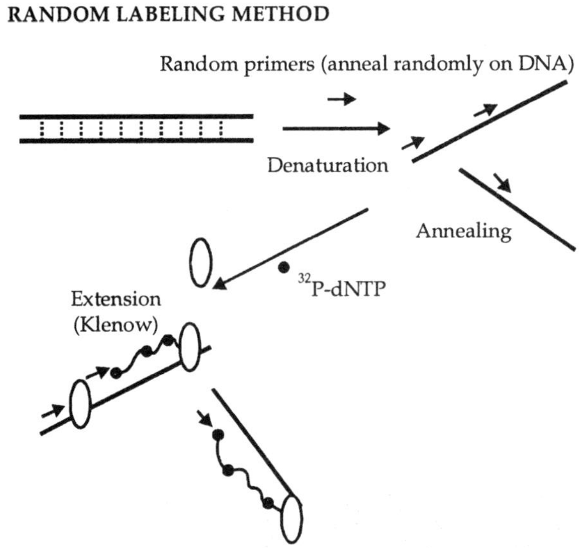

Fig. 4.22 : Labeling the probe DNA by random priming.

4.4.1.3 *Southern Blot—Its Use in Confirming the Identity of a Clone*

Once positive colonies have been identified by colony of plaque hybridization, and the DNA isolated and purified (Sec. 4.4.8), it is often a good practice to confirm the identity of a given piece of DNA by restriction map (see Sec. 4.2.3) followed by Southern blot analyses. Southern blot is a method developed for the transfer of single-stranded and denatured double-stranded DNA from an agarose gel to a membrane. The blotted DNA is immobilized on the membrane but is still able to hybridize to labeled probes.

The method involves is as shown in Fig. 4.23.

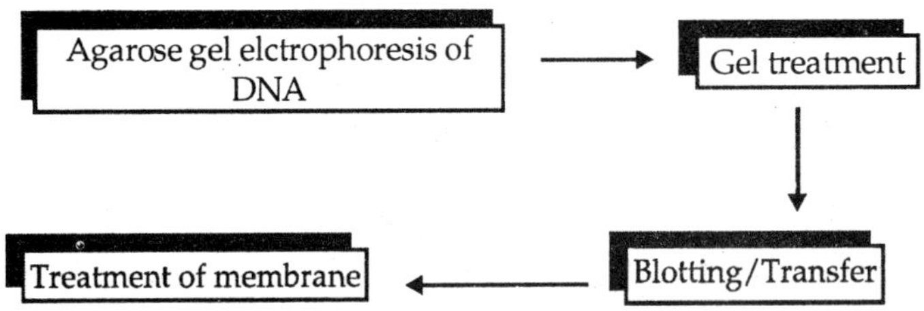

Fig. 4.23 : Steps in Southern blotting.

The aim of treating the gels before the blotting step is to aid the transfer of large fragments of DNA onto the membrane. The rate of DNA transfer from the gel to the membrane depends on the size of the DNA. The larger the DNA, the less efficient is the transfer to the filter. Using this procedure, the DNA in the gel is depurinated, denatured and neutralized using it series of solutions.

Capillary blotting is the most commonly used transfer method in many laboratories, as it does not require expensive instrumentation. There are two ways to carry this out. Traditionally, Southern blotting is carried out in the "upward transfer" fashion (Fig. 4.24). Quite often, the use of too heavy a weight at the top of the assembly will result in the gel getting squashed. This can result in the smearing of the DNA during transfer. An alternative to this mode of transfer is the "downward transfer". Basically, it shares the same principle, except that the arrangements of the assembly is opposite to that of the "upward transfer". The advantage is that it does not deform the gel during transfer. It is recommended as a rapid method for transfer, using minimal amounts of transfer solution. However, the transfer for a defined short period of time may not result in complete transfer.

After the restricted DNA fragments have been immobilized onto the membrane, the position and identity of the desired piece of DNA can be detected and visualized by nucleic acid hybridization, using its complementary labeled probes.

A. Upward transfer of DNA from gel to membrane.

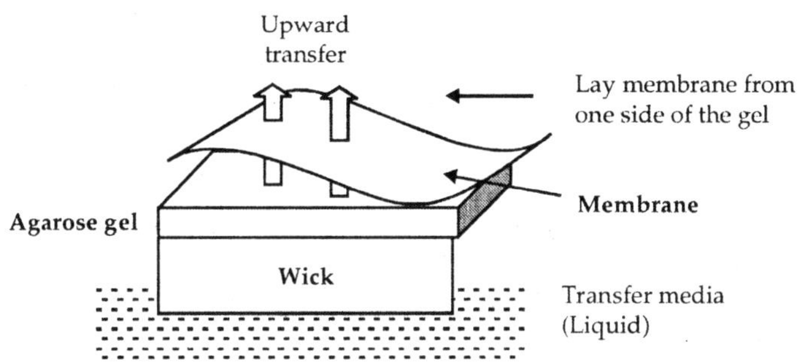

B. Assembly of an "Upward" transfer southern blot

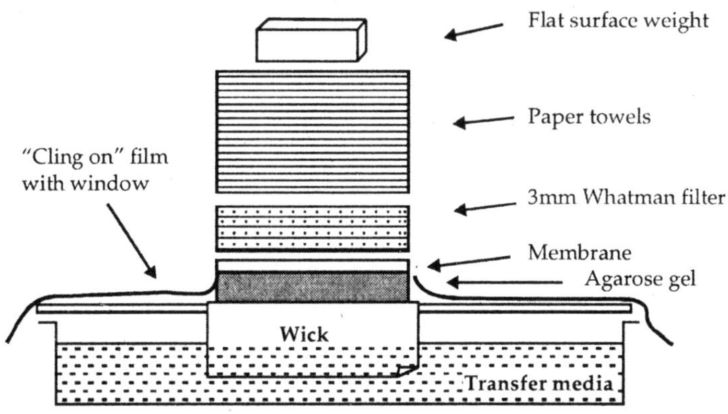

Fig. 4.24 : Southern blotting using the "upward transfer" method.

4.5.2 Polymerase Chain Reaction

As mentioned earlier, a successful cloning experiment relies mainly on the ability to identify and select the particular clone of interest. In this respect, the polymerase chain reaction (PCR) has proven to be an invaluable technique for the amplification of a specific copy of a gene of interest, and thus enhancing its successful cloning and subsequent selection and identification. Using a pair of gene-specific primers (GSPs), it is possible to selectively amplify millions of copies of a gene of interest from the microbial genome. In addition, by the incorporation of restriction endonuclease sites into these GSPs, PCR cloning will also enhance the efficiency of ligation between vector and insert DNA, or to allow in frame insertion into an expression vector for the production of recombinant proteins.

4.5.2.1 Basic Principles and Optimizing PCR

In PCR, short oligonucleotides act as specific primers hybridizing to a target DNA molecule. Under optimal conditions, two oligonucleotide primers complementary to the opposite strands of the template, can then be used to initiate DNA synthesis *in vitro* using thermostable DNA polymerases (Fig. 4.25). The hybridization conditions are carefully controlled to limit *in vitro* DNA synthesis due primarily to template sequences located between the two oligonucleotide primers. After annealing, the DNA polymerase will then extend the primers, generating an additional new copy of the template sequences located between the two primers. On completion of the primer extension, the sample is heated to denature the DNA duplex, regenerating approximately twice as much single-stranded template for primer annealing in the next cycle. A series of cycles is then performed, where the DNA template is denatured, oligonucleotide primers annealed specifically, and the primers extended by the thermostable DNA polymerase in each cycle. The repeated cycling results in a geometric accumulation of the products flanked by the two primers.

In slightly more detail, the initial extension products resulting from the original template do not have a distinct length because the polymerase will continue to synthesize new DNA until it either stops or is interrupted by the start of the next cycle. In the second cycle, the extension of products also result in variable lengths. However, by the third cycle, the segment between the primers is synthesized with a defined length. From the fourth cycle onwards the target sequence is amplified exponentially. Potentially, after 20 cycles of PCR, the products should be amplified 2^{20}-fold, assuming 100% efficiency during each cycle. However, in practice, this is seldom achieved, largely because the efficiency is highly dependent on the template and primers used, as well as the cycling conditions and reactants involved.

In later cycles, the accumulation of amplified products does not expand exponentially indefinitely but levels off ("plateau effect"). There are many reasons to account for this. Some of these factors include the gradual loss of polymerase activity and hence becoming a rate-limiting step. Another factor is the reannealing of target strands as their concentration increases. The reannealing of target strands then competes with primer annealing. Practically, the process levels out after microgram quantities of products are obtained and this is sufficient for many downstream procedures, e.g. subcloning into vectors, sequencing etc.

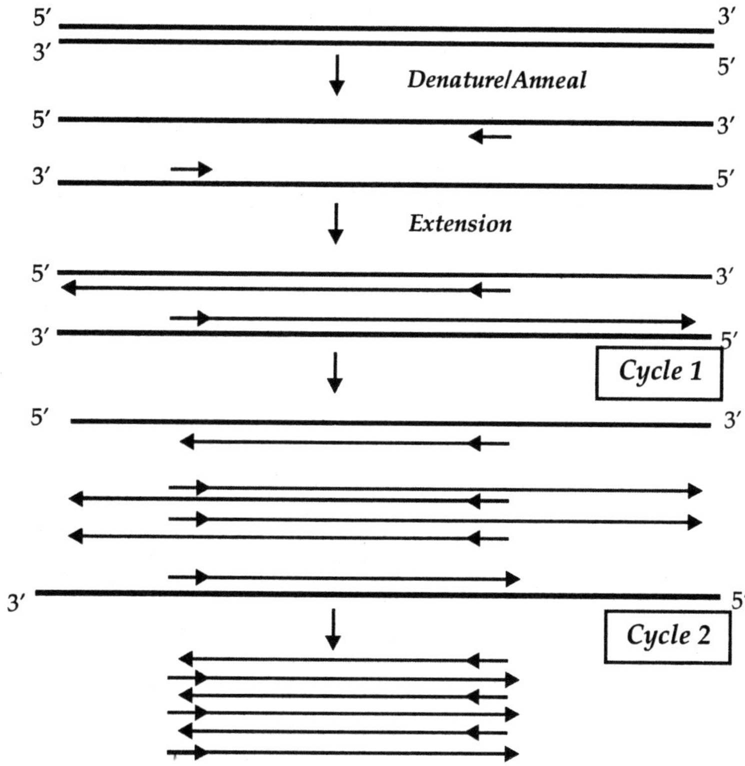

Fig. 4.25 : Steps in the polymerase chain reaction.

4.5.2.2 Cloning of PCR Products

PCR has allowed the isolation and amplification of any region within a complex DNA in a few hours. Cloning the PCR products into appropriate vectors allows the generation of the amplified products without having to repeat the reaction every time the products are needed. Therefore it is more convenient to amplify the gene of interest from its recombinant clone than from its original source, which could range from RNA to a DNA genome. In addition, if the PCR products are to be used in further studies, e.g. expression or the generation of single-stranded RNA probes, then it is necessary to clone the amplicons. In this section, some strategies to clone PCR products will be discussed. It is important to note that a well-planned cloning strategy is needed before starting the actual PCR reaction, as many of these strategies involve modification of the primers that are used in the PCR reaction.

(a) Blunt-end and T-A cloning—The type of DNA polymerase used in PCR will determine the type of vectors that can be used to clone the PCR products. For example, if Taq, Amplitaq or Tth DNA polymerase had been used in the reaction, the resulting amplified products will have a single A residue overhang at the 3′ end. This overhang nucleotide will facilitate ligation if the vectors used to clone these PCR fragments contain a complementary T nucleotide at the 3′ ends of the vector. This is called T-A cloning and is illustrated in Fig. 4.26. An example of this type of cloning vector is pGem-T.

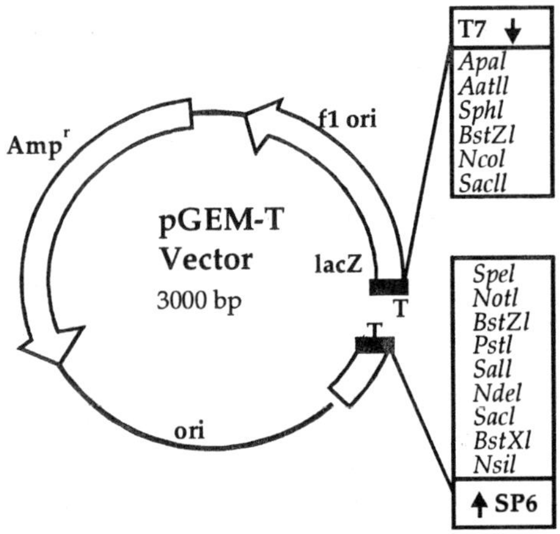

Fig. 4.26 : A plasmid vector for TA cloning of PCR products.

The orientation of the DNA insert cannot be controlled, as the overhang nucleotides on the inserts and vectors are bilateral. Precautions have to be taken to avoid non-recombinants due to religation of vector molecules, which have lost the T-overhang nucleotides due to a contaminating exonuclease activity. This problem is easily overcome with cloning vectors that has the multiple cloning sites within the *lacZ* gene, as in pGem-T. The presence of the DNA insert will result in the formation of a white colony, in contrast to a non-recombinant blue colony.

The use of other DNA polymerases, for example *Pfu* or *Ultma*, will however result in the formation of blunt-ended amplified products. As in the case of blunt-end ligations (Sec. 4.2.4), the ligation of blunt-ended PCR product to a vector is not as efficient as T-A cloning. However, this ligation reaction can be improved by using high amounts of starting inserts and vector. The choice of vectors that can be used for blunt-end ligation is broad, as long that the vector contains restriction sites that will result in blunt-ends, e.g. *Sma*I, making it terminally compatible with the amplified products. Alternatively, there are commercial vectors available for blunt-end cloning, for example, pCRScript and pCRScript™ Direct (Stratagene).

(b) Introduction of restriction sites for cloning—To aid in efficient and unidirectional PCR cloning, restriction sites can be added to the 5′ ends of one or both nucleotide primers that are to be used for PCR (Fig. 4.27). The sequences of these modified primers will be incorporated into the amplified PCR products. These amplified DNA can then be digested with the appropriate restriction endonucleases to generate either blunt or cohesive ends, as required by the scientists. By the use of different restriction sites at either ends of the DNA, it is thus possible to clone a DNA insert unidirectionally into a vector.

The addition of a few extra sequences to the 5′ end will not significantly affect the primer specificity, since this is determined mainly by the sequences at its 3′ end. There are, however, a few points that one needs to keep in mind when using this method:

1. The restriction site(s) used in the primer sequences should not have recognition sites within the DNA of interest.
2. The restriction sites that are added to the final product via the primers often need extra bases at the 5′ end to allow for efficient DNA digestion.

3. It may be essential that sequential digestions be performed on the amplified products if the enzymes chosen do not have optimal activity in a common buffer.

The techniques described above are widely used in recombinant DNA technology and provide one of the most reliable and simplest methods for PCR cloning.

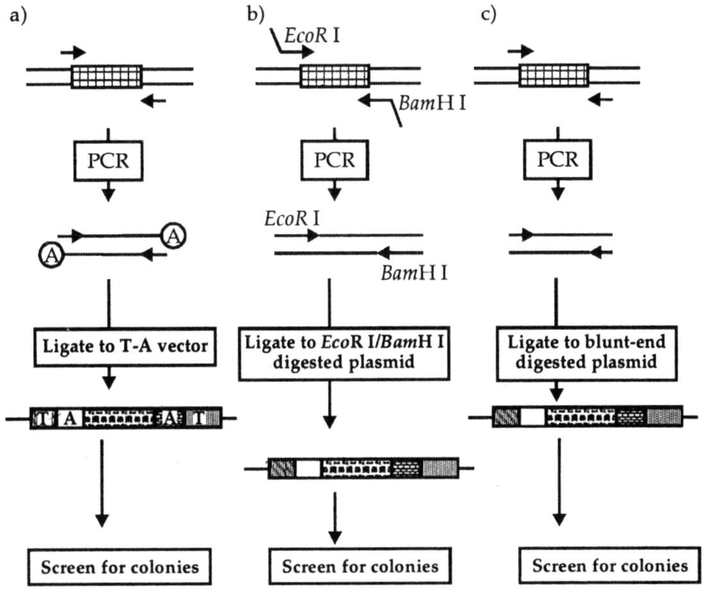

Fig. 4.27 : Restriction sites can be added to the 5′ ends of one or both nucleotide primers that are to be used for PCR.

4.5.3 DNA Sequencing

Probably the most important technique available in recombinant technology is DNA sequencing, which enables the scientist to decode the precise order of the nucleotides within a piece of DNA. The two of the most commonly used sequencing techniques are:

- Dideoxy chain termination method (Sanger), and
- Chemical degradation method (Maxam-Gilbert).

The former is the method of choice for the high throughput sequencing. Although the two methods differ in their strategies (Sanger method is enzymatic and Maxam-Gilbert method is purely chemical-based), the fundamental principle is the same. The principle is that two single-stranded DNA molecules differing in length by just one base can be resolved into distinct bands by some form of separations. The usual way of separation is the use of denaturing polyacrylamide gel electrophoresis.

4.5.3.1 Dideoxy Chain Termination Sequencing (Sanger Sequencing)

This method is conceptually rather simple. It uses polymerases to synthesize a new template by extending the primer and occasionally incorporates dideoxynucleotides (ddNTPs) instead of the usual deoxynucleotides (dNTP) during chain elongation at the 3' end (Fig. 4.28). When the dideoxynucleotide is added to the 3′ end of an elongating chain, no further elongation is possible as the 3′ end nucleotide lacks a 3′OH that is essential for forming the phosphodiester bond between the primer and additional dNTPs for elongation. Thus, the chain is terminated.

Structures of adenosine analogues

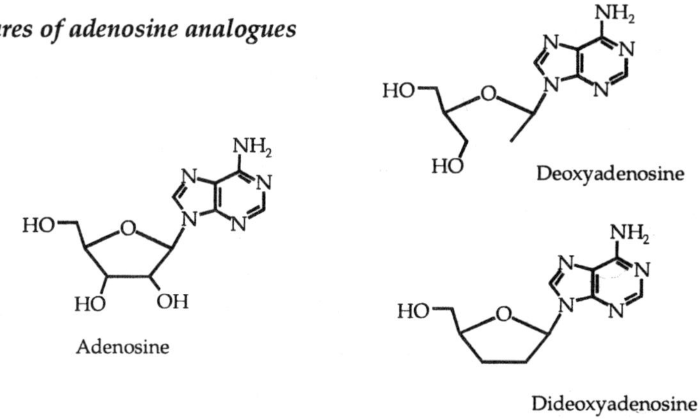

Fig. 4.28 : Dideoxynucleotides are used as chain terminators during DNA synthesis.

In manual chain-termination sequencing, each of the 4 dNTPs (dATP, dCTP, dGTP and dTTP) is added separately to one of the 4 primer extension reactions. In addition, a radiolabeled ddNTP is included in the reaction mix. The reaction with ddGTP is called the "G" reaction; the one with the ddATP is called "A" reaction, etc. Chain termination will occur in a base-specific fashion depending on which ddNTP is present during the extension reaction. For example, termination will occur only at the "G" position (complementary base of C) when ddGTP is present, or exclusively at the A when ddATP is present (Fig. 4.29). The concentration of ddNTP is kept low relative to the corresponding dNTP. Hence, base-specific chain termination will only occur in a small percentage of the time at a given position in the sequence. As the reaction proceeds, a mixture of all possible termination products for G, A, T, or C will be generated in the 4 primer-extension reaction tubes.

Termination with ddATP

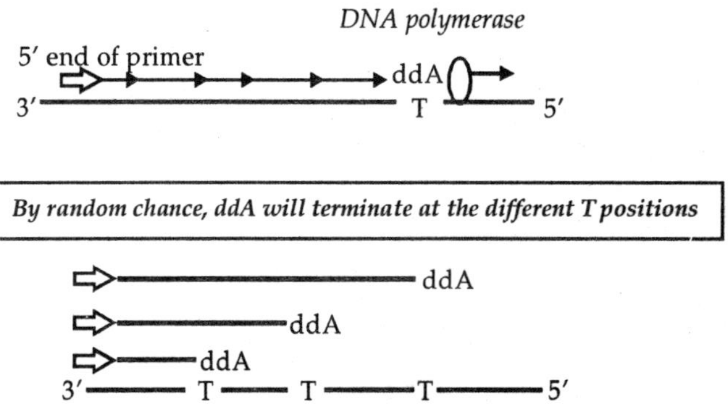

Fig. 4.29 : Incorporation of a dideoxynucleotide results in chain termination.

The radiolabeled products from the 4 base-specific termination reactions are then resolved on a high-resolution polyacrylamide gel. Up to 300-500 bases in length can be separated. Each terminated species will be visualized as a separate band on the gel with the shortest primer-extended products running fastest down the gel. The base sequence of the primer-extended products can then be read from the 5′ to 3′ ends by identifying which reaction mix (G, A, T, C) contains the next longest species, beginning with the shortest fastest migrating species as seen on the autoradiogram of the gel (Figs. 4.30 and 4.31).

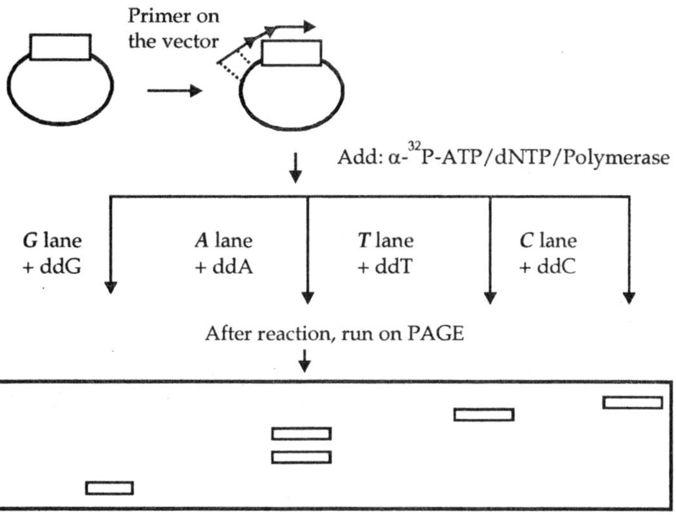

The sequence is read from the bottom up: **5′-GAATC**

Fig. 4.30 : Separation of DNA fragments during DNA sequencing.

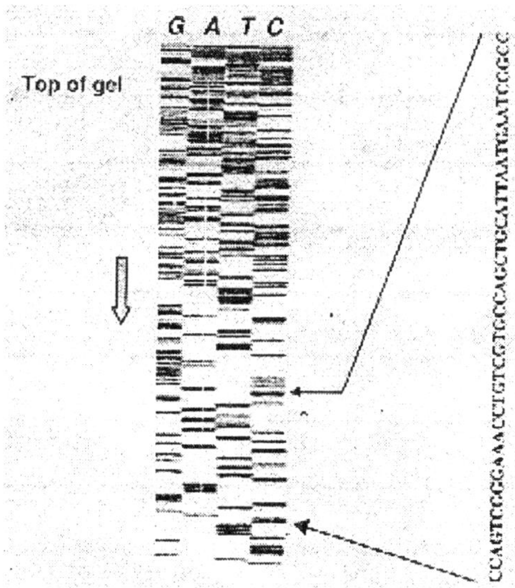

Fig. 4.31 : An example of a DNA sequencing gel after Sanger dideoxy chain termination sequencing. Each lane represents a chain termination reaction with a specific dideoxynucleotide.

4.5.3.2 DNA Polymerases

The properties of these polymerases are well-known. It suffices to say that the majority of those used in manual and automated sequencing are genetically modified for longer extension (high processivity) and diminished exonuclease activities. For manual sequencing, some examples of the commonly used polymerases include Sequenase, *Bst* and Klenow *exo*⁻.

Sequenase is a genetically modified version of the bacteriophage T7 DNA polymerase. The unmodified polymerase (wild type) is highly processive, with an average of 2000-3000 nucleotides polymer formed (before dissociating from the template) with an elongation rate (extension) of about 300 nucleotide/second.

Before the availability of Sequenase, Klenow polymerase was the enzyme of choice for manual sequencing (Sec. 4.2). Klenow polymerase is a modified enzyme derived from *E. coli* DNA polymerase I. The unmodified version is unsuitable for sequencing because of its 5′ to 3′ exonuclease. Klenow polymerase was originally prepared by proteolytic cleavage of the parent enzyme, but now as genetically modified enzyme.

4.5.3.3 *Alternative Labeling/Detection Strategies*

There are other strategies available besides incorporating radiolabeled nucleotides:

(a) End labeling of primer—Here, the primer is labeled instead of incorporating labeled nucleotides into the strands. T4 polynucleotide kinase is used, which catalyze the transfer of a ^{32}p or ^{35}S (in the γ-position) to the dephosphorylated 5′ end (not the 3′ end as it is where extension occurs). This method of labeling result in a chain-terminated molecule carrying just a single-labeled atom rather than the multiple labeled molecule as described above.

(b) Labeling with tags for chemiluminescence—This is an alternative to radiolabeled studies. Chemiluminescence can be as sensitive as radiolabeled approach, however, this involve more manipulations and can result in low sensitivity if mishandled. Here the 5′ end of the primer can be modified to contain a tag (e.g. biotin). Extension and termination is carried out as usual. At the end of the gel run, Southern transfer is carried out and the membrane is then processed for chemiluminescence.

(c) Fluorescent labels—This is extensively used in automated sequencing. Special imaging is required including a laser for monochromatic excitation and a detector with different wavelength discrimination. Fluorescent-labeled fragments with 4 different colours can be analysed in the same lane. As the different sizes pass the detector, one of the four colours will be recorded and deciphered using a software program.

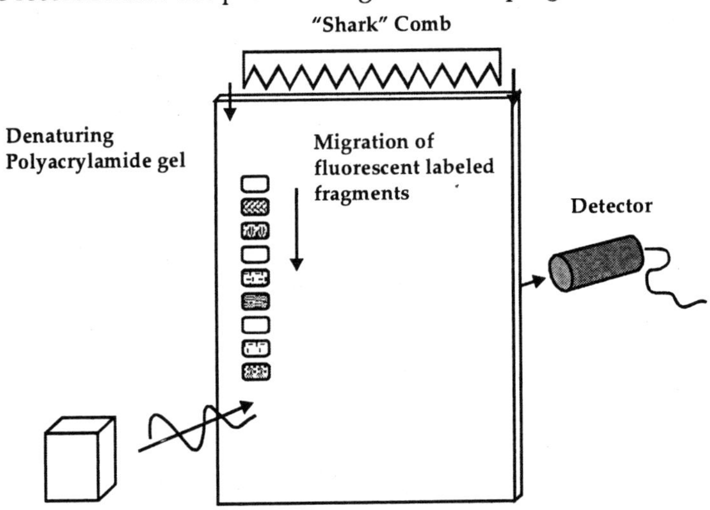

Fig. 4.32 : An automated DNA sequencer which uses 4 labels, single-lane separation.

Two main types of automated sequencers are commercially available. The first type uses single-label, 4 -lane separations, similar to the above manual sequencing, whilst the second type uses 4 labels, single-lane separation (Fig. 4.32). The latter type is often the instrument of choice for large-scale genome projects.

Sequencing is carried out with dye-terminators (similar to the use of radiolabeled ddNTPs) and the samples electrophoresed on a denaturing polyacrylamide gel. Fluorescent signals are detected after excitation by a laser positioned directly in front of the detector. Automated DNA sequencing has now become a routine in many laboratories because of its ease of use and the greater amount of sequencing data that can be obtained.

4.5.3.4 *Sequencing Strategies*

In sequencing a large piece of DNA (10 kb), the amount of sequences that can be read from a single template is often in the 500-700 bases region. Thus, various methods have been devised to obtain the rest of the sequences. These methods can be broadly divided into two types: **random and ordered sequencing.**

Random sequencing is a simple method to obtain large portion of the template quickly. The template is first fragmented physically by shearing the DNA into smaller pieces, subcloned into vectors and the recombinants sequenced all at one go. This is an example of a "shot-gun" method. Another is the use of transposon-mediated sequencing method where a transposon randomly integrates into and excise parts of the DNA. Sequencing of a collection of these recombinant transposons will also yield large amounts of data at a go. An advantage is that some difficult regions (G-C rich) may be broken up during the fragmentation step, thus, facilitating the ease of sequencing. However, both methods suffer from one major drawback. They both can obtain most of the sequences quickly but the last 10% of the sequence will pose some difficulty with the presence of similar clones. Usually, to obtain the last bit of the template sequence, ordered approaches are used.

Ordered sequencing comes in many forms. The simplest but most tedious and expensive method is to "walk" the whole template. Here, the sequencing data is used to synthesize new primers that can then be used to sequence further down. Moving the template closer to the primer progressively is another approach. One of these methods is to generate "collapse clones" where internal restriction sites and those on the multiple cloning site of the vector can be used to delete off sequences. The unidirectional deletion (nested deletion) and religation of template sequences is yet another approach which is easy to carry out. This uses enzymes, e.g. *Exo* III and *Bal*31, which can delete the template in a specific direction.

Obviously, directional sequencing is more convenient and once a fragment can be localized, the rest of the fragments produced can be zoomed in and overlapping clones used for the rest of the project. Furthermore, the deleted (shorter clones) can be used for expression or mutagenesis studies.

4.6 MUTAGENESIS

Recombinant DNA and gene-cloning technologies allow scientists to isolate and characterize any given gene from any living organism. In addition, these genes may also be modified in vitro in order to produce new restriction enzyme cleavage sites, to change a particular codon, or to introduce site-specific mutations by insertion and deletion of nucleotides. This procedure has allowed scientists to dissect genes at the nucleotide level by changing one nucleotide at a time and examining the effect of each change on the function of the gene or its product. *In vivo* mutagenesis protocols, on the other hand, either makes use of bacterial transposons to either mediate mutation via random insertion of the transposon into the host genome, or the replacement of the normal gene with an altered gene sequence via homologous recombination.

4.6.1 Random Insertional Mutagenesis

A gene that has been marked with a transposon is relatively easy to map, either by genetic crosses or by physical mapping using restriction endonucleases. A typical strategy is to construct a plasmid carrying the transposon, using temperature-sensitive vector, i.e. a plasmid that is unable to replicate at an elevated temperature. When this plasmid is introduced into bacterial cells, and the cells are grown at a temperature permissive for

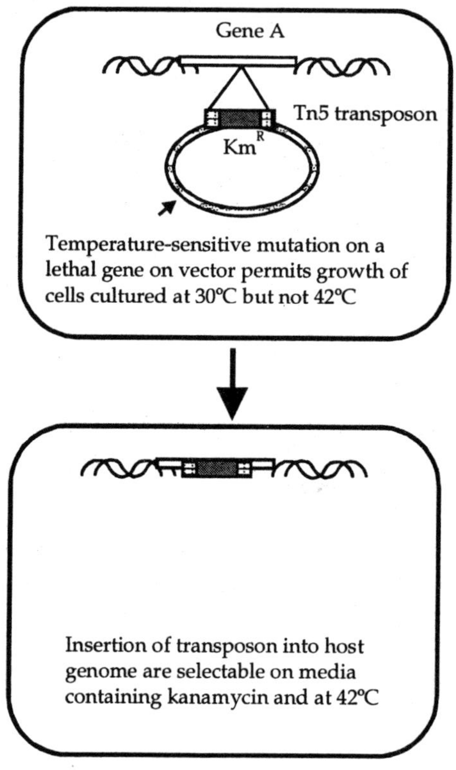

Fig. 4.33 : Gene disruption using a transposon carrying a kanamycin resistance marker.

plasmid replication (say 30°C), all the cells will carry the plasmid. When the culture is then shifted to the non-permissive temperature (say 42°C) at which the plasmid is unable to replicate, the only cells which retain the transposon (usually selectable by an antibiotic resistance marker) are those that have incorporated the transposon in its chromosome. Thus, using a transposon, such as Tn5, which carries the kanamycin resistance gene, will allow selection of transposon insertions by plating the culture on agar containing, kanamycin at 42°C (Fig. 4.33).

This generates a library of transposon mutants, with the transposon inserted at a variety of chromosomal positions, and therefore causing a range of mutations. After the desired mutants have been selected using an appropriate phenotype screen, e.g. the loss of bacterial motility, the genes into which the transposon has inserted may be identified. This is achieved by cutting the total genomic DNA of the mutant with a restriction enzyme, which

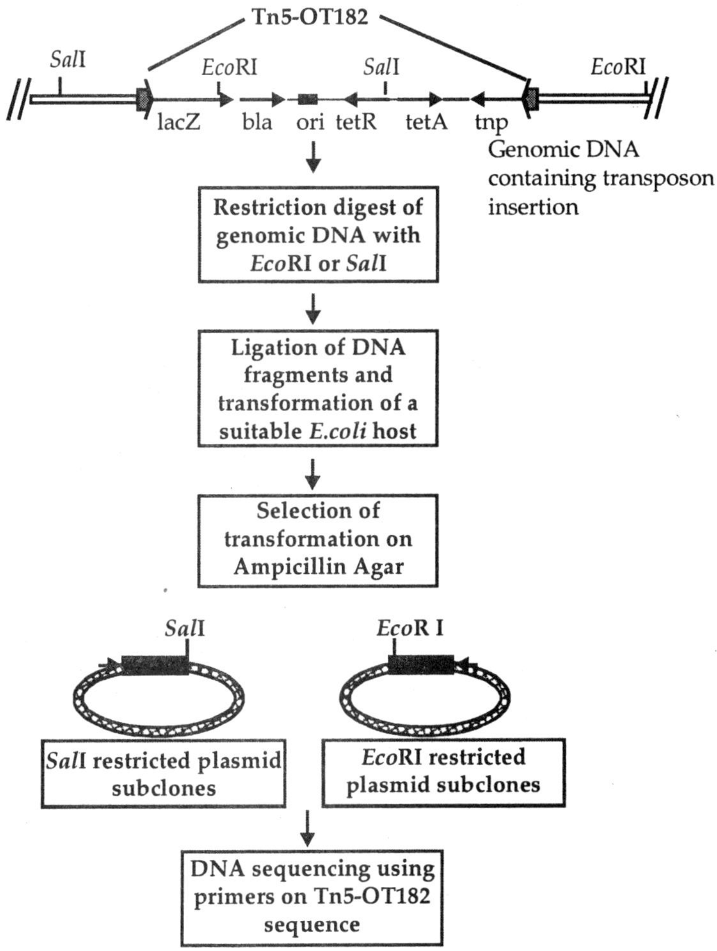

Fig. 4.34 : Strategy for the identification of DNA sequences flanking the Tn5-OT182 transposon insertion site.

does not cut the transposon itself, and then cloning the collection of fragments to produce a gene library. If the library is then plated on kanamycin agar, only those clones containing Tn5 will be able to grow. These clones will contain not only Tn5 but also some of the host sequence flanking the transposon. The cloning of DNA sequences flanking the transposon insertion is made easier when a transposon with self-cloning capabilities, e.g. Tn5-pOTI82, is used. In this example, the genomic DNA is digested with a restriction enzyme that cuts once within the transposon (e.g. *EcoRI* or *Sal*I), leaving its plasmid origin of replication and an antibiotic selection marker (ampicillin resistance) intact. The other *EcoRI* or *Sal*I restriction site would be in the host genomic DNA sequence flanking the transposon insertion. Self-ligation of the digested DNA fragments results in recircularization of the digested genomic DNA fragments. Upon transformation of *E. coli* host cells and selection on ampicillin-containing agar, only the recircularlized DNA containing origin on the transposon and the antibiotic resistance marker would result in ampicillin-resistant transformants. The disrupted host gene is then identified by isolating the plasmid DNA from such clones and sequencing the DNA flanking either ends of the known transposon sequence (Fig. 4.34).

4.6.2 Gene Replacement Mutagenesis

Having identified a gene of potential interest through DNA sequence analysis, one of the best ways of determining its function and significance is to deactivate it in the host organism. This may not be easy to achieve by the classical procedures of mutagenesis and selection, especially if the phenotype associated with that gene is not known, or not readily selectable. However, if the cloned gene is inactivated *in vitro*, it can be used to replace the wild-type chromosomal gene by homologous recombination, a process also known as gene knock-out (Fig. 4.35). In this case, the central part of the cloned gene has been removed and replaced by an antibiotic resistance gene, such as kanamycin resistance. The inactivated gene is constructed on a suicide vector, which cannot replicate in the host cell. In this example, the mobilizable suicide vector, pJQ200mp18, contained the *sac*B gene. Cells carrying the intact *sac*B gene, i.e. those that had not undergone a gene replacement event, will be lethal when plated on media containing 5% sucrose. The mutated DNA is introduced into recipient cells by conjugation. After conjugation, selection on kanamycin and streptomycin agar containing 5% sucrose will identify cells in which the kanamycin resistance gene has been inserted into the target gene by homologous recombination, involving a double, crossover. The addition of streptomycin selects against donor cells, which are streptomycin-sensitive, and the kanamycin selects against non-recombinant recipient cells. The target gene is thus deactivated. The phenotype of the knock-out mutant can then be studied using appropriate assays. A confirmation that the changed phenotype is actually due to inactivation of the gene done by introduction of the wild-type gene on a plasmid (complementation). Complementation is an important step, as the phenotypic changes could be due to polar effects on other genes, rather than directly caused by inactivation of the target gene. This not only provides a useful way of assessing the function of a cloned gene, but can also be used for producing attenuated strains of a pathogenic bacterium by inactivating a gene coding for toxin production or another virulence factor.

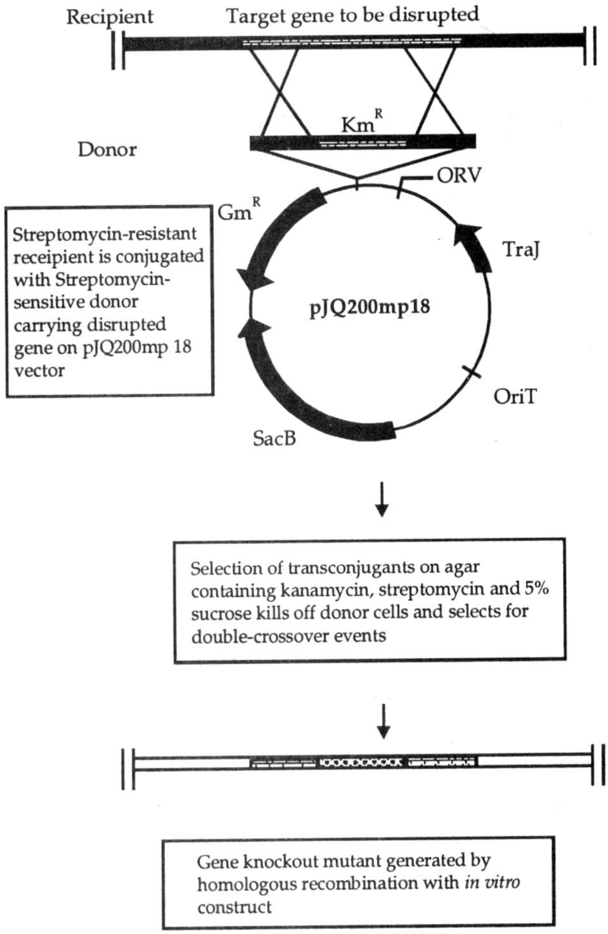

Fig. 4.35 : Targeted mutagenesis using a suicide vector, pJQ200mp18, for gene replacement by homologous recombination.

4.6.3 Site-directed Mutagenesis

It is sometimes desirable to replace a gene with one with specific alterations. This requires a combination of the techniques of gene replacement and site-directed mutagenesis. The knocked-out gene can be complemented with a plasmid carrying a specifically mutated gene copy. This enables us to test, for example, the significance of a specific amino acid in the activity of an enzyme, by replacing it with a series of different residues and determining the activity of the product. Or we could look at the precise requirements for the function of a regulatory sequence on the DNA such as a promoter site, by introducing defined mutations into that region.

A number of protocols have been established for site-directed mutagenesis using hybridization of a mismatched oligonucleotide to a DNA template, followed by second-strand synthesis by a DNA polymerase. One of these is a site-directed mutagenesis procedure developed by Deng and Nickoloff that eliminated the need for subcloning and generating single-stranded DNA templates by employing double-stranded plasmid DNA (Fig. 4.36).

The procedure involves simultaneously annealing two oligonucleotide primers to the same strand of heat-denatured double-stranded plasmid DNA. One of the primers (the selection primer) anneals to a region that contains a unique, nonessential restriction site. The other primer (the mutagenic primer) introduces a chosen mutation into the plasmid. Extension of these primers by T4 DNA polymerase followed by ligation of the resulting molecules with T4 DNA ligase results in a population of plasmid molecules, some of which contain the desired mutation but no longer contain the unique restriction site. The plasmid is treated with a restriction enzyme that will linearize the parental plasmid but not the heteroduplex. Since transformation of circular DNA is many folds more efficient than that of linearized DNA, the circular heteroduplex would be preferentially transformed. Using a mismatch repair-defective strain of *E. coli* (e.g. XL*mut*S) that cannot distinguish the original unmutated strand of DNA from the newly created strand containing the desired mutation, both strands of the heteroduplex would be equally present in the newly replicated DNA. The *mut*S-deficient strain randomly selects one of these strands as "correct" and changes the other strand to be complementary to the chosen strand. DNA is isolated from the plasmids and subjected to a second round of restriction digest, which will enrich the population of mutant plasmids after transformation into an *E. coli* strain (e.g. XL1-Blue).

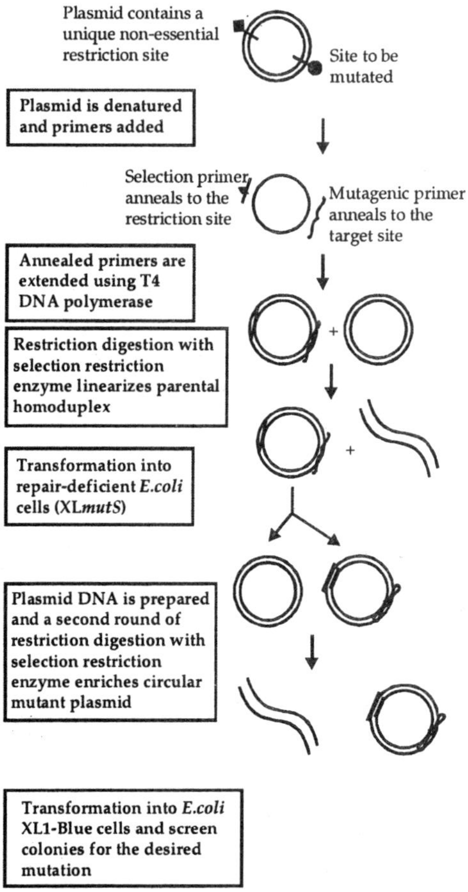

Fig. 4.36 : Protocol for double-stranded, site-directed mutagenesis.

4.6.4 PCR-based Mutagenesis

Various mutagenesis procedures based on polymerase chain reaction are available. A versatile and relatively easy method is the "megaprimer" procedure (Fig. 4.37). This method uses three oligonucleotide primers and two rounds of PCR performed on a DNA template containing the cloned gene to be mutated. Primers A and B represent the "flanking" primers that can map either within or outside the cloned gene. M represents the internal "mutant" primer containing the desired mutation, which may be a point mutation, a deletion or an insertion. In the first round of PCR, primer pair A and M1 is used. The PCR product, A-M1, is purified and used as one of the primers for the second round PCR together with the flanking primer B, hence the term "megaprimer". Prior separation of the two strands in the PCR product A-Ml is unnecessary since this will be achieved during the denaturation step in the second round of PCR. The wild-type cloned gene is used as the template for both PCR reactions. The final PCR product (A-M1-B) containing the mutation can be used in a variety of standard applications, such as sequencing and cloning in expression vectors. The inclusion of restriction sites to primers A and B would facilitate the cloning of the final mutant PCR product. This is especially important when sequences upstream of A or downstream of B are essential for gene expression or regulation, and therefore, must remain unaltered.

This protocol may also be applied to the mutagenesis of protein-coding genes and the study of structure-function relationship. In this case, primers A and B can be kept constant and a variety of mutant primers (M1, M2, M3 ...) can be used to produce the various mutants.

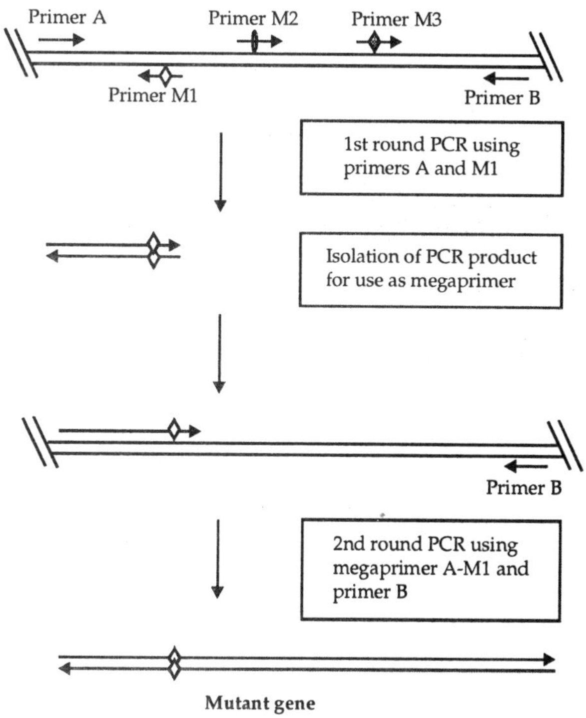

Fig. 4.37 : Site-directed mutagenesis using megaprimer PCR.

4.6.5 Ordered Deletions using Exonuclease III

Ordered deletions can be used in delineating sequences that are important for the function of a gene, such as those required for transcription. Deletions are generated by limiting, unidirectional digestion of DNA with *E. coli* exonuclease III (*Exo* III). *Exo* III preferentially digests one strand of a double-stranded DNA by removing nucleotides from 3′ ends if the end is blunt or has a 5′ protrusion, but not if the end has a 3′ protrusion of 4 or more bases (Fig. 4.38). The plasmid clone is first digested with restriction endonuclease A, which generates a 5′ protrusion or blunt end next to the target sequence, and restriction endonuclease B that generates a 3′ protrusion next to the sequencing primer site. The linearized plasmid DNA is then digested with *Exo* III, and aliquots of the digest are taken at a number of time points so as to yield a set of deletions of the desired lengths. These DNA fragments are then subjected to S1 nuclease digestion to remove the single-stranded DNA overhangs. The deleted DNA fragments may then be ligated to a vector with a promoterless reporter gene, such as the green fluorescent protein (GFP) or *lac*Z, to analyse the promotor activity. The cloned fragments can also be sequenced to determine the extent of the deletion.

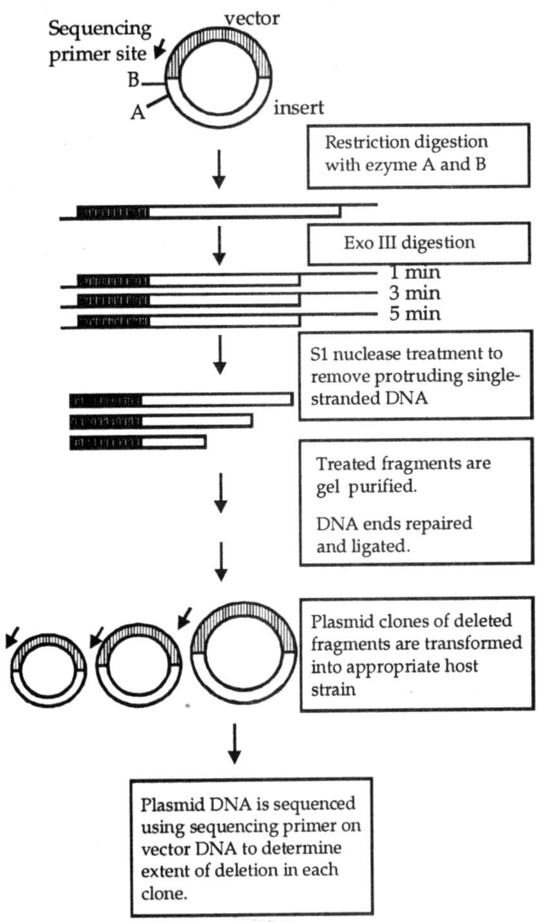

Fig. 4.38 : Generation of deletion mutants using *Exo* III.

4.7 RECOMBINANT DNA APPROACHES TO STRAIN IMPROVEMENT FOR LOW-AND MEDIUM-VALUE PRODUCTS

The advent of recombinant DNA techniques, first devised for *E. coli* in the 1970s, has meant that new strategies can be applied to strain improvement for low- and medium-value microbial products. It has taken some time for recombinant techniques to be developed and applied to the commercial strains, such as the filamentous bacteria and filamentous fungi which are the mainstay of this sector of the fermentation industry.

In addition, regulatory hurdles have to be crossed to gain approval to undertake fermentations at the production scale with these 'genetically-engineered microorganisms'. Gaining such approval is time-consuming and costly. Thus, there have to be good long-term economic reasons for adopting a recombinant DNA strategy for strain improvement.

In the case of antibiotic fermentations, temporal expression of antibiotic biosynthesis is regulated tightly as part of cellular differentiation pathway. Production of antibiotics is costly to the cell in terms of carbon and energy. Therefore, it is hardly surprising that tight control of expression has evolved. The genes for antibiotic biosynthesis are invariably clustered together, irrespective of whether the microbe is a prokaryote or eukaryote. In *Streptomyces*, regulation of expression of such gene clusters is controlled by a positive activator - a master gene which, when it is switched on, makes a protein that targets itself to the various promoters of the gene cluster and switches them on in concert (Figure 4.39).

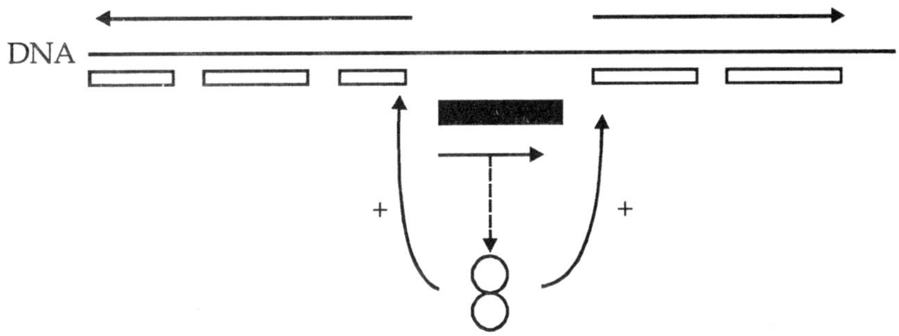

Fig. 4.39 : Schematic diagram to illustrate how antibiotic gene clusters are controlled. The DNA encodes a number of genes clustered together on the chromosome. The genes (rectangles) are usually transcribed as polycistronic mRNAs (shown by the unbroken straight arrows). Transcription of the production genes (unshaded rectangles) must be dependent on transcription of the positive activator (master gene, shown by the shaded rectangle and the protein by the two circles) that migrates and binds to the DNA (curved arrows) to allow transcription of the production genes to take place. Without the activator protein, there is no expression of production genes.

In this way, the cell ensures that the full complement of enzymes necessary to make the antibiotic are produced at the same time, at the correct levels.

By cloning this master gene regulator and then expressing it at unnaturally high levels, the cellular complement of the entire biosynthetic machinery for production of an antibiotic may be enhanced. Of course, there has to be sufficient fuel (metabolic precursors and energy) to realize the full potential available from this boosted level of biosynthetic machinery. The roles of metabolic flux analysis and process optimisation in assuring this advantage are very important.

It is also possible to force expression of the master gene during the growth phase and so to produce antibiotics during growth. In some instances, this may be to the advantage of the overall process, but often it is better to mimic the situation in Nature: firstly to focus the design of the process on maximizing the acquisition of biomass, and then to turn that biomass to best advantage by using the cells as a factory (with the maximal level of installed biosynthetic capability) to make the product at a fast rate and to achieve a high titre. Therefore, controlled expression of the master gene so that it is switched on decisively at the end of growth is often preferred.

Molecular genetic analysis has also shown that some of the strains from improvement programmes, isolated over the years by random mutagenesis and selection, have increased dosages of the biosynthetic genes. Thus, some of the best strains for penicillin production have multiple copies of the critical part of the biosynthetic gene cluster arranged in tandem array. This effect can also be achieved by gene cloning strategies.

The biochemical pathways for antibiotic biosynthesis are long linear series of enzyme reactions. The flux through the entire pathway is governed by the pace of the slowest catalytic step. If the flow of metabolic intermediates through that step can be improved, then there is a good chance that the productivity of the overall process will be improved. Thus for the antibiotic tylosin, produced by *Streptomyces fradiae*, it was established that the rate of the last step in the pathway, conversion of macrocin to tylosin by macrocin-O-methyltransferase (encoded by the gene *tylF*) limited the overall productivity. As the strain improvement programme had developed with time and the production strain lineage was reviewed, it was apparent that mutants with higher titres of tylosin also displayed extremely high levels of macrocin which was excreted as a shunt metabolite (Figure 4.40) because conversion of macrocin was limiting. Strains which had the methyltransferase gene cloned and expressed at high levels showed improved titres of tylosin.

tylf-limits conversion

(a) → → tylactone ——→ tylosin

closed *tylf*-no longer limitation

(b) → → tylactone ━━➤ tylosin

Fig. 4.40 : Schematic diagram highlighting (a) the limitation of tylosin biosynthesis by *Streptomyces fradiae* and (b) the elevation of such limitation by the use of a recombinant strain.

Often, process analysis shows that some metabolites on the main biosynthetic pathway are being diverted to other shunt products (Figure 4.41). This not only represents a waste of valuable carbon source, but also presents a problem for the ultimate purification of the desired product, as the second metabolite has to be purified away. Genetic manipulation can be used to specifically inactivate the gene for the enzyme that diverts the metabolite to the shunt metabolite. This precise 'genetic surgery' enhances the flow through the pathway to the desired product.

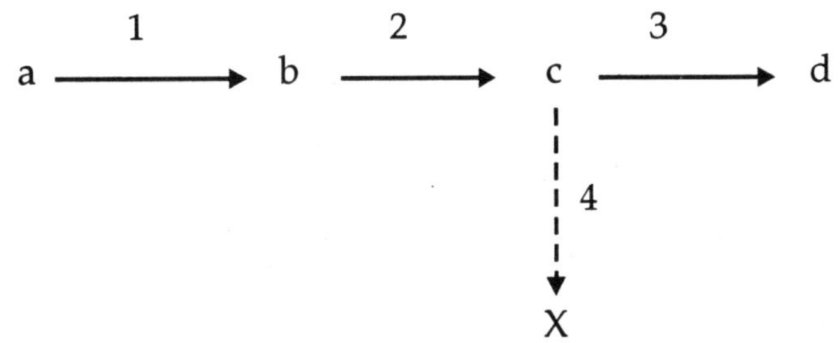

Fig. 4.41 : Schematic diagram highlighting the problem of shunt products and how this can be prevented. The metabolic pathway (substrate a to product d) is composed of three enzymes (1, 2, 3) with intermediates b and c. Because the carbon flow through steps 1 and 2 is greater than through step 3 (shown by the width of the arrows), the shunt metabolite, X, is formed through the action of enzyme 4. The presence of shunt metabolite, X, may complicate the recovery and purification of the desired product, d. By making a mutant devoid of enzyme 4, this is prevented.

One last important contribution that genetic manipulation makes is in providing flexibility of utilization of a carbon source. If, for example, there is a plentiful supply of cheap lactose available (from the milk whey industry) but the strain does not use lactose naturally, then lactose utilization genes can be cloned from another species and introduced into it.

At times, the world market-place has a glut of hydrolysed sucrose.. This cheap carbon source is problematic for fermentation because one of the hydrolysis products (glucose) is capable of inhibiting the utilization of the other (fructose). It is possible to overcome this problem by the cloning of the fructose utilization genes which, in turn, ensures that the two monomeric carbon sources are used concurrently. The overall economics of such a process benefits from being able to use a cheap and plentiful carbon source in an efficient manner, without excess fructose carbon being left in the broth at the end of the fermentation.

4.8 STRAIN IMPROVEMENT FOR HIGH-VALUE RECOMBINANT PRODUCTS

Strain improvement programmes for products derived from cloned genes follow a completely different strategy from those undertaken for their lower value counterparts. The techniques used to clone a gene (or cDNA) usually place it precisely in a vector (most often, a plasmid), in a context that allows continuous high level expression or, more advisedly, expression of the gene is controlled and induced at a high level only after growth has ceased—in the same way as antibiotic fermentations are performed. The biology of the host strain is an equally critical factor in the overall performance of a fermentation process for a recombinant product, and strain improvement programmes can have a significant impact on the economics of such a process.

In the early days of the 'new biotechnology' industry, considerable difficulty was experienced in 'scaling up' recombinant processes from the laboratory to the pilot plant (i.e. from a scale of around 100 ml to 400 l). This was because, at the larger volume, a greater proportion of cells had lost the recombinant plasmid from the cells. Plasmid-deficient cells do not make the recombinant protein, and this reduces the overall productivity. The root cause of the problem is the number of cell generations needed to attain significant growth of the culture at the larger volume, coupled to the inherent instability of engineered plasmids in cells. The growth rate of a cell carrying a plasmid is invariably slower than its counterpart that has shed its plasmid load. Therefore, plasmid deficient cells will outgrow the plasmid-containing cells in a population, a phenomenon which becomes more significant as the number of generations in a culture increases. The larger the volume of the fully-grown culture, the greater the likely proportion of plasmid-deficient cells in its population.

Resistance to an antibiotic, encoded by a plasmid-borne gene, is the usual selection strategy for the presence of a plasmid in a recombinant culture. The most commonly used selection is resistance to ampicillin, encoded, by β-lactamase. Ampicillin-resistant cells survive because they are protected by the β-lactamase, which breaks down the chemical backbone of the antibiotic. Eventually, all of the ampicillin in the fermentation broth is broken down and the selective pressure is lost—the longer the fermentation (i.e. the more generations that take place), the more likely it will be that the antibiotic will become inactivated and that plasmid-deficient cells will form.

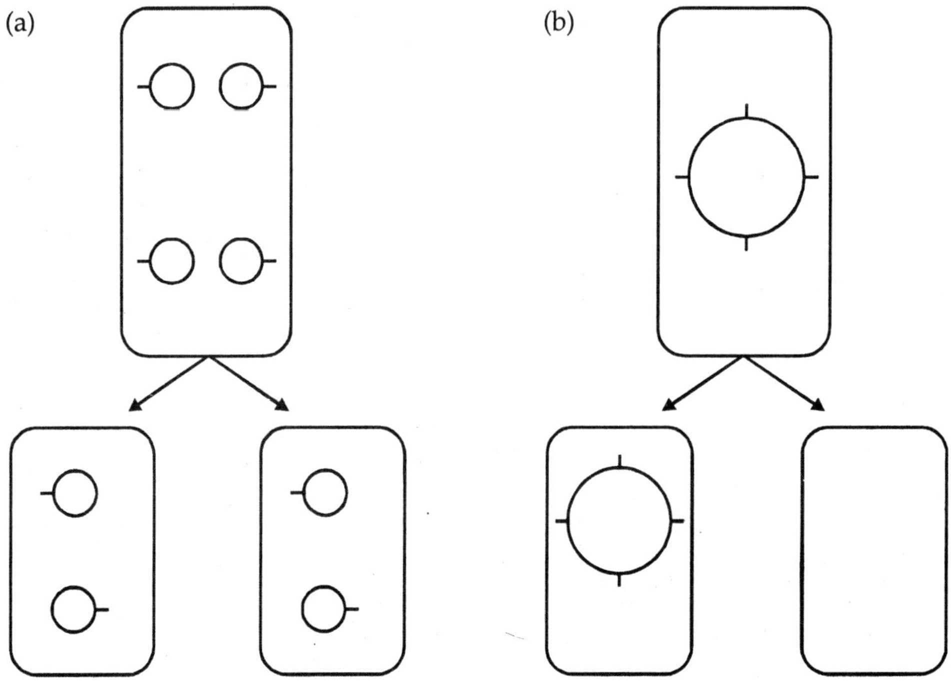

Fig. 4.42 : Plasmid segregation—the complication of plasmid oligomers. In case (a) the plasmids are present as four monomers and two segregate (on average) into each of the daughter cells. By contrast, in case (b), the plasmids are present as a tetramer. During growth, one daughter cell receives the tetramer, whereas the other does not—and becomes plasmid-deficient.

The frequency at which plasmid-deficient cells are formed, in the absence of antibiotic, is biased because of their natural tendency to form oligomers within the cell.

Consider a theoretical situation in which a cell has a plasmid copy number of four. This may be conceptualized as four separate plasmids, two of which segregate into each daughter cell at division (Figure 4.42a). However, because of oligomerization, an equally common situation is that there, will be a single tetramer of plasmids within the parent cell (i.e. the copy number will still be four). When the daughter cells are formed, one will receive the plasmid, which consists of four monomers, and the other will not (Figure 4.42b). In the absence of selective pressure (i.e. after all of the antibiotic has been exhausted) both will survive, and the plasmid-deficient daughter will outgrow its plasmid-containing sibling because of the advantage in growth rate.

Engineered plasmids undergo such oligomerization events at alarming rates. However, natural plasmids do not undergo such 'segregational instability', because they have a natural tendency to break back down from oligomers to monomers again. This ability has been lost during the process of conversion of natural plasmids into genetically-engineered vectors. It was discovered that a small piece of DNA, called *cer*, was the missing factor in the engineered vectors. When *cer* was reintroduced into vectors from the natural plasmids, the segregational stability of the plasmid construct containing the cloned gene to be expressed was improved. However, operation of *cer* is dependent on the strain and it is this aspect of the biology of plasmid instability that can be addressed by strain improvement programmes. The cellular machinery which allows *cer* to operate is rather complex and involves at least three different genetic loci. If any of the three is defective or missing, the result is that plasmids (even those containing a fully-competent *cer*) are unstable. Strain improvement of such cell lineages is best achieved using a targeted genetic approach, i.e. to use the power of genetic recombination to introduce the relevant machinery back into the chromosome of the strain, an extremely focused and targeted task.

Recombinant cultures are very prone to infection with bacterial viruses called **bacteriophages.** At a scale of fermentation above a litre, in a production environment, the recombinant cultures are at risk. There are two strain improvement strategies that can be brought into play. Firstly, it is possible to mutate the strain to become resistant to viral infection. This approach works for known viruses, but there is always the risk that infection from a new source will take place. A second, more secure strategy is to introduce a restriction/modification system into the cell line. The modification enzyme will alter the host's DNA so that it is no longer a target for the resident restriction system. However, viral DNA which enters the cell will be recognized by the restriction system and degraded, thus preventing the infection. The strategy can be enhanced further by introduction of a second restriction/modification system as a 'back up', should some of the viral DNA escape restriction by the first enzyme system. Again this is a very targeted approach to strain improvement.

Hosts for foreign gene expression often recognize the foreign protein as 'not natural' and selectively degrade it, thus reducing the overall productivity. The enzyme system which undertakes this function, the *lon* protease, works in a similar way in ordinary cells, selectively degrading ordinary cellular proteins which have been made defectively with altered amino acids. By knocking out the *lon* protease gene, degradation of the foreign protein is prevented, albeit to the detriment of the host cell which now has a slight growth rate disadvantage. Methods which generate *lon*-deficient cell lines are well established and extremely targeted.

In each of the examples in this section, there has been no need to use random mutagenesis and screening as the blueprint to derive the improved strain is direct and straightforward. Subsequent to this phase, strain improvement programmes use, the random approach to 'fine tune' the genetic content of the production recombinant strain to gain further advantages in productivity. Thus, the order of events is the reverse of that for the lower value counterparts.

❑❑❑

5

CLASSICAL STRAIN IMPROVEMENT BY MUTATION

5.1 INTRODUCTION

Naturally a 'wild-type' bacterium or yeast cell has become closely adapted through natural selection to its environment and to competition with other species. The industrial strain produces the desired commercial metabolite in vast quantities relative to its other metabolic end products.

The wild-type organism isolated from natural environment, besides optimising the cultural condition, requires genetic programming to make a normal metabolic product in an amount required at commercial level. Such a requirement calls for changing its genetic information in a way that eliminates undesirable properties or introduces entirely new ones. Such an attempt may result into shunting of more carbon and possibly other metabolites through metabolic sequences to a desirable end product formation. The genetic improvement of industrial strains may be achieved by mutation and/or recombination.

In this chapter, the brief introduction of mutations is given in section 5.2 and role of mutation in strain improvement programme is discussed in latter part of the chapter.

5.2. MUTATIONS

Inheritance is based on genes that are faithfully transmitted from generation to generation in microorganisms (in all organisms). Despite the biochemical mechanisms that facilitate such transmissions faithfully, sudden change in the sequence of nucleotide bases in genes can and do occur. These changes, called **"mutations"**, introduce variability into the gene pool and are heritable. Therefore, a mutation can be defined as *"a sudden and heritable change in the nucleotide sequence of a gene"*. Mutations always change the **genotype** (the genetic constitution of the cell) and thus the **phenotype** (the expression of the genotype in observable properties) of the microorganisms. A microbial cell exhibiting change in its phenotype due to the effect of a mutation is a **mutant** (mutant strain). Any physical or chemical agent that increases the mutation-rate is called **mutagen** or **mutagenic agent**.

Mutation is an important phenomenon because it is the ultimate source of genetic variation and provides the raw material for evolution. All the genes would exist in only one form and no mutants (alleles) would be produced without mutation. Thus the mutations are essential to provide new genetic variability to allow microorganisms to evolve and adapt to environmental changes. At the same time, if mutations occurred too frequently they would totally disrupt the transmission of genetic information from generation to generation. However, mutations occur only rarely in nature; a microbial population typically contains about one in 10^8 cells that carries a detectable mutation in any given gene. But, exposure of microorganisms to certain powerful mutagens increases dramatically the frequency of their mutant cells, by as much as 10^5-fold.

5.2.1 Occurrence of Mutations

Mutations occur in one of the two ways—spontaneous or induced.

5.2.1.1 Spontaneous Mutations

Spontaneous mutations are those that arise occasionally in the absence of a known cause, i.e., without exposure to external agents. These mutations may result from errors in DNA replication, or from the action of transposons, or even from the effect of some mutagenic agents present in the environment. Some most commonly operating spontaneous mutation mechanisms are as follows—

(i) Transition and Transversion Mutations (Errors in DNA Replication)

The bases of nucleotides in DNA strands exist in tautomeric form, i.e., they exist as structural isomers that are in chemical equilibrium and readily change into one another. Bases typically exist in keto form and seldom they can take on either an imino or enol form by error resulting in a rare tautomeric form (Fig. 5.1A), These tautomeric shifts alter the hydrogen bonding characteristics of the bases allowing purine for purine or pyrimidine for pyrimidine substitutions which can eventually result in a stable alteration of the nucleotide sequence (Fig. 5.1B). Such substitutions are called **transition mutations.** In **transversion mutation**, a pyrimidine substitutes a purine or a purine substitutes a pyrimidine. These mutations are rare due to the steric problems of pairing pyrimidine with pyrimidine and purine with purine.

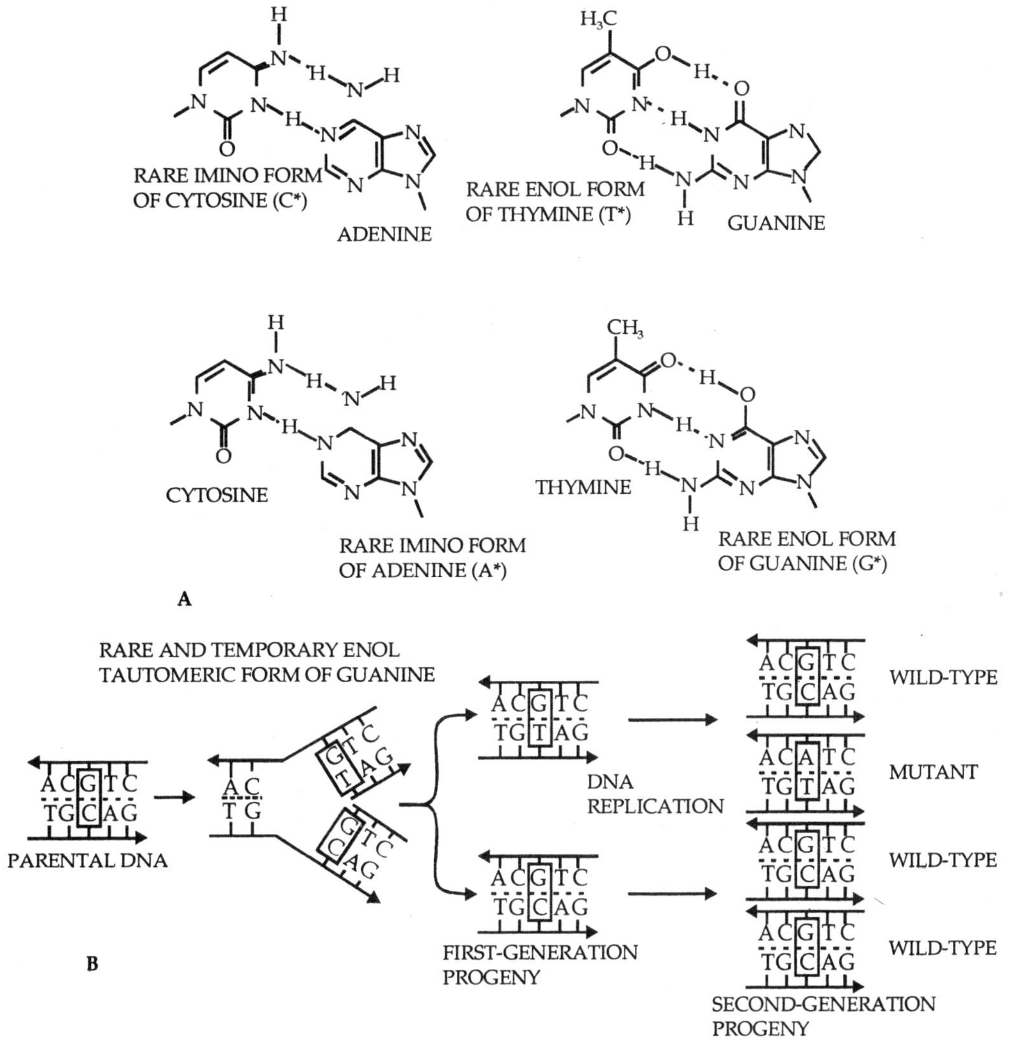

Fig. 5.1 : Transition and Transversion mutations. Errors in replication of DNA creap in due to base tautomerization. A. Usually AT and GC base pairs are formed when ketogroups participate in hydrogen bonding. Contrary to it, enol tautomers result in AC and GT base pairs. B. Mutation as a consequence of tautomerization during replication of DNA. The temporary enolization of guanine results in the formation of an AT base pair in the mutant, and a GC to AT transition mutation takes place.

(ii) Frameshift Mutations

Spontaneous mutations also arise from frameshifts, usually caused by the deletion of DNA segments resulting in an altered codon reading frame. It occurs where there is a short repeated nucleotide sequence. In such a location, the pairing of template and its complementary strand (new strand) can be displaced by the distance of the repeated sequence leading to additions or deletions of bases in the complementary strand (Fig. 5.2). These spontaneous frameshift mutations originate from lesions in DNA as well as from replication errors. For example, it is possible for purine nucleotides to be depurinated, i.e., to lose their base. This results in the formation of an **aprinic-site**, which will not base pair normally and may cause a transition type mutation after the next round of replication. Cytosine can be deaminated to uracil, which is then removed to form an **apyrimidinic site**.

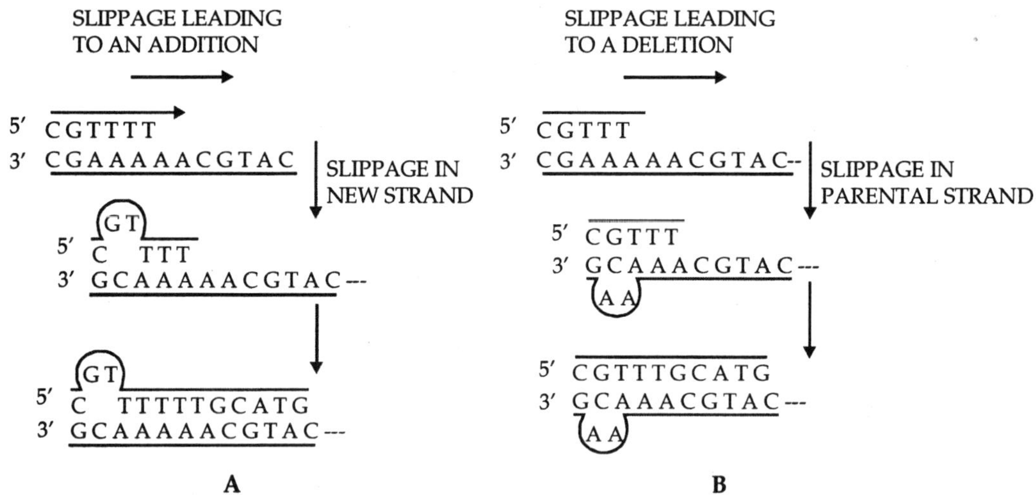

Fig. 5.2 : Additions and deletions in the complementary strand (new strand) of DNA. A hypothetical mechanism for the occurrence of additions and deletions during DNA replication is given. The large arrow indicates the direction of replication. In each case there is strand slippage leading to the formation of a small loop stabilized by the hydrogen bonding in the repetitive sequence, the AT stretch in this example. A. If the complementary strand (new strand) slips, an addition of one T results. B. Slippage of the parental strand results in a deletion, the loss of two Ts is example in this case.

5.2.1.2. Induced Mutations

Any agent which directly damages DNA, changes its chemistry, or interferes with repairs mechanisms will induce mutations and the agents which cause such actions are known as mutagens. They can be conveniently classified according to their mechanism of action. Following are the modes of mutagens actions:

(i) Base Analogs—These are similar to normal nitrogenous bases and can be incorporated into the growing polynucleotide chain during replication. Once in place, these compounds typically exhibit base pairing properties different from the bases they replace and can eventually cause a stable mutation. 5-bromouracil (5-BU), is a widely used base-analog of thymine (Fig. 5.3). It undergoes a tautomeric shift from the normal keto form to an enol much more frequently than does a normal base.

The enol forms hydrogen bonds like cytosine and directs the incorporation of guanine rather than adenine. The mechanism of action of other base analogs is similar to that of 5-bromouracil.

(ii) Specific Mispairing—This action is caused when a mutagen changes a base's structure and, therefore, alters its base pairing characteristics. Certain mutagens in this category are selective. They preferentially react with some bases and produce a specific kind of DNA damage. Methyl-nitrosoguanidine an alkylating agent that adds methyl groups to guanine, causing it to mispair with thymine.

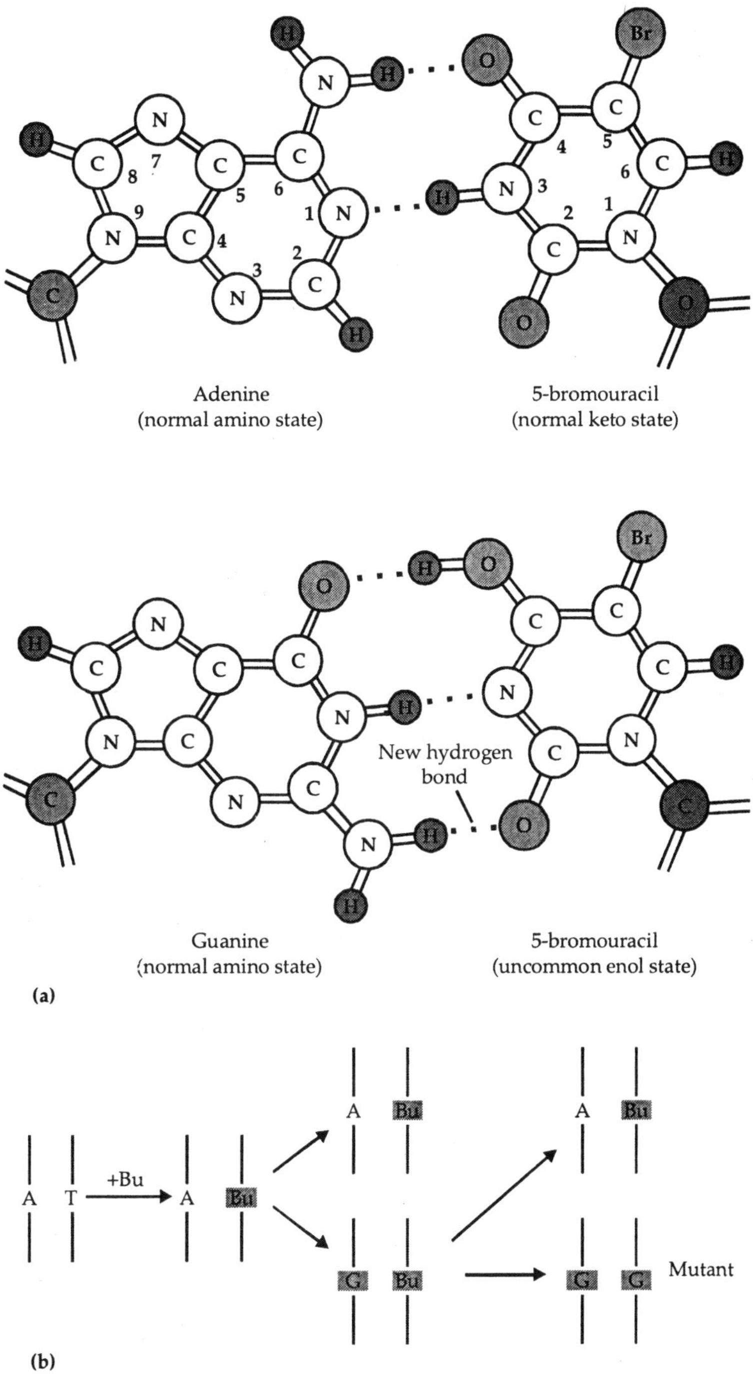

Fig. 5.3 : Mutagenesis by the Base Analog 5-Bromouracil. (a) The enol form of 5-BU base pairs with guanine rather than with adenine as might be expected for a thymine analog. (b) If the keto form of 5-BU is incorporated in place of thymine, its occasional tautomerization to the enol form (Bu_e) will produce an AT to GC transition mutation.

A subsequent round of replication could then result in a GC-AT transition. Ethylmethanesulfonate, and hydroxylamine are other examples of mutagens with similar mode of action.

(iii) Intercalating Agents—These agents distort DNA to induce single nucleotide pair-insertions and deletions. These mutagens are planar and insert themselves (intercalate) between the stacked bases of the helix. This results in a mutation, possibly through the formation of a loop in DNA. Acridines (proflavin) and acridine orange are important intercalating agents.,

(iv) Other Mutagenes—Several carcinogens directly damage DNA bases severely (hydrogen bonding between base pairs is impaired or prevented and the damaged DNA can no longer act as a template). **UV radiation** generates *cyclobutane type dimers (thymine dimers)*, between adjacent primidines. **Aflatoxin B1** and other **benzo (a) pyrene derivatives** cause damage to DNA which is generally lethal but may trigger a repair mechanism that restores much of the damaged genetic material.

5.2.2 The Expression of Mutations

The expression of a mutation can be observed if it produces a detectable, altered phenotype. A mutation from the most prevalent gene form, the wild-type, to a mutant form is called a forward mutation. Later, a second mutation may make the mutant appear to be a wild-type organism again. Such a mutation is called a **reversion mutation or back mutation.** In this the organism revertes back to its original phenotype. A true mutation converts the mutant nucleotide sequence back to the wild-type arrangement.

The wild-type phenotype also can be regained by a second mutation in a different gene, a **suppressor mutation,** which overcomes the effect of the first mutation. Sometimes, however, mutations may take place and do not alter the phenotype for a variety of reasons. When a large deletion and insertion mutations exist and affect only one base pair in a given location, they are called **point mutations**. Following are important types of point mutations.

5.2.2.1 Silent Mutation

It is the kind of point mutation that could not be detected without use of nucleic acid sequencing techniques. If a mutation is an alteration of the nucleotide sequence of DNA, mutations can occur and have no visible effect because of code degeneracy. When there is more than one codon for a given amino acid, a single base substitution could result in the formation of a new codon for the same amino acid. For example, if the codon CGU were changed to CGC, it would still code for arginine even though a mutation had occurred. The expression of this mutation would not be detected except at the level of the DNA or mRNA. Since there would be no change in the protein, there would be no change in the phenotype of the organism:

5.2.2.2 Missense Mutation

It is a second type of point mutation that involves a single base substitution in the DNA that changes a codon for one amino acid into a codon for another. For example, the codon GAG, which specifies glutamic acid, could be changed to GUG, which codes for valine.

5.2.2.3 Nonsense Mutations

It is a third type of point mutation and involves the early termination of translation and therefore results in a shortened polypeptide. Such mutations involve for conversion of a sense codon to a nonsense or stop codon. Depending on the relative location of mutation, the phenotypic expression may be more or less severely affected.

5.2.2.4 *Frameshift Mutation*

It is a fourth type of point mutation that arise from the insertion or deletion of one or two base pairs within the coding region of the gene. These mutations normally are very deleterious and yield mutant phenotypes resulting from the synthesis of non-functional proteins (Fig. 5.4).

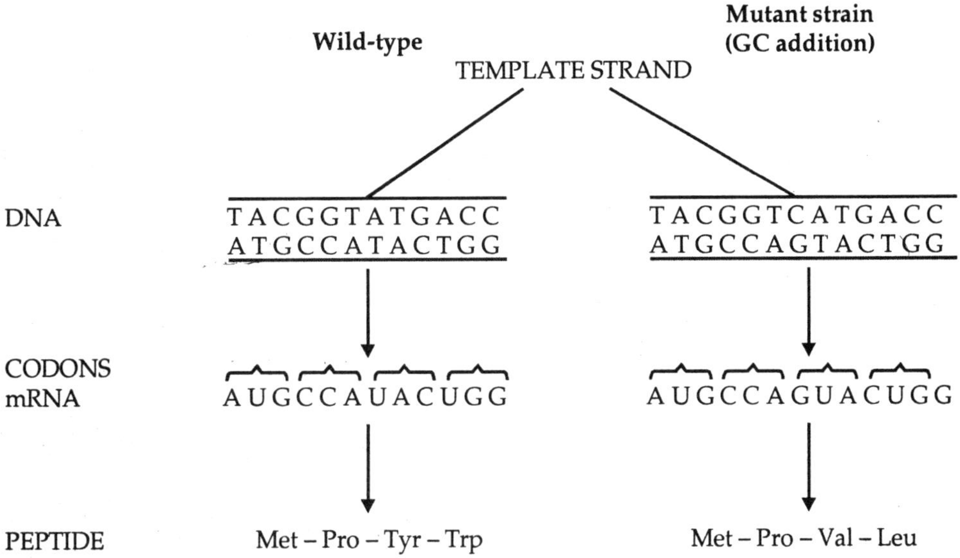

Fig. 5.4 : A frameshift mutation resulting in by the insertion of a GC base pair. The reading frameshift gives rise to a different peptide after deletion.

5.2.3 Types of Microbial Mutants

A large variety of microbial mutants have been isolated and studied intensively by microbiologists. Some important ones are as follows—

(i) Auxotrophic mutants. Auxotrophic mutants are those that are nutritionally deficient, i.e., they are unable to synthesize essential metabolites or growth factors (amino acid, purine, pyrimidine, vitamins, etc.) that are synthesized *de novo* by the parental or wild-type strains **(prototroph)** of the species. An auxotrophic mutant will not grow when the essential metabolite (growth factor) is absent in culture medium; it grows and reproduces only when the required metabolite is supplied in the medium.

(ii) Resistant Mutants. Resistant mutants exhibit an increased tolerance to inhibitory agents, particularly antibiotics and phages. Microbes may develop such mutants spontaneously through a range of mechanisms, e.g., acquisition of enzymes able to metabolize antibiotics, and loss of cell surface components which act as phage-receptors.

(iii) Cryptic Mutants. Cryptic mutants are those that have lost a specific function but retain the intracellular activities to catalyze the reactions of the function, for example, loss of a permease.

(iv) Conditionally Expressed Mutants. These mutants remain as wild-type phenotype under one set of conditions but may express as mutant phenotype when these conditions alter.

(v) Antigenic Mutants. Antigenic mutants show a change in the surface structure and composition of the microbial cell. Some antigenic mutants exhibit altered colonial shape due to effect on cell wall synthesis, and others lose the capacity to make specific cell components, e.g., capsular polysaccharides.

(vi) Metabolic Mutants. These mutants exhibit altered metabolic ability particularly the fermentation-ability of decreased or increased capacity to produce some end-product.

(vii) Regulatory Mutants. Regulatory mutants are those, in which mutation affects either the regulatory region of the promoter gene or the activity of a regulatory protein. Mutants of *Bacillus subtilis* are grossly deficient in the enzymes needed to separate daughter cell after division and they grow at normal rates as very long chains or unseparated cells.

(viii) Temperature-Sensitive Mutants. Temperature-sensitive mutants grow at one temperature but not at another. Most of them are heat-sensitive, some, however, are cold-sensitive. The temperature sensitivity usually results from increasing heat or cold liability of the mutant gene-product. For example, a temperature-sensitive mutant of *E. coli* fails to provide stability to enzymes involved in lactose utilization and does not grow on lactose at that temperature, while it retains the ability to grow on lactose at low temperature.

5.2.4 Practical Implications of Microbial Mutants

Microbial mutants, particularly the induced-mutants, are very much important because they are practically implicated in exploring various biochemical pathways and in achieving large yields of many end-products that prove to be extremely important in human affairs. Following some examples substantiate the practical implacability of microbial mutants

1. Induced mutants have been used extensively in the investigation of various biochemical processes, particularly metabolic reactions. Metabolic processes occur via sequences of enzyme-catalyzed reactions and by developing mutants in which the genes coding for enzymes are mutated (altered), the sequence of steps in a pathway can often be determined. Consider, for example, the following simple pathway

   Precursor A
   ↓ ← enzyme X ← gene X
   Intermediate B
   ↓ ← enzyme Y ← gene Y
   Product C

   Intermediate B is produced from precursor A by the action of enzyme X, the product of gene X. Intermediate B, however, may rapidly be converted to product C by enzyme Y, the product of gene Y. If so, intermediate B may be present in very low quantities and be very difficult to identify biochemically. However, in a mutant that has a mutation in gene Y resulting in the synthesis of either an inactive form of enzyme Y or no enzyme Y, intermediate B will often accumulate to much higher concentration facilitating its isolation and identification.

2. Induced mutants of microbes are used in producing large yields of many end-products important in various industries. For example, when penicillin was first discovered the yield was low and production was seriously limited. Then millions of spores were irradiated and mutants of *Penicillium* were selected. Such mutants dramatically increased the yield of penicillin in commercial production and proven invaluable for antibiotic industry.
3. Some microorganisms are known to develop resistance against certain antibiotics because of mutation. This fact is of great significance in the treatment of diseases because the antibiotics are originally effective for the control of a bacterial infection but they become less effective or ineffective as the antibiotic resistant mutants develop.
4. The maintenance of pure cultures of typical microbial species requires that occurrence of mutations be prevented. Otherwise, the cultures will no longer remain pure.

5.2.5 Isolation of Microbial Mutants—An Outline

As stated earlier, particular mutants are relatively rare in microbial population. So, in order to obtain a pure culture of such mutants, the latter are isolated from the population employing different techniques. Some important isolation techniques are as follows:

5.2.5.1 Direct Observation

In some cases, a colony growing on an agar plate can easily be seen to be different from the normal parental type (wild-type). For example, if the parental strain is pigmented, the observation of non-pigmented colonies may indicate the presence of mutants. Indicators can also be incorporated into the medium to detect microorganisms with and without particular metabolic capabilities. For instance, pH indicators can be used into the medium to detect the production of acidic products. The indication of acid production by one microbial strain and not by other of the same microorganisms growing under identical conditions would show the presence of a mutant.

5.2.5.2 Enrichment Technique

This technique is employed especially in isolating mutants resistant against phages, antibiotics, or toxic chemicals. Phage-resistant mutants can be isolated simply by plating the mutagenized, phenotypically expressed microbial population on plates containing phage particles. Cells expressing the parental wild-type phenotype are killed; only phage-resistant mutants develop into colonies. Such colonies are isolated. Similarly, mutants resistant to an antibiotic or a toxic chemical can be isolated by plating the microbial population with the antibiotic or the chemical.

5.2.5.3 Replica-Plating Technique

This technique is often used to isolate nutritional mutants (*auxotrophs*) as well as various other type of mutants, e.g., antibiotic resistant mutants. For convenience, if one wants to isolate nutritional mutants employing replica-plating techniques (Fig. 5.5), he is required to follow the following steps—

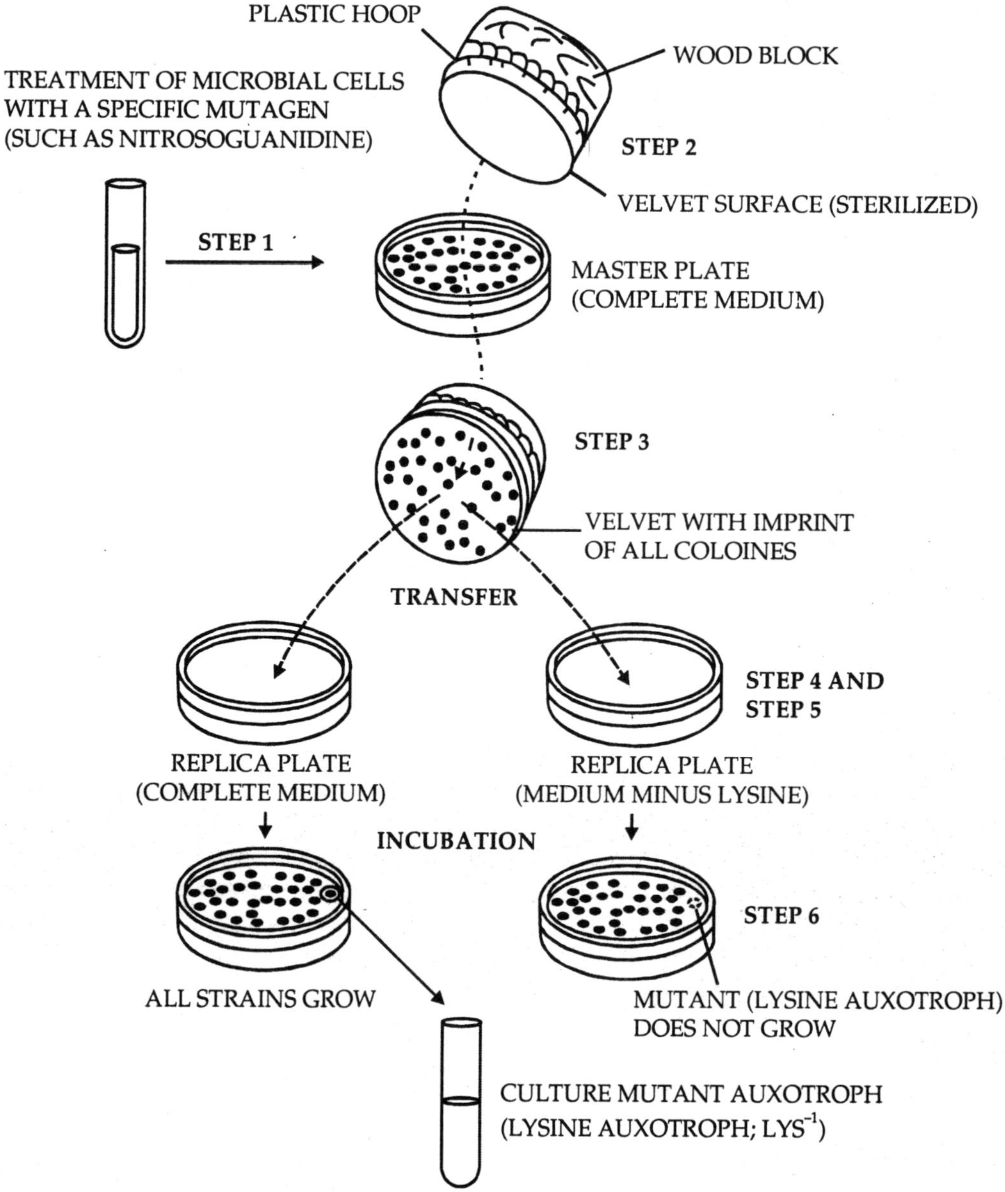

Fig. 5.5 : Replica plating technique used for selection of nutritional mutants.

1. Bacterial cultures are diluted, and the cells are spread on the surface of semisolid nutrient agar medium in a Petri dish (called **"master plate"**). The medium in the master plate is a complete medium i.e., containing all the nutritional components required by the bacterial population. After a sufficient incubation period, each bacterium produces a visible colony on the surface of the agar in the master plate.

2. A piece of sterile velvet cloth is stretched over a cylindrical block of wood or metal that is slightly smaller in diameter than the Petri dishes used in the process.
3. The master plate is now inverted and gently pressed onto sterile velvet. Since the fibres of the velvet act as fine inoculating needle, some cells from each colony of the master plate stick to the velvet.
4. Other Petri dish (called **"replica plate"**) is taken containing a minimal medium i.e., a medium deficient with specific nutritional component.
5. The replica-plate is now inverted and gently pressed onto the velvet thus stamping the bacterial cells onto the surface of its minimal medium. The replica plate is identically oriented at the application on the velvet with respect to mark placed on its rim so that the colonies that appear on the replica plate after incubation occupy positions congruent with those of their siblings on the master plate.
6. After sufficient incubation, it is observed that a colony that develops on the complete medium of the master plate fails to develop on the minimal medium of the master plate that lacks a specific nutritional component. Such colony is marked on the master plate and is isolated; it represents mutant for that specific nutritional component not used in the minimal medium of the replica plate.

The replica-plating technique was developed by Joshua and Esther Lederberg in 1952 in order to provide direct evidence for the existence of pre-existing mutations originated spontaneously in microorganisms.

5.3 DIRECTED SELECTION

The selection of a number of industrial microorganisms has been traditionally performed by plating the parent strain, after a previous mutagenic treatment or without it, on solid media and testing individual isolates colonies - for the required properties in liquid media in shake flasks. The acquisition of a new mutant requires the appraisal of a considerable number of isolates since the number of individuals with desirable properties in a population is low. The next step, i.e. the identification of the mutant in the population, is equally time-consuming. Appropriate identification and accumulation procedures have been elaborated for some types of mutants (auxotrophic, regulation mutants) while no such procedures exist for many other important processes such as secondary metabolite production.

Accumulation of the desired mutants may be strongly affected by the composition of the nutrient medium containing specific substrates, analogues and inducers. Thus, constitutive mutants can be isolated that will synthesize a given enzyme even in the absence of a specific substrate or inducer in the medium. If a source of carbon or the inducer is too expensive for industrial use, one can use an alternative and cheap carbon source with the constitutive mutants and the inducer may be omitted altogether. One of the cheaper carbon sources is glucose, but glucose will repress the synthesis of a number of enzymes. However, if one isolates a mutant that is resistant to catabolite repression, glucose may be used without repressing enzyme synthesis.

The possibilities for the selection of high-production strains forming secondary metabolites are quite restricted in this context. In the selection process one can choose conditions that may not guarantee the isolation of high-production strains but one may obtain variants, the application of which will have some other bearing on production economy. By using selective conditions for the evaluation of such mutants, as for example the nutrient medium composition, the mode of sterilisation, concentration of dissolved oxygen, time and temperature of culture, one can produce strains with new, more suitable properties. The examples shown below illustrate these possibilities.

One of the raw materials added to nutrient media in tetracycline biosynthesis is corn-steep liquor. Due to fluctuations in the content of phosphorus in individual batches of the liquor, the final phosphorus level in the medium may vary in the range 20-100 $\mu g\ ml^{-1}$. Strains of *Streptomyces aureofaciens* have the highest antibiotic production at inorganic phosphate concentrations of 30-40 $\mu g\ ml^{-1}$; at higher concentrations production is inhibited. Since it is extremely difficult to ensure a constant level of inorganic phosphate in the medium, some mutants were isolated in which the production of the antibiotic is not inhibited by high phosphate concentrations.

The production of tetracycline antibiotics by strains of *Streptomyces aureofaciens* is also considerably reduced when the aeration is interrupted., even for a very short period of time, especially between the 4th and 12th hour of culture. Selection of strains insensitive to interruptions in aeration was performed by interrupting aeration at regular intervals between the 4th and 12th hour of culture and collecting the surviving production variants.

The effect of the method of medium sterilisation as a selective agent is illustrated in the following example. The improvement of streptomycin-producing strains yielded variants with productivity in laboratory flasks 20 - 30% higher than that of the parent strain. However, in industrial fermenters the parent strain and the new strains exhibited the same productivity. Modification of the sterilisation procedure to achieve conditions closely resembling those used in the laboratory increased again the productivity of the new strains by 20-30% whereas no such increase was observed in the parent strain.

The isolation of new types of microorganism has gained a new impetus with the introduction of directed or programmed selection of microorganisms; the term was coined to show that this technique is the opposite to the standard, random selection (Table : 5.1).

Table 5.1 : Standard and programmed selction

		Selection	
		Standard	Programmed
1.	Mutation – spontaneous	Yes	Yes
	– Induced	Mutagenesis	Mutagenesis
2.	Determination of a mutant in a population	Plating, evaluation of isolates in flasks	Plating, colour reaction, auxotrophic mutants
3.	Accumulation of the mutant	None	Selection pressure in batch and continuous cultures

Table 5.2 lists types of mutants which can be obtained by this procedure, and the properties which make them favourable for industrial application. The mutants arise either spontaneously or after a mutagenic treatment; they comprise individuals which lack

completely or partially one of the regulatory mechanisms and are thus capable of accumulating large amounts of certain products and are only rarely found under natural conditions. The mutant types are not classified according to the metabolites they accumulate or the methods used for their isolation but according to their general applicability in microbial processes.

Table 5.2 : Types of mutants and their possible industrial applications

	Mutant type	Industrial application
1.	Resistance to bacteriostatic or bactericidal agents	Use of raw materials contaminated with the two types of agents
2.	Ability to utilise a smaller number of substrates than parent strain	Better utilisation of substrate and higher economic coefficient
3.	Change in transport of nutrients	Assimilation of a compound the parent strain could not assimilate
4.	Defects in anabolic pathways (auxotrophy)	Accumulation of important compounds (amino acids, nucleotides)
5.	Temperature resistance	Fermentation at higher temperatures and saving on cooling water
6.	Damaged control mechanisms at the level of enzyme synthesis	
-	Damaged system of induction and repression (constitutive and hyperinducible mutants)	Production of the desired substance (enzyme) without addition of inducer
-	Resistance to catabolite repression	Use of rapidly metabolised substrates
-	Resistance to oxygen effect at the level of enzyme activity	Insensitivity to fluctuation in oxygen supply
-	Resistance to inhibition and repression by feedback	Increased production of important compounds
7.	Altered or extended substrate specificity	Use of unusual substrates or of a greater number of substrates

An example of programmed selection which takes into account a number of technological and economical conditions and aspects is the evolution of an endoenzyme-producing strain shown in Table : 5.3.

Table 5.3 : Evolution of an endoenzyme-producing strain

	Wild strain	Mutant
1.	Requires an inducer (expensive, not available)	Isolation of a constitutive mutant (requires no inducer)
2.	Complex medium (increased raw material costs, complicated product isolation)	Isolation of a mutant resistant to catabolite repression (synthetic medium and cheap sources)

contd. ...

... contd.

	Wild strain	Mutant
3.	Low production of desired substance	Isolation of a hyperconstitutive or a magnoconstitutive mutant (increased production of the substance)
4.	Sensitivity to phages	Isolation of a phage-resistant mutant (elimination of cell lysis)
5.	Substrate is not available for economical or other (stability) reasons	Isolation of a mutant with an extended or changed specificity (possibility of using the most suitable substrates)

An example of directed selection is the improvement of *Bacillus licheniformis* which produces bacitracin. The limiting factor in the biosynthesis of peptide antibiotics are non-protein amino acids, in this case ornithine. A genetic block at the site of ornithine degradation by ornithine-δ-transaminase causes a rise in bacitracin production and a drop in the concentration of glutamate and proline in the cell. These two amino acids are formed from ornithine during the catabolic metabolism of arginine (Fig. 5.6).

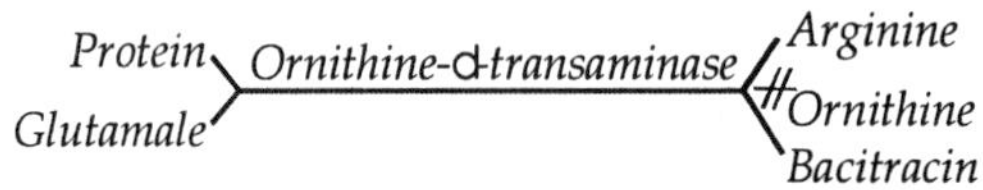

Fig. 5.6 : Scheme of genetic block.

The industrial production of a number of amino acids was made possible by a directed isolation of auxotrophic mutants and regulation mutants. As an example we may mention the genealogy of a tryptophan producer. The parent strain was a tyrosine- and phenyl-alaine-dependent auxotrophic mutant of *Corynebacterium glutamicum* which accumulated 0.15 g tryptophan per 1 litre medium. The mutant gradually acquired resistance to 5-methyltryptophan, tryptophan hydroxamate, 6-fluorotryptophan and 4-methyltryptophan. This multiple mutation provided a mutant producing about 5 g tryptophan per 1 litre medium. Further increase in production was achieved with mutants resistant to phenylalanine and tyrosine analogues such as p-fluorophenylalanine, p-aminophenylalanine, tyrosine hydroxamate and phenylaianine hydroxamate. The mutant isolated finally accumulated 15 g tryptophan per 1 litre medium.

5.3.1 Isolation of Auxotrophic Mutants

Auxotrophic mutants are those which need for their growth certain factors (vitamins, amino acids) in contrast to the prototrophic parent strains which do not require these factors. For this reason prototrophic strains can grow in the so-called minimal medium whereas auxotrophic mutants cannot. Thus, a mutant requiring methionine or thiamine does not grow in a medium lacking these substances.

The isolation of prototrophic mutants from a population of auxotrophs is relatively easily performed by inoculating a sufficiently large population into a minimal medium. The parent auxotiophic cells do not grow while the prototrophic mutants present form visible colonies.

Isolation of auxotrophic mutants from prototrophic populations requires more complicated methods. One of the most efficient is the penicillin method illustrated in Fig. 5.7.

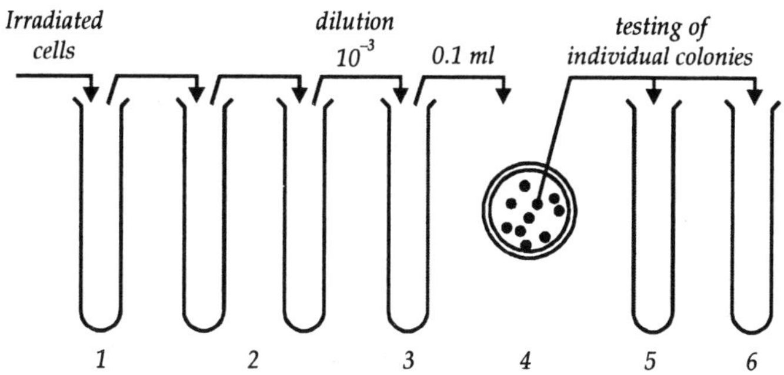

Fig. 5.7 Scheme or procedure used in the penicillin method 1—irradiated suspension of prototrophic cells is cultured for 24 h in a complete medium; 2—cells grown in the medium are repeatedly washed in physiological saline; 3—diluted inoculum of washed cells is introduced into a minimal medium containing 100-300units of penicillin per 1 ml. Only auxotrophic mutants survive: 4—cells from minimal medium are plated on a complete medium; 5—single colonies are inoculated into a complete medium; 6—the same colonies are inoculated into a minimal medium; if the colony grows in the complete medium only, it is taken to represent an auxotrophic mutant clone

The method is based on the fact that penicillin only kills actively growing microorganisms and it may be used for all penicillin-sensitive microorganisms. When the microorganisms are inoculated into a minimal medium containing a bactericidal concentration of penicillin, prototrophic cells begin to grow and are mostly killed. Auxotrophic mutants do not grow in this medium and consequently survive the penicillin treatment. Transfer of the population to a complex medium with agar yields a number of auxotrophic colonies. The auxotrophic character of these isolates can be verified by demonstrating their inability to grow on a minimal medium. An improved procedure of isolating auxotrophs is performed using exclusively solid media. Petri dishes are first filled with a layer of minimal agar with a microbial population (Fig. 5.8), incubated for several hours and a layer of minimal agar with penicillin is added. After a 24-h incubation penicillin is inactivated by adding another layer of minimal agar containing penicillinase and incubating. Next a layer of complete agar is added and the dishes are again incubated to obtain auxotrophic colonies. This technique yields a higher number of auxotrophic mutants than that using liquid media.

Fig.5.8 : Isolation of auxotrophic mutants on solid media.

The growth requirements of the isolated auxotrophic mutants can be determined in several ways. One method consists of inoculating the mutants into a minimal medium supplemented by the following complex substances:

a) casein hydrolysate (vitamin-free)— for mutants requiring a certain amino acid;

b) vitamin mixture—for mutants requiring a certain vitamin;

c) hydrolysed yeast nucleic acid—for mutants requiring purines, pyrimidines, nucleotides and nucleosides.

When a mutant has been found to grow on one of these substances, its specific requirements are ascertained by testing its growth on a minimal medium enriched in specific amino acids, vitamins, etc. The specific requirements of auxotrophic mutants can also be determined by the template method or replica plating technique (Fig. 5.5).

The isolation of auxotrophic mutants was instrumental in promoting the industrial production of amino acids, nucleotides and nucleosides.

The relationship between the accumulation of, and dependence on, a certain amino acid is illustrated in the following list. Each amino acid in the first column (Dependence on) corresponds to an amino acid in the second column (Accumulation of).

Dependence on:	**Accumulation of:**
homoserine	lysine
lysine	diaminopimelic acid
threonine	homoserine
diaminopimelic acid	threonine
arginine	ornithine
arginine	citrulline
isoleucine	valine
tyrosine	phenylalanine
phenylalanine	tyrosine

The accumulation of amino acids by auxotrophic mutants is conditional on the accumulation of the metabolite before the genetic block, such as in the case in the production of ornithine by mutants dependent on arginine or citrulline, or in the accumulation of citrulline by mutants requiring arginine. The accumulation of lysine by homoserine-requiring mutants includes the block of the metabolic pathway leading to homoserine due to a drop in the activity of homoserine dehydrogenase; analogously, the biosynthesis of valine by isoleucine-dependent mutants includes the inhibition of the parallel biosynthetic pathway.

The prototroph-auxotroph-prototroph route has been used in an attempt to increase production of a low chlortetracycline-producing strain of *Streptomyces viridifaciens*. About a hundred revertants from each of five auxotrops were tested. In the case of revertants from four of these auxotrophs, the productivities were similar to those of randomly tested survivors of the grandparent culture. However, in the case of a methionine auxotroph, 88% of the resultant revertants produced between 1.2 and 3.2 times as much chlortetracycline as the original prototrophic culture. Since methionine is involved in the synthesis of the antibiotic, these results indicate that reversion of mutants auxotrophic for a primary metabolite which is a precursor of a secondary product may be a useful means of obtaining high yielding producers of secondary metabolites.

5.3.2 Isolation of Mutants Requiring no Inducer (constitutive mutants)

The following methods can be used for the isolation of constitutive mutants:

a) Growth in a chemostat under substrate limitation. If an inducible population grows in a chemostat under a limitation by a substrate which induces the synthesis of an enzyme, mutants which require no inducer are selected and accumulate.

b) Cyclic growth of microorganisms in a medium with and without an inducer. The microorganisms are first cultured in an inducer-free medium; the constitutive and inducible populations grow under these conditions at the same rate. On transfer into a medium with the inducer, constitutive mutants grow at a higher rate since the inducible population requires a certain induction time before growth commencement. Repetition of these growth cycles results in an accumulation of constitutive mutants.

c) Application of a weakly inducing substrate. When a microbial population is cultured in a medium containing, as a source of carbon, a substance which is a good substrate but a mediocre inducer, constitutive mutants are selected.

d) Application of induction inhibitors. Microorganisms are cultured in a medium containing a substrate-inducer and an induction inhibitor. If the enzyme induction is inhibited, only mutants requiring no inducer grow.

e) Visual detection without accumulation. Microorganisms are plated on an inducer-free agar medium. When the colonies have grown, they are sprayed with a solution of a reagent detecting enzyme synthesis; colonies of constitutive mutants are stained. Since only constitutive mutants synthesise the enzyme during growth without inducer, they can be readily detected in the population.

5.3.3 Isolation of Mutants Resistant to End Product Repression

The isolation of these mutants can be performed as follows—

a) Application of antimetabolites—analogues. Analogue-resistant mutants are isolated by inoculating a microbial suspension previously treated with a mutagen on plates with a minimal medium containing the analogue. Resistant colonies appearing on the plates are transferred to plates with minimal medium and overlaid with a suspension of an indicator organism requiring for growth, e.g., the amino acid the analogue of which was used. Mutants resistant to these analogues possess altered enzymes which are insensitive to the effects of metabolic regulatory mechanisms preventing metabolite overproduction. Such resistant mutants can be divided into those possessing enzymes insensitive to end product inhibition, and those with enzymes production of which is insensitive to repression by this product. The first group includes, e.g., mutants of *Corynebacterium glutamicum* resistant to S-(-aminoethyl)-L-cysteine; the aspartate kinase of these mutants, i.e. the first enzyme of lysine biosynthesis, is insensitive to inhibition by threonine and lysine. Consequently, these mutants overproduce lysine. A similar mechanism is responsible for the accumulation of threonine by *Escherichia coli* mutants resistant to α-amino-β-hydroxyvaleric acid. Representatives of the second group are, e.g., *Escherichia coli* mutants resistant to kanavanin. Production of the arginine biosynthesis enzymes by these mutants is insensitive to arginine repression and the amino acid is consequently accumulated in the medium. *Escherichia coli* mutants resistant to 5-fluorophenylalanine, which accumulate tyrosin in the medium, possess enzymes production of which is not repressed by tyrosine.

b) Visual detection. Mutants resistant to end products can be detected directly without previous accumulation steps. Thus, mutants with constitutive phosphatase synthesis can be detected among colonies grown in the presence of a high concentration of inorganic phosphate by spraying the colonies with o-nitrophenylphosphate; constitutive mutants turn yellow while repressible colonies remain white. Alternatively, the colonies can be overlaid by a suspension of bacteria requiring for growth the substance which is assumed to be synthesised by a metabolic pathway repressed by the end product. Constitutive mutants synthesise the product in large amounts, excrete it and thus make possible the growth of the indicator culture.

c) Application of sporulation mutants. The formation of proteases and sporulation of bacilli is inhibited by a mixture of amino acids. Mutants capable of sporulating in the presence of the inhibitory amino acid mixture are selected by heating and inoculating onto a medium with agar. The resulting colonies exhibit a substantial increase in enzyme synthesis.

d) Reversion of auxotrophs to prototrophs. The mutation of prototrophic strains to auxotrophy and reversion of the auxotrophs back to prototrophy can be used to modify the regulatory properties of the enzyme responsible for the particular dependence in the auxotrophs. Thus the reversion of mutants incapable of synthesising homoserine dehydrogenase resulted in a 3-fold increase in the activity of the enzyme as compared with the original prototrophic population.

5.3.4 Isolation of Mutants Resistant to Catabolite Repression

Isolation of these mutants is important, particularly in enzyme production, since the enzymes of choice are synthesised even in media containing cheap carbon sources which cause catabolite repression in wild strains. Moreover, some mutants resistant to catabolite repression produce considerably higher amounts of enzymes; the reasons for this phenomenon are not yet quite clear. For instance, a yeast mutant with no catabolite repression of invertase production can produce almost 2% invertase referred to total cell protein.

There are several methods for isolating such mutants—

a) Use of the substrate of the repressed enzyme as a sole nitrogen source. Gradual transfers of a culture of *Aerobacter aerogenes* in a medium containing glucose and histidine as the only C and N sources lead to the selection of mutants resistant to catabolite repression. Under conventional conditions the enzymes dissimilating histidine are repressed by glucose; in this case, however, the enzymes are necessary for bacterial growth. The composition of the medium thus makes it possible to select mutants producing histidinase in the presence of glucose. During growth on glucose and histidine, these mutants exhibit derepressed levels of histidinase, urokinase and β-galactosidase.

b) Culture of microorganisms in media with and without glucose. The cells of *Escherichia coli* exhibit a growth lag on transfer front a glucose containing medium into another containing lactose, maltose, acetate or succinate. The lag is due to the repression of enzymes necessary for consumption of these substrates during growth on glucose. The change in growth rate on transfer from glucose to, e.g. succinate positively affects the selection of mutants resistant to catabolite repression. The selection is further facilitated by the fact that the generation times of the bacteria in a medium with maltose, acetate or succinate are much shorter in the mutants than in the parent strains.

c) Visual detection. The methods used for visual detection of constitutive mutants on solid media are also used for the detection of mutants resistant to catabolite repression. Thus, for instance, colonies grown on agar with glucose are sprayed with *o*-nitrophenylgalactoside to isolate mutants synthesising β-galactosidase and resistant to glucose repression. If the parent cells are inducible, the inducer has to be added to the agar; if the cells require no inducer, the agar is inducer-free. Mutants resistant to catabolite repression will be yellow due to the hydrolysis of o-nitro-phenylgalactoside. A similar technique has been used for the isoloation of mutants resistant to glucose catabolite repression during the production of extracellular polygalacturonic acid transeliminase. If the colonies are grown on polygalacturonic acid, the enzyme can be detected by spraying the dish with hydrochloric acid. Clear patches indicate the production of the enzyme. If glucose is introduced into the solid medium together with polygalacturonic acid, mutants resistant to glucose repression can be readily detected.

d) Streptornycin dependence. Couckell and Polglase found that in *Escherichia coli* mutants dependent on streptomycin the rate of glucose consumption decreased; this was associated with a non-specific resistance to catabolite repression.

5.4 THE SELECTION OF MUTANTS PRODUCING HIGH YIELD OF PRIMARY METABOLITES

Amino acids are primary metabolites, most commonly produced by *Corynebacterium glutamicum* and *Bravibacterium flavum*. Lysine is an important amino acid produced by *Corynebacterium glutamicum* and *Bravibacterium flavum*. In these wild-type organisms, the production of lysine is regulated in such a way that, it is produced to a level to meet the needs. The fermentation is based on understanding the pathways leading to bacterial biosynthesis of the amino acid and genetic regulation. In these bacteria lysine is one end product of a branched pathway that also leads to the synthesis of enough of these amino acids methionine and threonine.

The regulatory mechanisms involved are the inhibition of the activity of key enzymes in the biosynthetic route and the repression of their synthesis by the end products when present in the cell at required concentration to meet the organisms need. Such mechanisms are known as feedback control. That is, accumulation of these two amino acids in excess of the organisms requirements tend to inhibit their synthesis by inhibiting the enzymes activity; conversely, a shortage of threonine or lysine increases their rate of synthesis.

Overproduction of lysine (Fig 5.9) far beyond the bacterium's own needs, has been achieved by isolating two kinds of mutant. In one kind a mutation in the gene coding for the enzyme homoserine dehydrogenase abolishes the enzyme activity and thereby prevents the bacterium from making threonine (one of the inhibitory products) and methionine. In this pathway lysine does not exert any control over the biosynthetic route from aspartic semialdehyde to lysine. A mutant which could not catalyse the conversion of aspartic semialdehyde to homoserine would be capable of growth only in homoserine supplemented medium and the organisms are homoserine auxotrophs. When this auxotroph, or nutritionally deficient mutant, is cultured in a medium with just enough threonine and methionine to support growth but not enough threonine to cooperate with lysine in shutting of asparate kinase's activity, the pathway to lysine continues to operate at full speed. Auxotrophic mutants are selected by testing thousands of colonies that have been treated with a mutagen; auxotrophs grow only if particular growth factors (i.e., methionine and threonine) are supplied in the medium. Auxotrophs can be isolated more efficiently by inoculating a mutanized culture into a medium that lacks the appropriate growth factors and contain penicillin, which kills only proliferating bacteria; auxotrophs can not grow and so they survive, whereas the unmutated bacteria are killed.

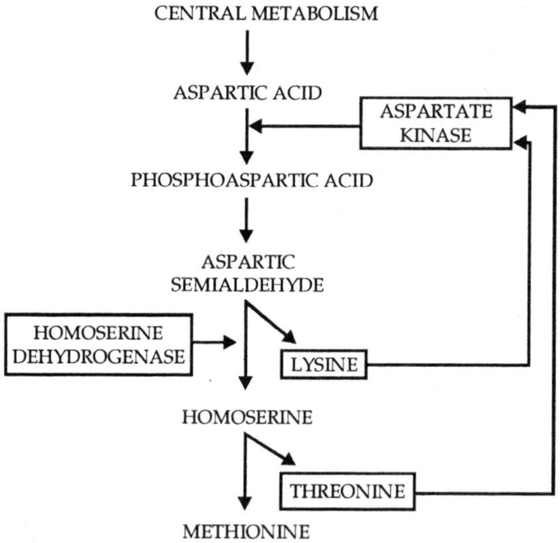

Fig. 5.9 : The control of lysine production in *Corynebacterium glutamicum* (Feedback inhibition).

A different kind of mutant which do not produce regulating end products, is to isolate a mutant which do not recognize the presence of controlling factors. Mutant has an altered form of aspartate kinase itself, which does not react with lysine and has therefore lost its sensitivity to feedback inhibition; again, high level of lysine accumulate in the fermenter. These mutants could be selected because they are resistant to lysine analogue S-(2-aminoethyl) cysteine (AEC), which resembles lysine so closely that it mimics its regulatory effect, inhibiting aspartate kinase even if no lysine in being syntnesized. In the presence of AEC wild strains die as a result of lysine starvation, where as mutants proliferate and form colonies. Lysine production is just one example of industrial process on the rational selection of mutants in which precise controls regulating amino acid production are disconnected.

For higher yielding capacity of a new mutant, the fermentation conditions are no longer the same, that were for wild type strains. A richer medium, or possibly, several media have to be tested to detect and evaluate a high yielding mutant.

5.5 THE SELECTION OF SECONDARY METABOLITE PRODUCING MUTANTS

Devising a procedure to isolate mutants, that over produce secondary metabolite is more difficult, as the information available on control of production is lacking, and also, such products are not required to the producing organisms for their growth. For example, antibiotics are synthesized only at particular stages in their life-cycle of certain molds, actinomycetes, and spore forming bacteria. Their genetic regulation is yet poorly understood, but the amount of antibiotic produced depends on many factors. Control system responds to various aspects of cells metabolism i.e., the availability of carbon, nitrogen and free phosphate, to the synthesis of chemical building blocks needed to make antibiotic and to the organisms resistance to its own potentially toxic antibiotic. Because the yield of an antibiotic depends on hundreds of genes, it is impossible to find individual mutations that can raise the yield from a wild strains low level to high yield at economic level.

Improved strains have been developed through many successive rounds of mutation and selection. In each round a culture is treated with a mutagen and thousands of resulting colonies are examined. When a mutant displaying significantly increased productivity is found, it serves as the starting point for a new round of mutagenesis and screening. In this way organisms evolution is channelled in an unnatural direction until a strain is developed that produces an improved economic yield of the antibiotic.

Such work is slow and labour intensive, and its results are unpredictable, because antibiotic levels are strongly influenced not by the genes of the producer but also by culture conditions. In the early stages of yield improving programme one can isolate the mutant simply by testing the yield of colonies growing on culture plates, but ultimately, evolution has to be performed under conditions mimicking those that will be encountered large fermenter tanks.

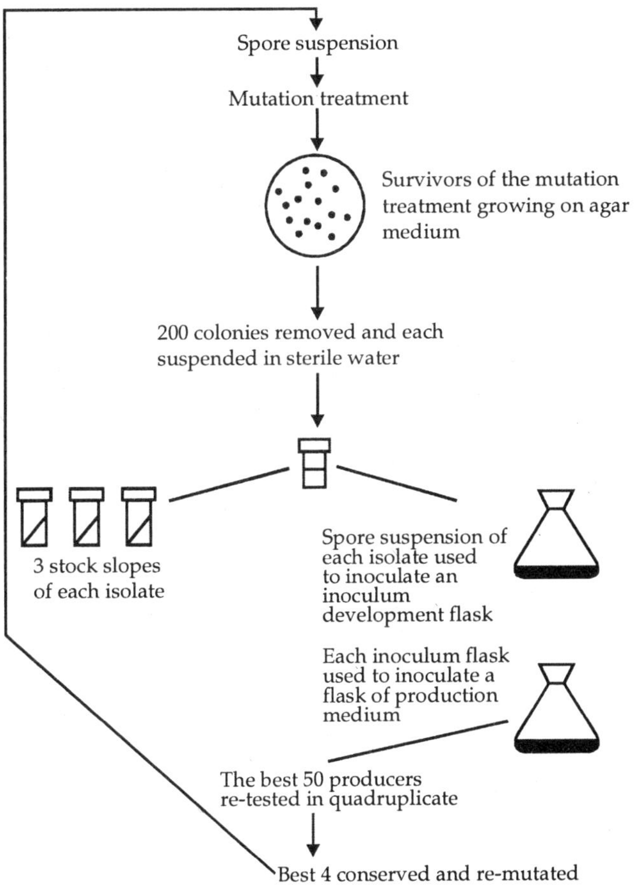

Fig. 5.10 : A strain improvement programme for a secondary metabolite producing culture.

A programme typical of the industry in the 1950s to 1970s is illustrated in Figure 5.10. The throughput of such programmes has been increased by miniaturizing the systems. Small volumes of liquid media in tubes or microtitre plates have been used coupled with the use of robots to automate the process. However, as more information has accumulated on the biosynthesis and control of secondary metabolites, directed selection approaches have also been used, thus reducing the empirical nature of the screens. Mutants capable of producing increased levels of secondary metabolite precursors have been isolated by techniques similar to those used for the improvement of primary metabolite producers. For example, analogue-resistant mutants have been isolated giving improved yields of pyrrolnitrin, candicidin and cephamycin. Relief of carbon repression has been achieved on of mutants resistant to 2-deoxyglucose, a glucose analogue.

❑❑❑

6

CLASSICAL STRAIN IMPROVEMENT BY RECOMBINATION

6.1 INTRODUCTION

Hopwood defined recombination as any process which helps to generate new combinations of genes that were originally present in different individuals. Compared with the use of mutation techniques for the improvement of industrial strains the use of recombination was fairly limited in the early years of improvement programmes. However, techniques are now widely available which allow the use of recombination as a system of strain improvement.

In vivo recombination may be achieved in the asexual fungi (e.g. *Penicillium chrysogenum*, used for the commercial production of pencillin) using the parasexual cycle. The technique of protoplast fusion has increased greatly the prospects of combining together characteristics found in different production strains. Such approaches for strain improvement are discussed in this chapter.

In vitro recombination has been achieved by the technique of *in vitro* recombinant DNA discussed elaborately in chapter 4 of this book. The efficiency of *Methylomonas methylotrophus*, the organism used in the ICI Pruteen process, was improved by the incorporation of a plasmid containing a glutamate dehydrogenase gene from *Escherichia coli*.

6.2 STRAIN IMPROVEMENT: THE POWER OF RECOMBINATION IN 'STRAIN CONSTRUCTION'

In the early stages of undertaking a random mutagenesis programme for a new product (i.e. starting from a new soil isolate which makes a small amount of product), there are likely to be rapid, concurrent advances in the performances of several strains. These strains, separately, will have individual desirable properties. For example in an antibiotic programme, one strain may give a higher titre, while another may use less carbon source to make a level of antibiotic equivalent to that of the parent (and so be more economical). Yet another strain may show no better performance in terms of titre of product or the amount of nutrients consumed to make it, but may carry out the fermentation at a lower viscosity, which will make it easier for the product to be extracted from the fermentation broth and purified. Different 'lineages' of the strain are developed very rapidly in these early days in the desire to improve the fermentation considerably. However it soon becomes desirable to construct

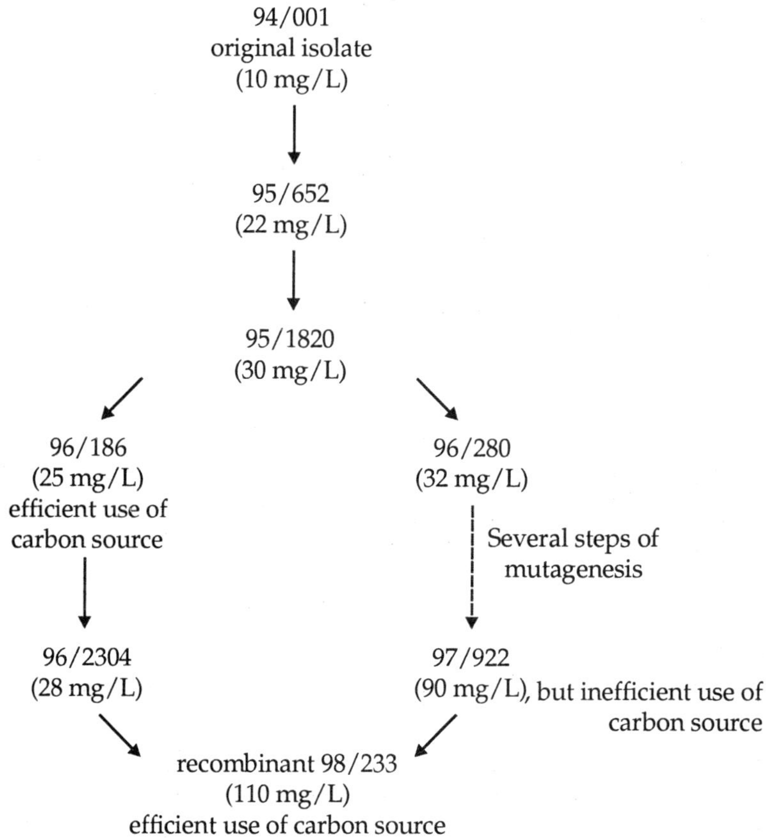

Fig. 6.1 : Strain construction by recombination. In this example two lineages have been derived from the first soil isolate. One lineage shows a steady improvement in titre with each new strain, but this is accompanied by wasteful utilization of carbon source. The other lineage makes very little antibiotic, but does so in a very efficient way. By genetic recombination of these 'traits', the attributes of both strands of the culture lineage can be combined together in a single strain.

strains with a combination of several of the desirable properties already present in the individual lineages. To achieve this, desirable traits from each lineage must be 'recombined' genetically into a single strain which possesses all of the desirable properties, or 'traits'.

Most microbial species are able to exchange genetic information with other members of their species by recombination. Construction of the genetic maps that appear in many textbooks is based on exploitation of these natural recombination processes. For *E. coli* and other enteric Gram-negative bacteria, protocols to undertake genetic recombination are well documented and depend on plasmid-based mobilization of the bacterial chromosome. However, the producers of many microbial metabolites, including the *Streptomyces*, have a life cycle that involves a sporulation stage (described above). With these species, the most effective method of constructing recombinants that have combinations of several desirable traits is to undertake "spore mating': to mix spores of the parents with the individual traits, germinate them together, allow them to go through an entire life cycle, and (from the spores formed at the end of this life cycle) to select those which have exchanged some genetic information (i.e. select those that show that they have undertaken some degree of recombination) and then to screen them to identify the individuals that have received the desired combination of traits (Figure 6.1).

Successful use of recombination for strain-improvement program in various industrial strains of actinomycetes has been mentioned in Table : 6.1.

Table 6.1 : Recombination in actinomycetes

Streptomyces spp.
S. coelicolor (Actinorhodin, Methylenomycin)
S. achromogenes var. *rubradiris* (Rubradirin)
S. acrimycini
S. aureofaciens (Chlortetracycline)
S. bikiniensis (Zorbamycin, Zorbonomycin)
S. erythreus (Erythromycin)
S. fradiae (Neomycin)
S. glaucescens
S. griseoflavus
S. griseus (Streptomycin)
S. olivaceus
S. rimosus (Oxytetracycline)
S. scabies
S. venezuelae (Chloramphenicol)
Nocardia spp.
N. erythropolis
N. mediterranei (Rifamycin)
Micromonospora spp.
M. chalcea
M. echinospora
M. purpurea (Gentamicin)
Mycobacterium smegmatis
Thermoactinomyces vulgaris

The antibiotic produced is indicated within the parenthesis. In *Thermoactinomyces*, transformation is responsible for the genetic exchange; in all other cases, a "conjugation" mechanism has been recognized, but fertility factors have not usually been detected (Elander *et al.*, 1977).

6.3 FUNGAL PARASEXUALITY

We all know that it is the process of sexual reproduction which, as a result of union of two compatible nuclei (karyogamy) followed by meiosis enjoying recombinations of genetic material, leads to the emergence of new individuals bearing some new hereditary characters. But, there are a great number of fungi which do not go through sexual reproduction nonetheless they produce new individuals bearing some new hereditary characters and, as a result, a large number of new species and genera are described frequently. How these new individuals of fungi evolve if sexual reproduction lacks? Mycologists have found the answer for this question. They claim that there is another device, called parasexual-reproduction or "parasexuality", which is responsible for genetic recombination resulting in the evolution of new species and genera in fungi that do not enter into true sexual cycle. Besides, reports are there that parasexuality operates in some such fungi which simultaneously enjoy sexual reproduction.

Parasexuality can be defined as a *phenomenon in which the three processes, e.g., plasmogamy, karyogamy and haploidization occur at an unspecified time at unspecified points in the life of a fungus.*

G. Pontecorvo and J.A. Ropper (1952) of **Glasgow** University, America, for the first time, discovered parasexuality in *Aspergillus nidulans*, the imperfect stage of *Emericella nidulans*. Later on, Pontecorvo (1956) continued his studies on parasexuality and discovered several steps involved in the process which are described below:

(a) Formation of heterokaryotic mycelium.
(b) Karyogamy and multiplication of diploid and haploid nuclei side by side.
(c) Mitotic crossing-over during multiplication or heterozygous diploid nuclei.
(d) Sorting out of diploid nuclei.
(e) Haploidization, and
(f) Sorting out of haploid nuclei.

(a) Formation of Heterokaryotic Mycelium—Existence of nuclei of different strains in the protoplasm of the cells of a hypha is termed 'heterokaryosis' and such hypha is termed 'heterokaryotic'. Heteroykaryosis generally occurs in the thallus of a fungus in the following ways:

(i) By ***anastomosis,*** i.e., plasmogamy between two genetically differing hyphae resulting in the entry of nuclei of different strains from one hypha into the cytoplasm of the other (Fig. 6.2A).

(ii) By ***mutation*** in homokaryotic multinucleate cells of a hypha followed by multiplication and spreading of the mutant nuclei from cell to cell in the hypha (Fig 6.2B).

(iii) By ***diploidization,*** i.e., fusion between two genetically similar nuclei of a homokaryon, the multiplication and spreading of the diploid nuclei among the haploid nuclei of the homokaryotic cells of the hypha (Fig. 6.2C).

(iv) By the ***germination*** of a heterokaryotic spore giving rise to a heterokaryotic hypha.

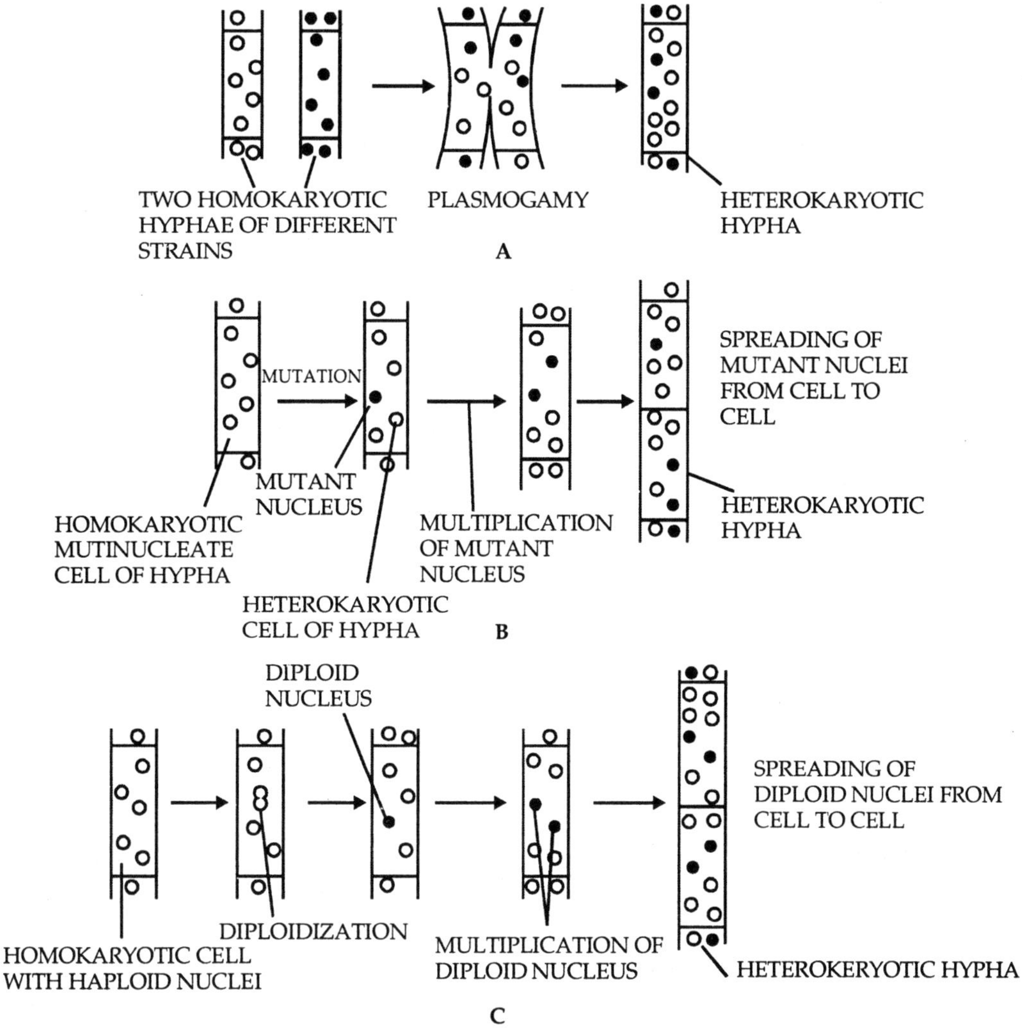

Fig. 6.2 : A Heterokaryosis by anastomoses. B. Heterokaryosis by mutation in homokaryotic multinucleate cells of a type. C. Heterokaryosis by diploidization.

b) Karyogamy and Multiplication of Diploid and Haploid Nuclei Side by Side— Roper (1952) first time studied karyogamy in somatic heterokaryotic hyphae of *Aspergillus nidulans*. When a hypha has become heterokaryotic, karyogamy occurs either between genetically similar nuclei or between nuclei of different genotypes resulting in the formation of homozygous or heterozygous diploid nuclei respectively. Thus, at this stage, hypha may possess five types of nuclei—two types of haploid nuclei (genetically different), two types of **homozygous diploid nuclei** (one type of homozygous diploid formed due to the fusion between two haploid nuclei of one genotype, and other type of homozygous diploid formed due to the fusion between two haploid nuclei of other genotype) and **heterozygous diploid**

nuclei. The latter case is seldom found in hyphae as Pontecorvo estimated a proportion of one heterozygous diploid nucleus to 1000 haploid nuclei.

(c) Mitotic Crossing-over During the Multiplication of Heterozygous Diploid Nuclei—Mitotic crossing over is an usual phenomenon occurring during the multiplication of heterozygous diploid nuclei resulting in new genetic recombinations in the parasexual cycle and hence is a most important phase. In *Penicillium chrysogenum* and *Aspergillus niger* diploidization and mitotic crossing-over take place frequently to produce many new genetic recombinations as the two species do not possess sexual reproduction. However, Pontecorvo calculated the rate of recombinant production through parasexual cycle 500 times lower than through its sexual cycle in an Ascomycete.

d) Sorting Out of Diploid Nuclei—The diploid nuclei formed due to the karyogamy and multiplication during parasexual cycle of an individual are sorted out by their incorporation in conidia thus forming uninucleate diploid conidia. Such conidia germinate and produce diploid thalli. Roper (1952) first time isolated diploid conidia in *Aspergillus nidulans*. Higher DNA content and bigger size differentiate the diploid conidia from those of the haploid ones.

(e) Haploidization—Haploidization is the process to bring the somatic chromosome number half. An important aspect in haploidization during parasexual cycle is its taking place through a process called **'aneuploidy'** in which the chromosomes are lost during mitotic division not during meiosis as it happens in a true sexual cycle. During anaphase stage of mitosis of the diploid nucleus some chromatids do not separate from each other resulting in one daughter nucleus getting one more chromosome (2N + 1) while the other daughter nucleus is less with one chromosome (2N - 1) than the normal set of chromosomes. These two daughter nuclei are referred to as *'aneuploids'*. An aneuploid nucleus with one chromosome less (2N – 1) may lose more chromosomes in the successive mitotic divisions and finally reduce to haploid condition (N).

(f) Sorting out of New Haploid Nuclei—While continuing his studies Pontecorvo marked the presence of sectors of haploid colonies in diploid colonies growing on the Petri plate. The hyphae of haploid sectors produce haploid conidia which may be isolated and grown into haploid colonies. Emergence of haploid sectors into the diploid colony point out that some diploid nuclei of hyphae undergo haploidization producing haploid nuclei which later get aggregated in the form of haploid conidia which germinate and produce haploid sectors. Mitotic crossing-over takes place during haploidization resulting in the formation of new recombinants. It is because some haploid nuclei are generally different from others.

Examples for strain improvement—Parasexual cycle has been successfully used to "breed" improved strains of *Penicillium chrysogenum, Aspergillus niger, Aspergillus nidulans*, some yeasts for commercial production of various primary as well as secondary metabolites.

6.4 PROTOPLAST FUSION

The protoplast fusion is one of the most important technique which enhances the potential for generating combination of desirable properties i.e., high yield, growth rate, degree of sporulation and rate of substrate consumption, that were important for fermentation processes has been accomplished in the improvement of industrial strains. Such industrially important filamentous fungal strain includes *Penicillium chrysogenum* and

Cephalosporium acremonium, and yeasts. Bacterial organisms include *Bacillus* spp., *Bravibacterium flavum* and *Streptomycetes*.

The principle of the protoplast fusion method is shown in figure 6.3 using an example the alkane yeast *Saccharomycopsis lipolytica*.

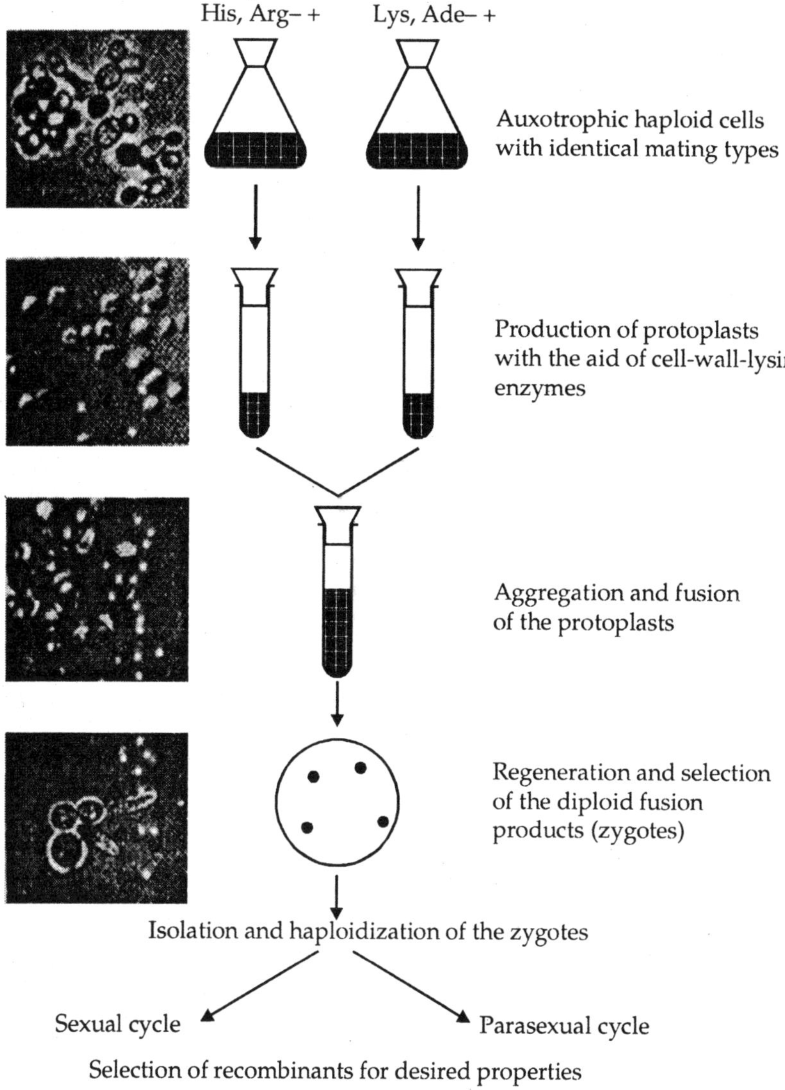

Fig. 6.3 : Production, fusion and regeneration of protoplasts as prerequisite for the recombination of genetic material shown for the case of the alkane yeast *Saccharomycopsis lipolytica*.

Protoplasts are cells from which the cell wall has been removed by enzyme treatment. For bacteria, the enzyme lysozyme is used, whereas for fungi, chitinase or cellulase is used. The protoplasts must be stabilized against lysis by suspension in a medium containing an osmotic stabilizing agent, such as sucrose. **Recombination by protoplast fusion or protoplast**

transformation is one of the most important developments in applied genetics in recent years. Protoplast fusion is normally rare because of the strong negative charge of the protoplast surface, but in the presence of polyethyleneglycol (PEG) the protoplasts aggregate and fusion occurs accompanied by DNA exchange. Besides the use of PEG to bring about fusion, the method of electric-field-induced fusion of protoplasts has been developed. This method results in a considerably higher frequency of protoplast fusion. After fusion, the cell wall is allowed to regenerate. In the regenerated progeny there is a significant number of recombinants.

Protoplast fusion can be used for the following—

- **Intraspecific recombination** of strains which lack sexual or parasexual systems or whose frequency of recombination is too low.
- **Interspecific hybridization** to obtain completely new organisms capable of synthesis of modified metabolites.

Engineered genes in plasmids or virus DNA can also be used to transform protoplasts.

An interesting alternative to PEG fusion is **electrofusion**, a technique developed by Zimmerman. When cells are placed in an alternating current electrical field, transient holes develop in the plasma membrane, promoting the process of membrane merging and cell fusion. With electrofusion, two or more protoplasts can be caused to fuse under microscopic control, or several cells can be fused into one giant cell. Further, protoplasts can be induced to fuse artificial phospholipid vesicles called *liposomes*. The fusion rate is around 80-90%, considerable higher than the 60% rate obtained using PEG as a fusogen. In addition to its use in the fusion of plant protoplasts, electrofusion can also be used with yeast and fungal protoplasts. However, the fusion of small bacterial protoplasts is more difficult to accomplish.

6.4.1 Intraspecific Recombination

Good genetic recombination systems exist for many industrial microorganisms: For various strains of *Bacillus, Lactobacillus, Streptococcus, Corynebacterium*, or *Brevibacterium*, in fungi such as *Aspergillus, Penicillium, Mucor, Claviceps*, and *Cephalosporium* strains, in yeasts such as *Candida, Saccharomyces, Kluyveromyces*, and in *actinomycetes* such as *Streptomyces, Micromonospora*, and *Nocardia*. A number of advantages arise from the use of these newer methods to achieve intraspecific recombination:

- Protoplast fusion is applicable in bacteria in which other recombination procedures have been successful. It is also applicable in *Bacillus strains*, which do not have a natural conjugation system. Gram-negative bacteria such as *E. coli* and *Providencia alcalifaciens* have also been used although the frequency of protoplast regeneration is rather low.
- The exchange of genetic material does not require the presence of fertility factors, and DNA can be recombined from up to four parental genotypes.
- The entire genome can be transferred, rather than only fragments as in conjugation, transduction, and transformation.
- The frequency of recombination is significantly increased.

An apparent increase in the formation of heterokaryons as a result of protoplast fusion has been found in some fungi. This is especially significant in *Cephalosporium acremonium*, in which the hyphal cells are mainly uninucleate, which hinders heterokaryon formation during anastomosis of hyphae. In this fungus, several recombinants with increased cephalosporin formation were isolated after protoplast fusion. Protoplast fusion has resulted in dramatic increases in frequency of recombination in some Streptomycetes, as shown in Table : 6.2.

Table 6.2 : Frequency of recombination after conjugation and protoplast fusion in streptomycetes

Strain	Mechanism of recombination	Recombination frequency
S. coelicolor A3(2)	Conjugation	1.7×10^{-6}
($SCP1-SCP2^+$	Protoplast fusion	4.7×10^{-2}
$\times$ $SCP1^--SCP2^+$)		
S. coelicolor A3(2)	Conjugation	1.7×10^{-7}
($SCP1^--SCP2^-$	Protoplast fusion	6.2×10^{-2}
$\times$ $SCP1-SCP2^{—}$)		
S. parvulus ATCC 12434	Conjugation	6.3×10^{-6}
	Protoplast fusion	0.6×10^{-2}
S. lividans 66	Conjugation	1×10^{-6}
	Protoplast fusion	6.0×10^{-2}
S. griseus CUB 94	Conjugation	2.3×10^{-6}
	Protoplast fusion	1.0×10^{-2}

After ultraviolet radiation (approximately 99% inactivation), a further enrichment of recombinants can be attained in *S. coelicolor,* because only viable cells remain when lethal lesions are eliminated through recombination. Thus the original frequency of recombination is increased from 10^{-2} to 10^{-1}.

At these high recombination rates, a population can be assayed for recombinants after protoplast fusion without the cumbersome use of selective markers. Hence, because of the development of high-efficiency protoplast fusion techniques, not only mutation and selection but also recombination and selection may be considered equally useful methods of strain development.

6.4.2 Interspecific Hybridization

This approach allows genetic information from different species to be combined *in vivo* in order to create new or modified products. The use of protoplast fusion for this objective is being examined by different antibiotic manufacturers.

Among the fungi, interspecific crosses have been attempted between *Aspergillus nidulans X A. rugulosus; A. nidulans X A. fumigatus; Penicillium chrysogenum X P. cyaneofulvum; P. cyaneofulvum X P. citrinum; Saccharomyces cerevisiae X S. diastaticus; Kluyveromyces lactis X K. fragilis.* Correct heterokaryons and recombinants have only been obtained in the cross between *P. cyaneofulvum X P. citrinum,* in which a fusion frequency of less than 10^{-5} was obtained as compared to an intraspecific frequency of 40-60%. Interspecific crosses among streptomycetes have given similar results. The low frequency of heterospecific fusion may because of genome inhomologies which prevent recombination from taking place or because of the presence of restriction/modification systems that lead to physiological incompatibility.

The fusion frequency of intergeneric crosses between *Candida tropicalis* X *Saccharomyces fibuligera* is also quite low, around 10^{-5}. Crosses between *Saccharomyces cerevisiae* X *Lipomyces kononkoae* showed several fusion products which exhibited metabolic properties of both parental strains, as well as the formation of nonparental segregants after haploidization.

6.4.3 Advantages of Protoplast Fusion Technique

(1) Technique yields a high number of recombinants.

(2) The whole genome is subjected to recombination in the process, and not just a protein of genome as occurs in conjugation process.

(3) This technique has given an enormous advantages in generating heterokaryons of the filamentous fungi—the most difficult task in the use of the parasexual cycle.

6.4.4 Transformation and Transfection of Protoplasts

Recombinant DNA technology made possible for the first time the development of methods for transformation of protoplasts with plasmid, chromosomal, or viral DNA. In the basic method, which is simple and generally applicable, protoplasts are treated with DNA in the presence of PEG and Ca^{2+}. In this way, all of the numerous techniques of genetic engineering become available for use with microorganisms of industrial interest (fungi, actinomycetes, other bacteria): amplification of genes, restoration of metabolism in deficient mutants, in vitro mutagenesis with protein engineering. Thus, even microorganisms for which conventional host-vector systems are not available can be handled. For example, excellent transformation systems have been developed for microorganisms of interest to the dairy industry such as *Streptococcus lactis* and various lactobacilli, for *Straphylococcus carnosus* and *Bacillus subtilis*, and vitamin B_{12}-producing *Propionibacterium freudenreichii*. Transformation systems have also been developed for β-lactam producing strains of *Streptomyces clavuligerus* and *S. wadayamensis*, vancomycin-producing *Nocardia orientalis*, erythromycin-producing *S. erythreus*, and gentamicin-producing *Micromonospora purpurea* subspecies *luridus*. Using virus vectors, transformation (transfection) systems have been described for *Streptomyces*, *Thermomonospora*, *Mycobacterium smegmatis*, and *Brevibacterium lactofermentum*.

6.4.5 Examples of Commercial Application

(1) Hamlyn and Ball fused strains of *Cephalosporium acromonium*, a cross between an asporulating, slow-growing strain, with a sporulating fast-growing strain which synthesized only one-third of the level of cephalosporin of the first strain; this resulted in the selection of a recombinant which sporulated well, had a high growth rate and synthesized higher level of cephalosporin.

(2) The use of protoplast fusion has resulted in the isolation of recombinants capable of synthesizing only Penicillin V with the desired yield.

(3) Recombinant formation of high-yielding lysine-producing strain of *Brevibacterium flavum*.

7

MEDIUM ENGINEERING AND METABOLIC ENGINEERING

7.1 INTRODUCTION

In fermentation or bioprocessing, microbial/enzymic or biochemical reactions need appropriate designing of the medium in which reactions occur. Many technological concerns are involved in medium design engineering (Fig. 7.1) in bioprocessing. For fermentative processing this need is to establish the most economic medium for any particular fermentation taking care of certain basic requirements to be met by any nutritional medium.

The medium design engineering for fermentation should take care of (a) source of energy, (b) source of carbon, (c) source of nitrogen, (d) source of minerals, and (e) source of growth factors in relation to technological concerns. On the other hand biocatalytic medium engineering take care of microenvironments consisting of a thin layer of water around the biocatalyst and an interfacial region, which forms the transition between the biocatalytic and continuous phases.

In recent years an increasing important development in the field of synthetic bioinorganic and bioorganic chemistry relates to application of biocatalysts to chemical reactions. This is not surprising. Biocatalysts are known to perform catalysis very effectively a wide variety of organic and inorganic reactions under relatively mild conditions in a stereo and/or region-selective way. So medium design engineering has many concerns. One of the most important concerns is optimization of a fermentation process in the design of the growth and production medium. Besides meeting nutritional and environmental requirements of the fermenting microorganism, the medium must also minimize several technoeconimic constraints.

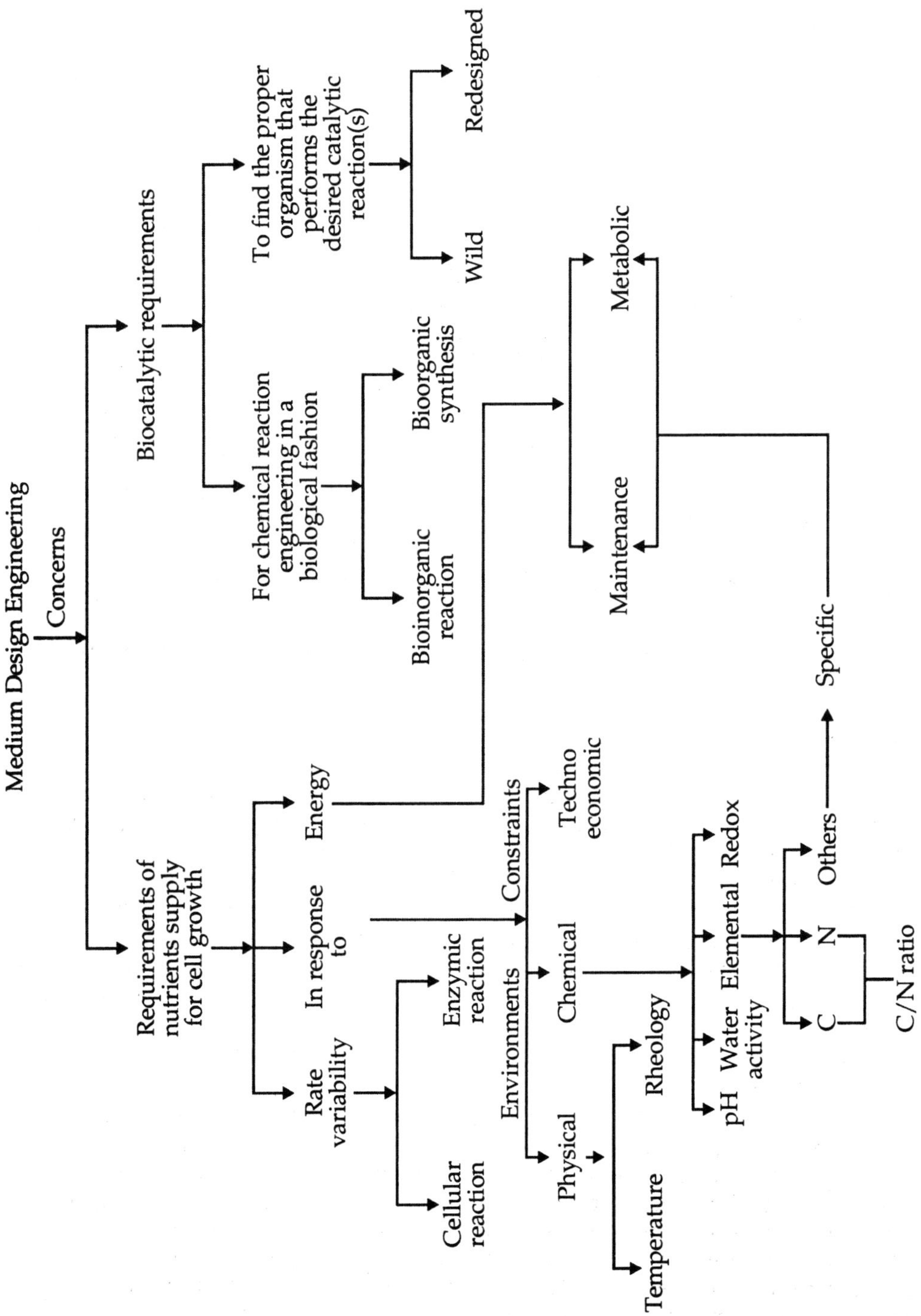

Fig. 7.1 : Technological Concerns of Medium Design Engineering in Bioprocessing

7.2 BASIC CONCEPT IN MEDIUM DESIGN

A. Fermentation

In meeting nutritional and environmental requirements the first approximation is the minimal medium design. Minimal medium design is based on the simple stoichiometry for growth and products formation as given below.

$$\text{C-source} + \text{N-source} + O_2 + \text{minerals} + \text{growth factors/specific nutrients}$$
$$\rightarrow \text{cell biomass} + \text{product} + CO_2 + 2H_20 + Q \text{ cals.}$$

The basic concept in such design is simply a material balance and energy balance.

B. Chemical reaction in Biological Fashion

In chemical reactions like bioorganic synthesis or bioinorganic transformations, medium engineering concerns designing microenvironments for the biocatalyst. The basic objective of this is to reach maximal activity, stability and productivity.

Reactant substrate containing liquid medium + biocatalyst
surrounded by ↓ microenvironments
products + coproducts + biocatalysts if any

Medium design engineering in such systems, therefore, involves aspects such as stabilizing the essential liquid layer around the biocatalyst, controlling electrostatic and hydrophobic interactions between substrate(s) and biocatalyst.

7.3 DESIGN PROCEDURE

Design of medium is carried out on the basis of the objective set for.

7.3.1 Growth and Production Medium

All growth processes are endergenic. In the course of cellular growth, organic and inorganic molecules like glucose, NH_3, water and CO_2 may be synthesised to form biomass product. Supply of energy is needed for growth. It must come from chemical energy of oxidation of medium nutrients or from light energy. In medium design procedure the first and foremost importance is to provide suitable and sufficient nutrients (reactants) and in appropriate proportions for a specified amount of biomass (products) to be synthesised.

The most suitable energy and/or carbonaceous nutrient source is selected on the basis of experimental results performed with different energy/carbon source. Present day industrial bioprocessing uses starch or molasses as the commonest carbon or energy source. For computing the necessary amounts of various substrates reaction stoichiometry is useful. It requires the knowledge of the elemental composition of the product biomass (e.g. microorganism). Besides biomass, if fermentative production of other materials occur, the chemical composition of the major products are also required in ascertaining nutrient medium composition. Design of nutrient medium composition is not only important in microbial cell culture system but also the nutrient regime is an important factor in mammalian and plant cell culture bio-processing. Its regime may affect both growth and product synthesis. For cultivations of plant cells most of the used medium has a defined design of chemical composition. The media design for the growth of plant tissue/cells are more complex and costly. This is due to the inclusion of plant growth regulators in the media. Optimization in media design in case of plant cell culture mostly confines to carbon and mineral nutrients. However, media supporting good biomass growth do not necessarily provide maximal yields of natural products. It necessitated defining 'maintenance media' and 'production media'. Maintenance media supports enhanced cell yield and cell division with reduced product yield. Production media on the other hand gives enhanced product yield at the expense of growth.

So in the medium engineering stoichiometric design based and experimental requirement based media formulation approaches have been used in bioprocessing.

7.3.2 Stoichiometric Design Approach: Material and Energy Balances

This approach is entirely based on laws of conservation of mass of each element taking part in the bioreaction. It is essential to balance the elemental requirements for growth and product formation. For example, in hetero-fermentative anaerobic bioreaction for production of propionic acid by *Propionibacter acidipropionici ATCC 25562* (at 30°C, pH 6.8 for 34 hours) the minimum carbon requirement in the medium may be determined stoichiometrically. In this bioprocessing the carbon content of substrate (glucose) was found to be distributed mainly in products propionic acid, acetic acid, CO_2 and cell mass. On elemental analysis the carbon content in the cell mass was found to be 44%. The stoichiometric carbon balance in this bioprocessing at the end of fermentation (34 hrs) period was as below. Propionic acid is rarely formed as a sole product in an acidogenic bioprocessing. It is usually accompanied by formation of acetic acid. This occurs for stoichiometric reasons and to maintain hydrogen and redox balance.

P. acidipropionici
30°C, pH 6.8
34 hrs

Glucose (S) ($5 gl^{-1}$) ⟶ Cell biomass ($2.24 gl^{-1}$ with 44% carbon)

⟵ CH_3CH_2COOH + CH_3COOH
($3.28 gl^{-1}$) ($1.15 gl^{-1}$)
+ CO_2 + Residual sugar
(1.095 g in fermenter) $0.1 gl^{-1}$

About 95% of the initial substrate carbon present could be accounted in various products. Trace amounts of succinic and isovaleric acids as reported by some workers requiring very sensitive methods of estimation could probably account for the rest of the 5% carbon. In anoxic bioprocessing every oxidation is accompanied by a reduction that yield product(s) formed from the substrate. Thus the balance between the oxidized and reduced product provides an important criterion for evaluation of the accuracy of the analysis. A oxidation-reduction (O/R) balance in propionic acid production by the above stated microorganism revealed the values as shown in Table : 7.1 and 7.2.

Table 7.1 : Substrate carbon distribution in propionic acid fermentation by *P. acidipropinici* (Substrate carbon distribution at the end of 34 h. gC in fermentor at zero hours = 4.75)

	g/l	g in fermentor	gC
CO_2	—	1.095	0.298
Propionic Acid	3.28	4.52	2.198
Acetic Acid	1.15	1.587	0.634
Cell mass	2.24	3.09	1.359
Residual Sugar	0.1	0.138	0.055
			4.544

Table 7.2 : Oxidation-reduction balance in propionic acid fermentation based on 100m Moles (30°C, initial pH 6.8)

Products	mMoles of product per 100 mMole of Glucose	O/R Number	Milli-equivalent of oxidized	Reduced
Acetic acid	32.95	0	—	—
Propionic acid	92.0	–1	—	92.00
Carbon dioxide	43.26	+2	86.52	—
			86.52	92.00

O/R Index = 0.94

The balance of hydrogen and oxygen were obtained by comparing their ratio in the substrate and product. When in a substrate 2H = 0, its oxidation value is zero. Every additional pair of hydrogen atom constitutes O/R number of —1 and every additional oxygen atom constitutes O/ R number of +1. From Table : 7.2 O/R index of 0.94 is observed.

An O/R index close to 1 indicated that all products were accounted for in the right proportions. The deviation from unity accounted for experimental error.

Cell composition and product composition data availability can enable one to assess the elemental requirements in medium design considering requirements and constraints (Table : 7.3).

Thus mass and energy balances associated with growing cultures play a great role in media design. The heat of reaction per electron transferred to oxygen for a wide variety of organic molecules, the number of available electrons per carbon atom in biomass and the weight fraction of carbon in biomass have been found to be relatively constant.

Table 7.3 : Nutritional and environmental requirements and techno-economic constrains in medium design

A. Nutritional and Environmental requirements
   1. Elemental requirements
   2. Specific nutrient requirements
   3. Energy requirements
   4. Environmental needs (culture)

contd. ....

... contd.

B. Fermentation Process Objectives
   1. Cell mass
   2. Product
   3. Maintenance
   4. Special growth conditions
      (a) avoid catabolite repression
      (b) avoid phosphate repression
      (c) limit a specific nutrient

C. Technoeconomic Constraints
   1. Cost
   2. Availability of raw materials
   3. Requirements of specific C— or N— sources
   4. Recovery
   5. Pollution control

7.4 BASICS OF MATERIAL BALANCE

In any process biotechnology consideration, material balance is essential for the sake of metabolic economy. Frameworks using material balances in metabolic pathways to study cellular metabolism has been investigated by various researchers. From material balances the concept of stoichiometric coefficients for the over all reaction representing the biological system has been introduced and were determined on line to provide information about yield factors, substrate consumption and product formation rates. Despite the excellent performance and the suitability of this approach for on line bioprocess monitoring and control, its application to infer cellular behaviour is limited, however. This is because the model does not contain any intracellular mechanisms and is therefore unable to provide essential information at the cellular level. Several researchers have paid attentions on this limitations. In their studies, material balances on the extracellular compounds were employed to deduce most probable metabolic model for citrate production from glucose. Also, fermentation balance equation for the acetone-butanol using metabolic biochemistry to describe intracellular reactions have been provided.

7.5 METABOLIC ENGINEERING

7.5.1 Introduction

One of the major aims in medium engineering is to make maximal conversion of the designed medium into useful product through cellular metabolic network.

In more recent years based on microbial biochemistry pathways of metabolism attempts have been laid to set up metabolic network stoichiometric models. For example, based on a review of the *Penicillium chrysogenum* biochemical pathways. Jorgensen (1995) and his associates have set up its stoichiometric models. Biochemical engineering analysis of these stoichiometric pathway models providing measures of individual step rates and metabolic fluxes have been defined by the name metabolic engineering. Recombinant methods have further enhanced the study of metabolic engineering. Application of recombinant DNA methods to restructure metabolic networks can improve production of metabolite and protein products by altering pathway distribution and rates. Cell functions can also be modified through precisely targeted alterations in normal cellular activities. Recruitment of heterologous proteins enables extension of existing pathways to obtain new chemical products, alter post-translational protein processing, and degrade recalcitrant wastes. The experimental and mathematical tool, required for rational metabolic engineering are becoming available more and more. However, complex cellular responses to genetic perturbations can complicate predictive designs. Thus, metabolic engineering naturally acquired design engineering (NADE)/redesigning of DNA that deals with metabolic pathway net work in a way that changes metabolic process of the cell. The cell manufactures more of the desired product as a result of the changes in several metabolic reactions.

Advantages of metabolic engineering has been noticed from cloning and expression of heterologous genes. These can serve several useful purpose, including extending an existing pathway to obtain a new product, creating arrays of enzymatic activities that synthesize a novel structure, shifting metabolic flow towards a desired product, and accelerating a rate determining step. So far the usefulness of metabolic engineering includes the following.

A. In recruiting heterologous activities for strain improvement

In such recruitment activities metabolic engineering is concerned with the following.

1. Synthesis of new products is enabled by completion of partial pathways
2. Transferring multistep pathways: Hybrid metabolic networks
3. Creating new products and new reactants
4. Transfer of promising natural motifs

B. Redirecting metabolic flow

The routes of reactions in a metabolic pathway is directed towards maximizing useful product formation by the following means—

1. Directing traffic towards the desired branch
2. Reducing competition for a limiting resource
3. Revising metabolic regulation

Thus consequence of completing metabolic engineering cycle may exhibit potentials and perils of rational design through:

1. cloning in industrial strains.
2. dissecting physiological responses.
3. design principles and cell models; coping with complexity and coupling.

7.5.2 Metabolic Engineering Case Examples

(A) Penicillin Fermentation

Through ages production of penicillin by *Penicillium chrysogenum* has been improved through the development of modified strains by repeated rounds of mutations and selections on the same parental strain *P. chrysogenum* NRRL 1951. The current estimated report of production of Penicillin from mutant strain from NRRL 1951 is around 40-50 g/L penicillin V after 200 h of cultivation. It is hoped that the present production strains of *P. chrysogenum* could be further improved with respect to penicillin production, and with the development of recombinant DNA technology it has become possible to use the more rational approach i.e. metabolic engineering to strain improvement than traditional mutation and selection.

In order to use metabolic engineering successfully the primary requirement of a map of the rate controlling step (s) in the metabolic pathway leading to penicillin production has been fulfilled.

(B) Lysine Fermentation

Lysine is an important essential amino acid. Its overproduction has been achieved using metabolic engineering principles. This was done by significant flux alterations in the primary metabolism of *Corynebacterium glutamicum* ATCC 21253 in its metabolic networks. In the new biotechnology area, chemical and biochemical engineers are more likely to find newer roles in quantification of fermentation processes based on metabolic engineering. It will enable them to alter genetic character of the amino acid or other product forming normal catabolic pathways of a cell so that new substrates can be used or new or enhanced chemical products can be formed by that cell.

(C) Plant Cell Culture Fermentation

In more recent years metabolic engineering has provided a new promising avenue for enhancing product formation in a plant or plant cell culture. Metabolic engineering of plant secondary metabolite pathways in the production of fine chemicals has been reported to be a possible approach to increase yields. Several ways of application of metabolic engineering to improve yields have been reported. Compartmentation strategy used for alkaloid biosynthesis in *Cantharanthus roseus* is one of the classical example of the same.

(D) Solventogenic Clostridial Fermentation

By developing suitable software appropriate experiments of *Clostridium acetobutylicum* ATCC 824 fermentation was designed for metabolic flux distribution analysis in the biochemical pathway of acetone-butanol solvent production. This software development served, therefore, as a tool for the metabolic engineering of solventogenic, clostridial fermentation and was useful in Stoichiometric modelling of the fermentation process.

7.5.3 Engineering Science Foundations of Metabolic Engineering

(i) Molecular stoichiometric model

For such a model a knowledge of the *in vivo* fluxes involved in associated reactions in metabolic pathway is essential. This is because metabolic engineering seeks to replace the shotgun random mutagenic approach with a more rational effort. It is an integrative procedure involving identification of metabolic bottlenecks or limitations followed by manipulation to alter metabolic fluxes. Many powerful molecular biology techniques have been developed in more recent years to manipulate the genetics of an organism. Also, these techniques enabled cloning and sequencing of genes for enzymes associated with the desired product. The resultant genetic repertoire served as an useful tool in redesigning of an organism to enhance product yields. The task of engineering science here is to devise process conditions so as to maintain stable cultures and to realize the full genetic potential of the used culture. Thus, thorough mapping of biosynthetic pathways is a prerequisite for any metabolic engineering programme. Such mapping is necessary to develop a so called fermentation equation to verify fermentation process data consistency, to develop gateway sensors and to predict maximum theoretical yields.

(ii) Appropriate computational software development

In metabolic engineering for assessing *in vivo* fluxes, various strategies are now available to resolve singularities. Many investigators have incorporated optimality concepts to promote linear programming methods of resolving singularities. This was done for general utility of reformulating any stoichiometric model as a nonlinear constrained minimization system. It is therefore, essential that appropriate computer software be constructed. Also, it should be able to process and/or act on data and make calculations which are beyond human comprehension. In metabolic engineering the calculation of *in vivo* fluxes becomes a nontrivial issue due to interactions of fermentation pathways.

7.6 BIOORGANIC REACTION MEDIUM ENGINEERING

In bioorganic reactions a biocatalyst in a liquid medium is surrounded by various microenvironments. The biocatalyst may be an enzyme, or more complex systems such as microorganisms, eucaryotic cells or parts thereof. The micro-environments consists of a thin layer of water around the biocatalyst and an interfacial region. This forms the transition between the biocatalytic and continuous phases. Medium engineering in such cases involves modification of the microenvironment of the biocatalyst either by introducing additives or solid matrices into an existing medium or by varying the composition of the liquid medium itself. Presently medium engineering is a rapidly expanding area of research in which the objectives of biocatalyst engineering viz. an increased biocatalytic reactivity, stability and hence productivity are pursued. The nature of different phases and their effects on biocatalysis are therefore investigated intensively. In bioorganic reactions generally each biocatalyst is surrounded by a thin layer of tightly bound water. This aqueous layer is believed to be essential for biocatalytic activity. Based on research findings following two important rules in aqueous medium engineering of bioorganic reactions have been laid down.

Rule 1. Do not distort the essential water-around the biocatalyst

Rule 2. Try to stabilize the biocatalyst conformation by strengthening the essential water-biocatalyst interactions

The first rule assuring the integrity of the biocatalyst provides expectation of full activity. Second rule on the other hand leads to possible way to achieve the goal in this rule.

In such reactions systems immediately adjacent to the essential water layer surrounding the biocatalyst is the interfacial region. Here the associated problem remains that it is difficult to predict for any given combination of properties of microenvironment what the effect of the matrix will be on the biocatalyst activity, stability and/or productivity. Also, electrostatic interaction effects of organic solvents on biocatalysts, the mutual hydrophobic interactions between matrix, reactants and biocatalyst are far less specific imposing another difficulty in medium design engineering. However for various biocatalytic systems a trend has been observed between biocatalytic activity and polarity of the interphase/bulk phase, when the latter was expressed as the logarithm of the partition coefficient (logp). For biocatalytic systems in organic solvents forming continuous phase like enzymes entrapped in reversed micelles the medium engineering principle should be such that optimal activities were reached when the concentration of substrate in the interphase is maximal. One way of achieving this is by matching the polarity of the interphase to that of the substrate, and/or vice versa, by increasing the difference in the polarity between the continuous phase and the substrate.

In medium engineering other considerations involved are solvent toxicity, viscosity, cost, reusability, ease of product isolation and bioreactor design.

Additives are also important in medium engineering. These are compounds capable of stabilizing the biocatalyst against denaturation without affecting the activity to a great extent. Few examples of additives are various sugars, glycerol, ethylene glycol and polyhydric polymers. They are important to stabilize various enzymes. However, practical use of additives seems to be restricted to biocatalyst.

The findings that enzymes can function in apolar solvents like heptand dramatically expanded the range of reactions which can be approached through biocatalytic medium engineering. The influencial factors for activity and stability of enzymes in organic solvents include the ionic state of the enzyme, support characteristics and the extent of hydration of biocatalyst and solvent. In bioprocessing of oils and fats through esterification and inter esterification for industrial application, medium design engineering served a potential role. Commercialization of cocoa butter is a classical example.

❑❑❑

8

FERMENTATION MEDIA FORMULATION

8.1 INTRODUCTION

Once a strain capable of producing a high yield of the desired product has been developed in the laboratory, it is necessary to design or formulate a fermentation medium economically usable on an industrial scale. All industrially important microorganisms require water, sources of energy, carbon, nitrogen, mineral elements, vitamins (plus oxygen if aerobic) as their nutrients. So, the choice of a good fermentation medium and raw materials is virtually as important to the success of an industrial fermentation as is the selection of an organism to carry out the fermentation.

The fermentation medium supplies nutrients for growth, energy, building of cell components, and biosynthesis of fermentation products. Of particular importance are the sources of carbon and nitrogen compounds in the fermentation medium, since microbial cells and fermentation products are composed largely of these elements. In addition to carbon and nitrogen compounds, a fermentation medium contains inorganic salts, water, vitamins and other growth factors, precursors of fermentation products, buffers, antifoam agents etc. All of these can be considered as an essential ingredients of fermentation media.

Microbial growth under industrial fermentation conditions usually utilizes a "luxury metabolism" of the organisms. Thus, good yields of fermentation products occur only if relatively large amounts of carbon, and possibly nitrogen, are channelled through the metabolic pathways of the microorganism.

8.2 CRITERIA FOR AN IDEAL FERMENTATION MEDIUM

(1) Fermentation medium should support good growth of fermenting organism. The constituents of a medium must satisfy the elemental requirements for cell biomass.

(2) Should produce the maximum yield of product or biomass per gram of substrate used or the medium should permit the maximum rate of product formation.

(3) There should be the minimum formation of undesired product.

(4) Nutrients of cheap cost like agricultural by-products should be utilized and they would be readily available throughout the year with consistent quality.

(5) Should provide the required quantities of micronutrients like S, P, Mg, K and trace elements like Fe, Zn, Cu, Mn, Co, Mo, B etc.

(6) Should cause minimal problems in other aspects of the production process particularly aeration and agitation, extraction, purification and waste treatment.

(7) Should have buffer capacity.

(8) Foam production should be minimum.

(9) Should inhibit the significant growth of contaminants.

(10) Should neutralize acidic or alkaline metabolites.

(11) Genetic stability should be maintained.

(12) There should be easy and cheap recovery of fermentation products.

(13) During sterilization, there should not be interactions between medium ingredients.

(14) The medium should provide precursors, if required.

(15) In certain cases, metabolic inhibitors are required to block the particular pathway for high yield which should be supplied by the medium.

(16) In certain cases, sporulation or suppression of sporulation is necessary for product formation which should be supported by fermentation medium.

8.3 TYPES AND COMPOSITION OF FERMENTATION MEDIA

The culture medium should contain all the essential elements required for growth of the microorganisms, in proportions similar to those occurring in the microbial cell (Table : 8.1). It is customary to distinguish between the macro-nutrient elements (required in large quantities, e.g. C, H, O, N), the meso-nutrient elements (required in smaller but significant quantities, e.g. Mg, P, S.) and the micronutrient (trace) elements required in only minute amounts such as Fe, Cu, Zn and Mo.

Table 8.1 : The main elements present in microorganisms (average composition)

Element	Physiological function	Dry weight (%)
Hydrogen	Constituent of organic compounds and water	8
Oxygen	Constituent of organic compounds and water	20
Carbon	Constituent of organic compounds	50
Nitrogen	Constituent of proteins, nucleic acids and coenzymes	14
Sulphur	Constituent of proteins and several conenzymes	1
Phosphorus	Constituent of nucleic acids, phospholipids and coenzymes	3
Magnesium	Cofactor of numerous enzyme reactions (ATP)	0.5
Manganese	Cofactor of several enzyme	0.1
Calcium	Cofactor of enzymes (proteases)	0.5
Iron	Constituent of cytochromes, non-haem proteins and cofactor of some enzymes	0.2
Cobalt	Constituent of vitamin B_{12}	0.03
Copper, Zinc, Molybdenum	Constituents of some enzymes	0.03

8.3.1 Simple and Complex Fermentation Media

The particular composition of a fermentation medium can be simple to complex depending on the particular microorganism and its fermentation. Autotrophic microorganisms require only the simplest of inorganic media. They require a few inorganic salts, water, and a nitrogen source; their carbon requirement is fulfilled by the carbon dioxide of the air or by carbonates. Thus, from simple inorganic nutrients, autotrophs are able to synthesize all of the complex organic compounds required to sustain life and to allow growth and multiplication of their cells, and they meet their total energy requirements by oxidation of some particular inorganic component of their medium. At the opposite end of the scale are the highly fastidious microorganisms, such as some of the lactic acid bacteria, which lack the ability to synthesize many of their sustenance and growth requirements. These organisms require the presence of many types of simple to complex preformed nutrients in the medium, and they must have an organic carbon supply to provide for synthesis of cell substance and release of metabolic energy. These are the two extremes and, obviously, microorganisms exist with requirements intermediate between these extremes. However, in addition to these considerations, fermentation growth conditions impose a metabolic stress on microorganisms, as for instance, the high aeration rates and high substrate levels commonly employed, so that additional nutrients and growth factors may be required as compared to the usual laboratory culture of the organisms. Thus, enzyme systems that normally are not limiting factors in metabolic sequences, because of a lack of sufficient levels of coenzymes or for other reasons, may become limiting under the stresses of fermentation growth: under these conditions more complex media are required than would normally be employed.

8.3.2 Synthetic and Crude Fermentation Media

Simple and complex media are further subdivided into two categories: **"synthetic"** and **"crude"**. A synthetic medium, at first glance, would appear to be the medium of choice. It is a medium which all of the constituents are specifically defined and known compounds. Every constituent is a relatively pure compound, and the exact amounts incorporated into the medium are known. Thus, an example of a synthetic medium would be one containing inorganic salts, water, purified sugar, and an ammonium or nitrate compound or an amino acid to supply combined nitrogen. Obviously, if the nitrogen requirement were to be met with something like dried blood, we would not know the exact components and their amounts in the medium.

8.3.2.1 Synthetic Media and Advantages

Synthetic media present distinct advantages for certain types of studies. Since the amount and chemical structure of each component are known, the concentration of any one component or of several components can easily be varied to determine its specific effect on cell growth and product yield. Also, individual components are easily deleted or added. These considerations allow the redesigning of a medium to obtain the greatest possible yield of product, and they also allow a study of the metabolic processes leading to or governing product formation. Studies of this type can yield additional information if one of the substrates or a part of the molecule of one of the substrates contains a radioactive or heavy isotope. Isotopes allow us to determine which nutrients or parts of nutrients end up as an integral part of the product. Thus, information of the latter type may allow us to predict alternate substrates or other means for further manipulation of the medium to increase yields. Synthetic media also provide other advantages. Because they are completely defined media, they aid in obtaining reproducibility of growth and product yield from one fermentation run to another so that error due to medium composition is held at a minimum. Foaming usually is not a problem with synthetic media, because they do not contain any protein or relatively high molecular weight peptides. The recovery and purification of fermentation products are relatively simple with these media, and because most of the compounds that might interfere with recovery are known. With advantages such as these, we might assume that synthetic media would find great use in industrial fermentations. However, this is not the case. These media can be expensive, because of the cost of the relatively pure ingredients used, and the yields derived from these media are relatively low. In regard to the latter point, the metabolic stress on the microorganism brought on by fermentation conditions may induce requirements for growth or biosynthesis factors which are not normally associated with the particular organism being employed, or for growth factors that are, as yet, unknown in relation to microbial metabolism.

8.3.2.2 Crude Media and Advantages

The alternative is the nonsynthetic or "crude" medium, which usually allows much higher yields of fermentation products than does a synthetic medium. An example of a crude medium is one containing soybean meal, blackstrap molasses, cornsteep liquor, $(NH_4)_2SO_4$, $CaCO_3$, and K_2HPO_4. It also may supply precursors for the biosynthesis of the fermentation product, and it can allow the use of crude antifoam agents without radically changing the nutritive balance of the medium. Thus, crude media contain crude or ill-defined sources of nutrients and, as such, they provide an excess of both nutrients and growth factors. However, it is assumed that a medium such as this does not contain metals, inorganic salts, or organic compounds toxic to the organism or to its product formation. Also, it is assumed that the crude carbon and nitrogen sources in such a medium should not provide a pentose sugar for an organism that can only use a hexose, protein for an organism that does not have proteolytic activity, or starch for an organism that lacks amylase.

Crude media, and to some extent synthetic media, must at times meet requirements in addition to providing nutrients for growth and product formation. These requirements include buffering capacity, lack of foam production, control of oxidation-reduction potential, inhibition or slowing of the growth of contaminating microorganisms, neutraliztion of acidic or alkaline growth products, contribution to maintenance of the genetic stability of the microorganism, promotion of vigorous aeration and agitation, allowing of recovery of the fermentation product without resort to complex recovery procedures, withstanding of sterilization without adverse interaction of nutritive constituents, allowing the growth of the organism in its proper morphological state, provision of precursors if required, allowing or preventing sporulation, provision of specific metabolic inhibitors if such are required, and provision of alternate substrates for a possible diauxic phenomenon. In addition, the medium must be economically feasible for the particular fermentation.

8.4 MEDIUM FORMULATION

Medium formulation is an essential stage in the design of successful laboratory experiments, pilot scale development and industrial manufacturing processes. The constituents of fermentation medium must satisfy the elemental requirements for cell biomass and metabolite production and there must be an adequate supply of energy for biosynthesis and cell maintenance.

The first step to consider is an equation based on the stoichiometry for growth and product formation, as shown under:

$$\begin{matrix}\text{Carbon and} \\ \text{energy source}\end{matrix} + \begin{matrix}\text{nitrogen} \\ \text{source}\end{matrix} + \begin{matrix}\text{other} \\ \text{requirements}\end{matrix} \rightarrow \begin{matrix}\text{cell} \\ \text{biomass}\end{matrix} + \text{products} + CO_2 + H_2O + \text{heat}$$

The above equation should be expressed in quantitative terms, which is important in the economical formulation of fermentation media if component wastage is to be minimal.

Following points should be considered during design or formulation of fermentation media:

(1) It should be possible to calculate the minimal quantities of nutrients which will be needed to produce a specific amount of biomass. Knowing that a certain amount of biomass is necessary to produce a defined amount of product, it should be possible to calculate substrate concentration necessary to produce required product yields. There may be medium components which are needed for product formation which are not required for biomass production.

(2) A knowledge of the elemental composition of a fermenting microorganism (which should include the content of C, H, O, N, S, P, Mg, K) is required in the solution of the elemental balance equation.

(3) Trace elements (Fe, Zn, Cu, Mn, Co, Mo, B) may also be needed in smaller quantities and needed to include in an initial medium recipe.

(4) Some microorganisms cannot synthesize specific nutrients like amino acids, vitamins, nucleotides etc. Once a specific growth factor has been identified, it can be incorporated into a fermentation medium in adequate amounts as a pure compound or as a component of a complex mixture.

(5) The carbon substrate has a dual role in biosynthesis and energy generation. So an adequate supply of the carbon source is very important and critical.

The carbon requirement under aerobic conditions may be estimated from the cellular yield coefficient (Y) which is defined as:

$$\frac{\text{Quantity of cell dry matter produced}}{\text{Quantity of carbon substrate utilized}} = Y$$

8.4.1 Importance of Carbon Sources in Medium Formulation

With the exception of the algae and autotrophic bacteria, which use carbon dioxide as a source of carbon, those microorganisms used in industry require organic compounds both as a source of carbon and of energy. Carbon is the most abundant elements, by weight, in microbial cells, comprising approximately 50% of the biomass; the concentration of nitrogen varies from 3-15%.

The carbon source is therefore the largest ingredient of the culture medium: if it is limiting, the total biomass X is proportional to the initial concentration *s* of the organic source of carbon:

$$X = Ys$$

The constant of proportionality Y in this equation is termed the yield constant for the substrate and the organism, and can be determined at the beginning of the stationary phase of a batch culture. The original definition of the yield constant by Monod arises from his observation that the rate of growth of a microorganism is proportional to the rate of consumption of the rate limiting substrate:

$$\frac{dX}{dr} = Y\frac{ds}{dr}$$

If *Y* were truly a constant it could be considered as the mass of biomass produced per unit mass of substrate consumed but, in fact, *Y* has been shown to vary with growth rate in experiments using the chemostat (Tables : 8.2)

Table 8.2 (a) : Yield constant Y of a variety of carbon and energy sources

Medium	Culture	cells/ml	DRY weight (g/1)	Y(%)
Nutrient broth with	*Escherichia*	5×10^9	1	12.5
8 g/l dry weight	*coli*			
Synthetic medium with	*Serratia*	2×10^{11}		

contd. ...

... contd.

Medium	Culture	cells/ml	DRY weight (g/1)	Y(%)
57 g glucose + 20g acetic acid/1	*marcescens*			
Mineral solution + 100 g glucose/1	*Candida utilis*		45	45

Table 8.2(b) : Cell yield from a variety of carbon and energy sources

Organism	Substrate	Cell yield Y (g dried cells/ mole substrate)
Saccharomyces ceteriside	glucose	20
Streptococcus faecalis	arginine	10.2
Aerobacter aerogenes	glycerol (m = 0.24 h^{-1})	36.8
Aerobacter aerogenes	glycerol (m = 0.004 h^{-1})	14.7
Pseudomonas $C_{12}B$	acetate	23.5

Data from Ribbons, D.W. (1970). Quantitative relationships between growth media constituents and cellular yields and composition in *Methods in Microbiology* Vol. 3A. ed J.R. Norris and D.W. Ribbons. Academic Press London.

The two part of Table : 8.2 indicate two of the different units in which cell yields may be expressed.

Amongst microorganisms a wide range of nutritional types is found. There are versatile species which can use as carbon source a large number of compounds ranging from simple C_2 compounds (acetic acid, ethanol) to complex compounds (polysaccharides and proteins) and even aromatic compounds. A good example is *Pseudomonas cepacia* (Table : 8.3) : it is worth recalling that another species of this genus (*Ps. testosteroni*) will degrade the aryl sylphonate group which terminates the alkyl chain of biodegradable detergents.

In contrast, there are microorganisms narrowly adapted to a single substrate which may well be unsuitable for the growth of others. Thus the methylobacteria *Methylomonas* and *Methylococcus* utilize only methane and methanol as sources of carbon and energy.

Table 8.3 : Organic compounds serving as carbon sources for *Pseudomonas cepacia*

Sugars and polyols		*Hydroxy-and Keto acids*	*Amino acids*	
Ribose	Acetate	Citrate	Alanine	Betaine
Xylose	Propionate	x-ketoglutarate	Serine	Sarcosine
Arabinose	Butyrate	Pyruvate	Threonine	Hippurate
Fucose	Isobutyrate	Aconitate	Asparate	Acetamide

contd. ...

... contd.

Rhamnose	Valerate	Citraconate	Glutamate	Nicotinate
Glucose	Isovalerate	Levulinate	Lysine	Trigonellin
Mannose	Caproate	Glycolate	Arginie	-
Galactose	Heptanoate	Malate	Histidine	*Non-*
Caprylate	Tartrate	Proline		nitrogenous
Sucrose	Pelargonate	Hydroxybutyrate	Tyrosine	aromatics
Trehalose	Caprate	Lactate	Phenylalanine	-
Cellobiose	-	Glycerate	Tryptophan	Benzoyl
Salicin	*Dicarboxylic*	Hydroxymethyl-	Cynurenin	formate
Gluconate	*acids*	glutarate	Cynurenate	Benzoate
2-ketogluconate	-	-		o-hydroxy-
Saccharate	Malonate	*Primary*	*Other*	benzoate
Mucate	Succinate	*alcohols*	*nitrogen*	m-hydroxy-
Mannitol	Fumarate	-	*compounds*	benzoate
Sorbitol	Glutarate	Ethanol	-	p-hydroxy-
Inositol	Adipate	Propanol	Anthranilate	benzoate
Adonitol	Pimelate	Butanol	Benzylamine	Phenyl-
Glycerol	Suberate	-	Putrescine	acetate
Butyleneglycol	Azelate	-	Spermine	Phenol
	Sebacate		Tryptamine	Quinate
			Butylamine	Testosterone
			Amylamine	

When the yield of ATP is known it is observed that the yield of biomass of a range of unicellular microorganisms (bacteria and yeasts) is proportional to the number of molecules of ATP formed from a number of different carbon and energy sources in defined media. It suffices to collect the biomass at the end of the growth phase and calculate the dry weight obtained as function of ATP produced during catabolism of the energy source (Table : 8.4).

Table 8.4 : Yield of anaerobic growth as a function of substrate and ATP production

			Dry weight of cells in grams	
Culture	**Energy source**	**Mole ATP/ mole substrate**	**(Per mole substrate)**	**(Per mole ATP)**
Saccharomyces cerevisiae	Glucose	2	21	10.5
Streptococcus faecalis	Glucose	2	22	11
	Arginine	1	10	10

contd. ...

... contd.

Culture	Energy source	Mole ATP/ mole substrate	Dry weight of cells in grams (Per mole substrate)	(Per mole ATP)
Lactobacillus delbrueckii	Glucose	2	21	10.5
Zymomonas mobilis	Glucose	1	8.6	8.6
Propionibacterium acidi-propionici	Glucose	4	37.5	9.4
	Glycerol	2	20	10
	Lactate	1	7.6	7.6

8.4.2 Modes of Energy Production

Energy is obtained in essentially two ways

(i) By substrate phosphorylation : In this case oxidation of the substrate, by loss of electrons, is accompanied by the synthesis of an 'energy-rich' phosphate which may be transferred to ADP with the formation of ATP. Thus, during glycolysis, microorganisms obtain 2 molecules ATP for every molecule of glucose consumed. Thus substrate level phosphorylation occurs in both aerobic and anaerobic situations because the initial reactions in the aerobic degradation of glucose do not, in fact, require free oxygen.

(ii) *By oxidative phosphorylation* In this case energy is conserved during the transfer of electrons along a respiratory chain. Three molecules of ATP are obtained for each pair of electrons transported form NADH to oxygen. Since there are 12 pairs of electrons available from each molecule of glucose metabolized via glycolysis followed by the Krebs cycle. 36 molecules of ATP may be obtained. Thus aerobic metabolism yields (36 + 2)/2 = 19 times more ATP than anaerobic. In general, aerobic microorganisms convert a much larger proportion (about 50%) of carbon substrates into biomass than do anaerobes, because they do not need to use so much for obtaining energy. It is possible to calculate the minimum quantity of a carbon substrate required in a medium to obtain a specified yield of biomass. For example, a preliminary experiment using a medium containing an excess of the substrate might show that it was possible to obtain a maximum of 40 g (dry weight) of bacterial cells per litre. The biomass is assumed to contain 50% of the carbon of the substrate into biomass carbon. The medium should therefore contain:

$$40 \times \frac{100}{50} \times \frac{50}{10} = 40 \text{ g carbon} / 1$$

If the carbon source were glucose it would be necessary to add to the medium:

$$\frac{40 \times 180}{72} = 100 \text{ g glucose} / 1$$

Pure sugars such as glucose and sucrose are generally too expensive to use industrially and it usually becomes necessary to find a cheap source of carbon, especially as this is quantitatively the most important medium constituent. There has thus been considerable interest in utilizing agricultural or industrial byproducts which may not be otherwise usable and may be stable in price. Moreover, a fermentation industry may choose between several primary materials depending on their cost, and contribute to regulating an essentially fluctuating market by using surpluses.

8.5 MEDIUM OPTIMIZATION USING STATISTICAL EXPERIMENTAL DESIGNS

8.5.1 Significance

In most instances, the microbiologist begins with some medium and set of conditions that allow for at least modest expression of the metabolite or activity of interest. The task then is to improve that expression to a level sufficient for isolation and characterization of the desired product. Preliminary experiments are usually done with shaken flasks or similar culture vessels to minimize time and cost involved in the study of medium optimization. Many of these nutritional factors will need to be evaluated to identify a fermentation process yielding an optimal result (response).

At this stage, it is important to consider the optimization of a medium such that it meets as many as possible of the criteria listed for an ideal fermentation medium in the beginning of the chapter. The meaning of optimization in this context does need careful consideration. When considering the biomass growth phase in isolation it must be recognized that efficiently grown biomass produced by an "optimised" high productivity growth phase is not necessarily best suited for its ultimate purpose, such as synthesizing the desired product. Different combinations and sequences of process conditions need to be investigated to determine the growth conditions which produce the biomass with the physiological state best constituted for product formation. There may be a sequence of phases each with a specific set of optimal conditions.

8.5.2 Need for Statistical Experimental Designs

Medium optimization by the classical method of changing one independent variable (nutrient, antifoam, pH, temperature, etc.) while fixing all the others at a certain level can be extremely time consuming and expensive for a large number of variables. To make a full factorial search which would examine each possible combination of independent variable at appropriate levels could require a large number of experiments, x^n, where x is the number of levels and n is the number of variables. This may be quite appropriate for three nutrients at two concentrations (2^3 trials) but not for six nutrients at three concentrations. In this instance 3^6 (729) trials would be needed. Industrially the aim is to perform the minimum number of experiments to determine optimal conditions. Other alternative strategies must therefore be considered which allow more than one variable to be changed at a time. These methods have been discussed by Stowe and Mayer (1966), McDaniel *et al.* (1976), Hendrix (1980), Nelson (1982), Greasham and Inarnine (1986), Bull *et al.* (1990) and Hicks (1993).

8.5.3 Plackett-Burman Design

When more than five independent variables are to be investigated, the Plackett—Burman design may be used to find the most important variables in a system, which are then optimized in further studies (Plackett and Burman, 1946). These authors give a series of designs for up to one hundred experiments using an experimental rationale known as balanced incomplete blocks. This technique allows for the evaluation of $X - 1$ variables by X experiments. X must be a multiple of 4, e.g. 8, 12, 16, 20, 24, etc. Normally one determines how many experimental variables need to be included in an investigation and then selects the Plackett—Burman design which meets that requirement most closely in multiples of 4. Any factors not assigned to a variable can be designated as a dummy variable. Alternatively, factors known to not have any effect may be included and designated as dummy variables. As will be shown shortly in a worked example (Table : 8.5), the incorporation of dummy variables into an experiment makes it possible to estimate the variance of an effect (experimental error).

Table 8.5 : Plackett-Burman design for seven variables (Nelson, 1982)

Trial	Variables							Yield
	A	*B*	*C*	*D*	*E*	*F*	*G*	
1	H	H	H	L	H	L	H	1.1
2	L	H	H	H	L	H	L	6.3
3	L	L	H	H	H	L	H	1.2
4	H	L	L	H.	H	H	L	0.8
5	L	H	L	L	H	H	H	6.0
6	H	L	H	L	L	H	H	0.9
7	H	H	L	H	L	L	H	1.1
8	L	L	L	L	L	L	L	1.4

H denotes a high level value; L denotes a low level value.

Table : 8.5 shows a Plackett-Burman design for seven variables *(A—G)* at high and low levels in which two factors, *E* and G, are designated as 'dummy' variables. These can then be used in the design to obtain an estimate of error. Normally three dummy variables will provide an adequate estimate of the error. However, more can be used if fewer real variables need to be studied in an investigation (Stowe and Mayer, 1966). Each horizontal row represents a trial and each vertical column represents the H (high) and L (low) values of one variable in all the trials. This design (Table : 8.5) requires that the frequency of each level of a variable in a given column should be equal and that in each test (horizontal row) the number of high and low variables should be equal. Consider the variable *A*; for the trials in which *A* is high, *B* is high in two of the trials and low in the other two. Similarly, *C* will be high in two trials and low in two, as will all the remaining variables. For those trials in which *A* is low, *B* will be high two times and low two times. This will also apply to all the other variables. Thus, the effects of changing the other variables cancel out when determining the effect of *A*. The same logic then applies to each variable. However, no changes are made to the high and low values for the *E* and G columns. Greasham and Inamine (1986) state that although the difference between the levels of each variable must be large enough to ensure that the optimum response will be included, caution must be taken when setting the level differential for sensitive variables, since a differential that is too large could mask the other variables. The trials are carried out in a randomized sequence.

The effects of the dummy variables are calculated in the same way as the effects of the experimental variables. If there are no interactions and no errors in measuring the response, the effect shown by a dummy variable should be 0. If the effect is not equal to 0, it is assumed to be a measure of the lack of experimental precision plus any analytical error in measuring the response (Stowe and Mayer, 1966).

This procedure will identify the important variables and allow them to be ranked in order of importance to decide which to investigate in a more detailed study to determine the optimum values to use.

The stages in analysing the data (Tables 8.5 and 8.6) using Nelson's (1982) example are as follows:

1. Determine the difference between the average of the H (high) and L (low) responses for each independent and dummy variable.

   Therefore the difference = ΣA (H) – ΣA (L)

   The effect of an independent variable on the response is the difference between the average response for the four experiments at the high level and the average value for four experiments at the low level.

   Thus the effect of

$$A = \frac{\Sigma A(H)}{4} - \frac{\Sigma A(L)}{4}$$

$$= \frac{2(\Sigma A(H) - \Sigma A(L))}{8}$$

   This value should be near zero for the dummy variables.

2. Estimate the mean square of each variable (the variance of effect).

   For A the mean square will be =

$$\frac{(\Sigma A(H) - \Sigma A(L))^2}{8}$$

3. The experimental error can be calculated by averaging the mean squares of the dummy effects of E and G.

   Thus, the mean square for error =

$$\frac{0.045 + 0.020}{2} = 0.0325$$

   This experimental error is not significant.

4. The final stage is to identify the factors which are showing large effects. In the example this was done using an F-test for

$$\frac{\text{Factor mean square.}}{\text{error mean square.}}$$

   This gives the following values:

$$A = \frac{15.125}{0.0325} = 465.4,$$

$$B = \frac{13.005}{0.0325} = 400.2,$$

$$C = \frac{0.0500}{0.0325} = 1.538$$

$$D = \frac{0.0000}{0.0325} = 0.00$$

$$F = \frac{10.580}{0.325} = 325.6.$$

When Probability Tables are examined *it* is found that Factors *A*, *B* and *F* show large effects which are very significant, whereas *C* shows a very low effect which is not significant and *D* shows no effect. *A*, *B* and *F* have been identified as the most important factors. The next stage would then be the optimization of the concentration of each factor, which wilt be discussed later.

Nelson (1982) has also referred to the possibility of two factor interactions which might occur when designing Table : 8.5. This technique has also been discussed by McDaniel *et al* (1976), Greasham and Inamine (1986), Bull *et al* (1990) and Hicks (1993).

Table 8.6 : Analysis of the yields shown in Table 8.5 (Nelson, 1982)

	Factor						
	A	B	C	D	E	F	G
Σ(H)	3.9	14.5	9.5	9.4	9.1	14.0	9.2
Σ (L)	14.9	4.3	9.3	9.4	9.7	4.8	9.6
Difference	-11.0	10.2	0.2	0.0	-0.6	9.2	-0.4
Effect	-2.75	2.55	0.05	0.00	-0.15	2.30	-0.10
Mean square	15.125	13.005	0.005	0.000	0.045	10.580	0.020

Mean square for 'error' $= \frac{0.045 + 0.020}{2} = 0.0325$

8.5.4 Response Surface Methodology [RSM]

The next stage in medium optimization would be to determine the optimum level of each key independent variable which has been identified by the Plackett—Burman design. This may be done using response surface optimization techniques which were introduced by Box and Wilson (1951). Hendrix (1980) has given a very readable account of this technique and the way in which it may be applied. Response surfaces are similar to contour plots or topographical maps. Whilst topographical maps show lines of constant elevation, contour plots show lines of constant value. Thus, the contours of a response surface optimization plot show lines of identical response. In this context, response means the result of an experiment carried out at particular values of the variables being investigated.

The axes of the contour plot are the experimental variables and the area within the axes is termed the response surface. To construct a contour plot, the results (responses) of a series of experiments employing different combinations of the variables are inserted on the surface of the plot at the points delineated by the experimental conditions. Points giving the same results (equal responses) are then joined together to make a contour line. In its simplest form two variables are examined and the plot is two dimensional. It is important to appreciate that both variables are changed in the experimental series, rather than one being maintained constant, to ensure that the data are distributed over the response surface. In Fig. 8.1, the profile generated by fixing X_1 and changing X_2 and then using the best X_2 value and changing X_1 constitutes a cross which may not encroach upon the area in which the optimum resides.

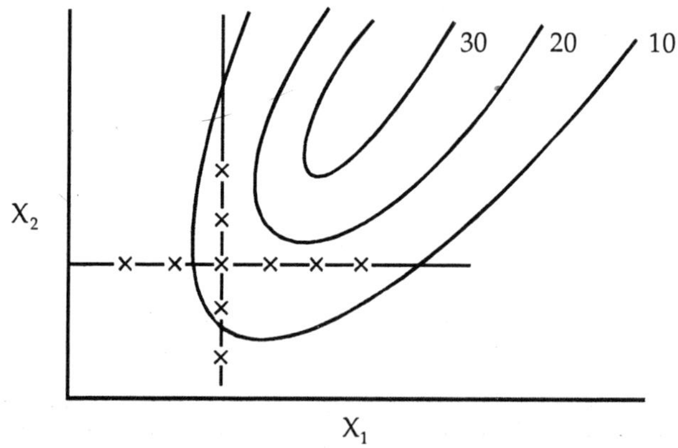

Fig. 8.1 : Optimal point of a response surface by one factor at a time.

The technique may be applied at different levels of sophistication. Hendrix applied the technique at its simplest level to predict the optimum combination of two variables. The values of the variables for the initial experiments are chosen randomly or with the guidance of previous experience of the process. There is little to be gained from using more than 15-20 experiments. The resulting contour map gives an indication of the area in which the optimum combination of variables resides. A new set of experiments may then be designed within the indicated zone. Hendrix proposed the following strategy to arrive at the optimum in an incremental fashion:

1. Define the space on the plot to be explored.
2. Run five random experiments in this space.
3. Define a new space centred upon the best of the five experiments and make the new space smaller than the previous one, perhaps by cutting each dimension by one half.
4. Run five more random experiments in this new space.
5. Continue doing this until no further improvement is observed, or until you cannot afford any more experiments.

The more sophisticated applications of the response surface technique use mathematical models to analyse the first round of experimental data and to predict the relationship between the response and the variables. These calculations then allow predictive contours to be drawn and facilitate a more rapid optimization with fewer experiments. If three or more variables are to be examined then several contour maps will have to be constructed. Hicks (1993) gives an excellent account of the development of equations to model the different interactions which may take place between the variables. Several computer software packages are now available which allow the operator to determine the equations underlying the responses and, thus, to determine the likely area on the surface in which the optimum resides. Some examples of the types of response surface profiles that may be generated are illustrated in Fig. 8.2.

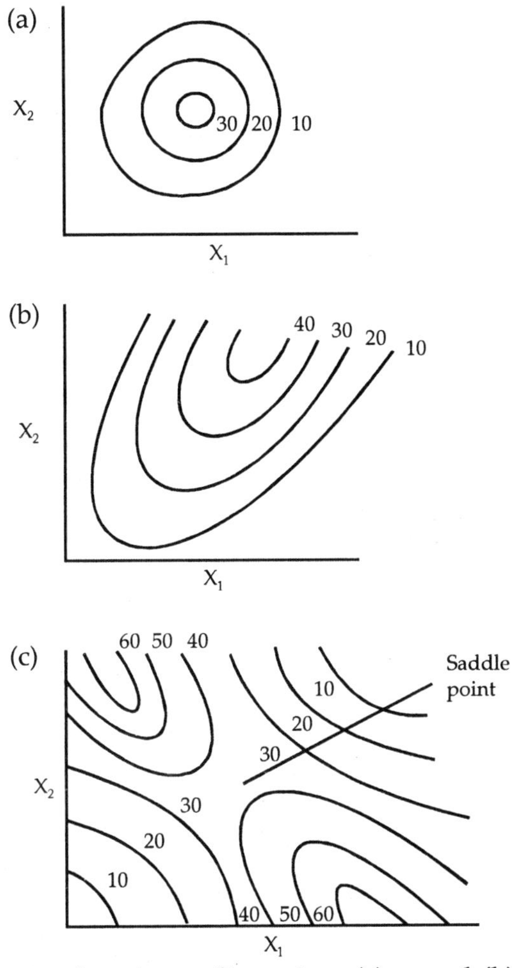

Fig. 8.2 : Typical response surfaces in two dimensions; (a) mound, (b) rising ridge, (c) saddle.

The following examples illustrate the application of the technique:

(i) McDaniel *et al.* (1976), Fig. 8.3. The variables under investigation were cerelose and soybean level, with the analysis indicating the optimum to be 6.2% cerelose and 3.2% soybean.

(ii) Saval *et al.* (1993). The medium for streptomycin production was optimized for four components resulting in a 52% increase in streptomycin yield, a 10% increase in mycelial dry weight and a 48% increase in specific growth rate (Table : 8.7).

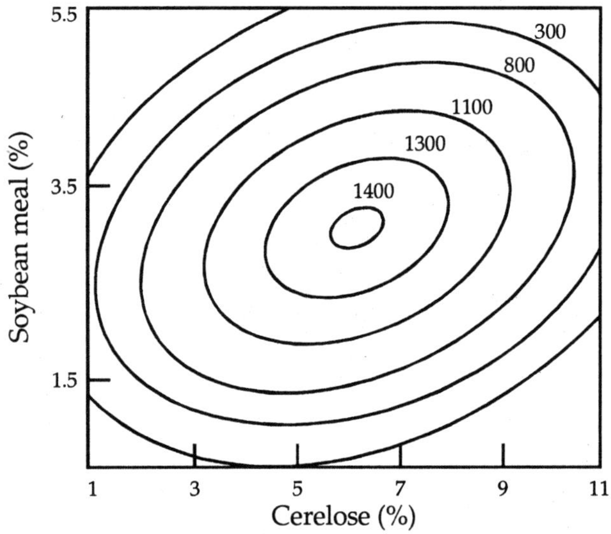

Fig. 8.3 : Contour plot of two independent variables, cerelose and soybean meal, for optimization of the Candidin fermentation (Redrawn from McDaniel *et al.*, 1976; Bull *et al.*, 1990).

Table 8.7 : Concentration of nutrients in an original and optimised medium for streptomycin production (Saval *et al.* 1993).

Nutrient (g dm^{-3})	Original medium	Optimized medium
Glucose	10	23
Beer-yeast autolysate	25	27
NaCl	10	8
K_2HPO_4	1	1

8.5.5 Simplex Search Method

When further optimization experiments are necessary for medium development in large vessels, the number of experiments will normally be restricted because of the cost and the lack of spare large vessels (Spendley *et al.*, 1962). The simplex search method attempts to optimize *n* variables by initially performing $n + 1$ experimental trials. The results of this initial set of trials are then used to predict the conditions of the next experiment and the situation is repeated until the optimum combination is attained. Thus, after the first set of trials the optimization proceeds as individual experiments. The prediction is achieved using a graphical representation of the trials where the experimental variables are the axes. Using this procedure, the experimental variables are plotted and not the results of the experiments. The initial experimental conditions are chosen such that the points on the graph are equidistant from one another and form the vertices of a polyhedron described as the simplex. Thus, with two variables the simplex will be an equilateral triangle. The results of the initial set of three experiments are then used to predict the next experiment enabling a new simplex to be constructed. The procedure will be explained using an example to optimize the concentrations of carbon and nitrogen sources in a medium for antibiotic production.

In our example a graph is constructed in which the x axis represents the concentration range of the carbon source (the first variable) and the y axis represents the concentration range of the nitrogen source (the second variable). The first vertex A (experimental point) of the simplex represents the current concentrations of the two variables which are producing the best yield of the antibiotic. The experiment for the second vertex B is planned using a new carbon-nitrogen mixture and the position of the third vertex C can now be plotted on the graph using lengths AC and BC equal to AB (the simplex equilateral triangle, Fig. 8.4a). The concentrations of the carbon and nitrogen sources to use in the third experiment can now be determined graphically and the experiment can be undertaken to determine the yield of antibiotic. The results of the three experiments are assessed and the worst response to antibiotic production identified. In our example, experiment A was the worst and B the best.

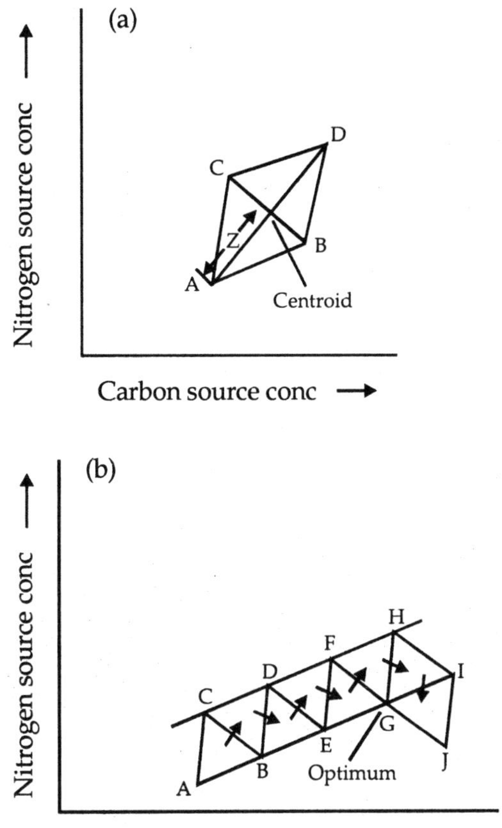

Fig. 8.4 : (a) Simplex optimization for a pair of independent variables (with reflection). (b) Simplex optimization of pair of independent variables which has reached the optimum.

The simplex design is now used to design the next experiment. A new simplex (equilateral triangle) BCD is constructed opposite the worst response (i.e. A) using the existing vertices B and C. A line is drawn from A through the centroid (mid point) of BC. D (the next experiment) will be on this line and the sides BD and CD will be the same length as BC.

This process of constructing the new simplex is described as reflection. Once the position of D is known, the concentrations of the carbon and nitrogen sources can be determined graphically, the experiment performed and the production of antibiotic assayed. Thus, a series of simplexes can be constructed moving in a crabwise way. The procedure is continued until the optimum is located. At this point the simplex begins to circle on its self, indicating the optimum concentration (Fig. 8.4b; Greasham and Inamine, 1986). However, if a new vertex exhibits the lowest response, the simplex would reflect back on to the previous one, halting movement towards the optimum. In this case the new simplex is constructed opposite the second least desirable response using the method previously described. If it is decided that the supposed optimum should be reached more rapidly then the distance z between the centroid and D may be increased (expanded) by a factor which is often 2. If the optimum is thought to have nearly been reached then the distance z may be decreased by a factor of 0.5 (contraction). This modified simplex optimization was first proposed by Nelder and Mead (1965) and has been discussed by Greasham and Inamine (1986).

The simplex method may also be used in small scale media development experiments to help identify the possible optimum concentration ranges to test in more extensive multifactorial experiments.

❑❑❑

9

Raw Materials for Fermentation Media

9.1 INTRODUCTION

The choice of a good fermentation medium and raw materials is virtually as important to the success of an industrial fermentation as is the selection of a fermentative organism to carry out the fermentation.

The fermentation medium supplies nutrients for growth, energy, building of cell components, and biosynthesis of fermentation products. Of particular importance are the sources of carbon and nitrogen compounds in the fermentation medium, since microbial cells and fermentation products are composed largely of these elements. In addition to carbon and nitrogen compounds, a fermentation medium contains inorganic salts, water, vitamins and other growth factors, precursors of fermentation products, buffers, antifoam agents etc.

All of these can be considered as an essential raw materials of fermentation media and these essential raw materials are elaborately discussed in this chapter.

9.2 CARBON SOURCES

Quantitatively carbon sources represent the main component of nutrient media.. They serve not only for the biosynthesis of cell matter, but also as the main source of energy. The earlier concept, according to which only sugars were the most suitable or even the sole carbon sources for biochemical processes, while other organic compounds were thought to be biologically ineffective, has been corrected in recent years owing to the increasing knowledge of microbial metabolism. Thus some branches of the microbial industry have recently begun using entirely new substrates such as synthetic alcohols and n-alkanes. The nature of the microbial process requirements for product purity and the metabolism of the production strain may dictate the use of chemically defined carbon sources such as glucose, sucrose, lactose or starch, as well as complex raw materials such as molasses, sulphite liquors, wood hydrolysates, cellulose, and raw materials of petrochemical origin.

9.2.1 Carbohydrates

Carbohydrates form a distinct group of organic compounds and, according to the number of building blocks, they are usually divided into three groups—monosaccharides, oligosaccharides and polysaccharides. In fermentation processes glucose is the most frequently used monosaccharide; the most often used oligosaccharides are sucrose and lactose, polysaccharides with wide application include starch and cellulose.

9.2.1.1 Monosaccharides

Glucose occurs in nature mainly in fruits of shrubs and trees. For instance, grapes contain 10-30% of glucose. Glucose is obtained industrially by hydrolysis of starch either chemically or by enzymes. It is rapidly assimilated by many microorganisms. Because of its rapid assimilation, glucose may therefore be added continuously to growing cultures of microorganisms, especially for the biosynthesis of more complex compounds. Glucose must be sterilised separately from other medium components because at neutral pH and high temperatures it reacts with amino acids, giving rise to dark-brown solutions. It is usually sterilised at pH 3 as a 30-50 % solution (w/v); under these conditions the solution remains colourless. In microbial processes the so-called hydrol, a waste product in the manufacture of pure glucose, can also be used.

9.2.1.2 Oligosaccharides

Lactose is contained in the milk of mammals at levels of 3-8%. It is prepared by evaporation of whey which remains after the processing of milk to butter. Only some microorganisms can assimilate lactose; yeasts, with a few exceptions, do not assimilate this oligosaccharide. The rate of assimilation by some moulds, such as *Penicillium chrysogenum,* is low but efficient and lactose was formerly used in the production of penicillin. Nowadays a continuous dosage of glucose or sucrose is used for the purpose instead of lactose. The main disadvantage of lactose is its varying quality caused by differences in milk processing.

Sucrose is obtained from sugar beet or sugar cane. It is available in different purities as refined sugar, raw sugar and molasses. Sucrose is now used as a major source of carbon in the biosynthesis of many antibiotics, amino acids and organic acids.

9.2.1.3 Polysaccharides

These substances represent a suitable carbon source for bacteria, yeasts and fungi and are used in the biosynthesis of many products. They are macromolecular, forming colloidal solutions or exhibiting negligible solubility in water; for this reason they must often be converted into a soluble form before being used in nutrient media.

(a) Starch is one of the most important polysaccharides both biologically and industrially. It is accumulated as a reserve substance in roots, fruits, seeds and bulbs of higher plants in the form of granules of different shapes and sizes. It is usually denoted according to its origin, i.e. according to the plant from which it is manufactured—potato, maize, wheat, rye, or barley starch. Starch granules are hygroscopic and do not change at low temperatures while at high temperatures they swell and change into a starch gel. The gelation is usually performed at 100°C in water (1: 4). At 120-130°C the gel liquefies and may be easily degraded by enzymes. Starch is a mixture of two related polysaccharides, amylose (linearly arrayed chain of α-glycosidically linked molecules of D-glucose with 1 → 4 bonds) and amylopectin (branched chain of D-glucose molecules bonded also α-glycosidically by 1 → 4 and, less often, 1 → 6 bonds; one molecule contains more than one thousand glucose residues). Acid hydrolysis of amylose yields first maltose and then glucose. Enzymic degradation by amylases proceeds *via* amylodextrin and low-molecular weight dextrins to maltose.

(b) Cellulose—Another important polysaccharide is cellulose, which forms a substantial part of all plant cells and is thus the most widespread biological substance in nature. Isolation of pure cellulose is difficult as it is naturally accompanied by other substances, especially lignin, hemicellulose and waxes. Industrially the most important source is the cellulose obtained from wood for the paper industry, and from the straw of various plants. However, its mass utilisation has not yet been solved because of economical factors.

9.2.2 Molasses

Beet and cane molasses are by-products of the sugar industry. These molasses are the concentrated syrups or mother liquors recovered at any one of several steps in the sugar-refining process, and different names are applied to the molasses depending on the particular step from which it was recovered.

(i) Black strap molasses : Of these, **blackstrap molasses** prepared from sugar cane normally is the cheapest and the most used sugar for industrial fermentations. In the commercial production of sugar, the juice from crushed sugar cane is concentrated to allow crystallization of its sucrose. The crystallized sugar is then separated from its mother liquor, and the mother liquor is further concentrated to allow recovery of additional crops of crystalline sugar. This procedure is repeated several times until crystallization inhibitors accumulate to such a concentration that further recovery of sucrose is not economical. At this point, the mother liquor still contains approximately 52 per cent total sugars calculated as sucrose (30 per cent sucrose, and 22 per cent invert sugars) and is known as black-strap molasses. When this molasses is used as a fermentation medium component, it is considered to contain 50 per cent fermentable sugars. **Refinery blackstrap** molasses is a similar product that differs from black-strap molasses only in that it is the residual mother liquor that has accumulated in the recrystallization refining of the crude sucrose.

(ii) High-test or invert molasses contains approximately 70 to 75 percent sugar, and it is produced in a manner different from that previously described. The whole cane juice is partially inverted to prevent sugar crystallization; that is, the sugar is partially hydrolysed to monosaccharides with heat and acid then neutralized and concentrated without removal of any of the sugar. Thus, high-test or invert molasses contains much of the original sugar of the cane juice, although it has been partially hydrolyzed to D-glucose and D-fructose. It is preferred to blackstrap molasses because of the lower shipping charges on a sugar concentration basis and because of its lower levels of nonfermentable solids including salts and unfermentable sugars. In blackstrap molasses, the unfermentable sugars result from the action of heat on the sugars, particularly fructose, during the refining process. High-test molasses is produced only during years of sugar cane overproduction and, hence, its availability at any one time may be somewhat questionable.

(iii) Beet molasses and Hydrol : Beet molasses are produced by procedures resembling those for sugar cane. However, beet molasses may be limiting in biotin for yeast growth so that a small amount of cane blackstrap or other source of —biotin should be added for growth of these microorganisms. "Hydrol" is a molasses resulting from the manufacture of crystallin dextrose from corn starch. It contains approximately 60 per cent sugar, but it also contains a relatively high salt concentration that must be considered if this molasses is to be used as a medium component.

In addition to sucrose, blackstrap molasses contains small amounts of complex polysaccharide and invert sugars. The presence of the invert sugars is attributed to the action of the "invertase" enzyme present in the original cane juce. Blackstrap molasses also contains various noncarbohydrate materials. Thus, dark coloured, nitrogen-containing polymeric substances result from "browning", a reaction of the sugars with amino acids because of the heat and alkali used in processing. Inorganic ions are present in high concentrations and include most of the ions of the original cane juice which were concentrated in the mother liquor during sugar crystallization. Calcium also is present, being added during processing. Organic-acid constituents include aconitic, malic, citric, lactic, formic, acetic, and propionic acids. The nitrogen-containing compounds (other than the polymeric forms) are mainly amino acids and, particularly, aspartic and glutamic acids resulting from the deamidation of the asparagine and glutamine of the cane juice. A few heat and alkali stable vitamins are present, such as myo-inositol, niacin, pantothenic acid, riboflavin, and small amounts of biotin. Also present are organic phosphorus compounds such as inositol hexaphosphate and its calcium-magnesium salt known as "phytin":

The overall compositions of the various molasses differ according to the specific geographic areas of production. This is particularly true for their contents of certain metal ions and, in fact, for certain fermentations, such as that for citric acid, the molasses is pretreated with cation exchange resins or potassium ferrocyanide before use so as to remove interfering cations. The composition of beet molasses and cane molasses are mentioned in Table : 9.1.

Table 9.1 : Composition of beet and cane molasses

Component	Molasses	
	Beet %	Cane %
Dry weight	78.0-85.0	78.0-85.0
Total carbohydrates	48.0-58.0	50.0-58.0
N	0.2-2.8	0.08-0.5
P_2O_5	0.01-0.02	0.009-0.07
CaO	0.15-0.7	0.15-0.8
MgO	0.01-0.1	0.25-0.8
K_20	2.2-4.5	0.8-2.2
$Si0_2$	0.1-0.5	0.05-0.3
Al_20_3	0.005-0.06	0.01-0.04
Fe_20_3	0.001-0.02	0.001-0.01
C	28.0-34.0	28.0-33.0
Ash	4.0-8.0	3.5-7.5

9.2.3 Barley and its Products

A traditional source of carbon, used especially in brewing and formerly also in yeast production, is barley and its products such as malt, wort and malt extract. The contents of individual carbohydrates in barley are given in Table : 9.2. Malt is prepared by germination of barley. During malting amylases and pectinases are only partially activated since the process takes place at low temperatures and in a medium containing only vegetable liquor. In the brewery malt is crushed, mixed with water and kept at a higher temperature; under these conditions the enzyme reactions proceed rapidly and completely as far as the degradation of malt starch is concerned. Wort is obtained from malt residue by filtration. It is used as a raw material for beer brewing. After the addition of hops it is boiled and filtered. The resulting solution is hopped wort; after cooling it is inoculated by a culture of brewer's yeast and fermented. Malt extract is prepared from wart by filtration and evaporation; it is a thick syrup containing 80% dry weight which may be vacuum-dried to a powder. A typical composition of dry malt extract is given in Table : 9.3.

Table 9.2 : Monosaccharides and oligosaccharides in barley

Carbohydrate	Dry weight mg g^{-1}
Glucose	0.2-0.93
Fructose	0.33-1.59
Maltose	0.0-1.35
Sucrose	3.43-16.90
Raffinose	1.44-8.32
Ketose	0.70-4.33
Fructosans	
soluble in ethanol	0.97-5.36
soluble in water	0.40-9.00

Table 9.3 : Composition of powdered malt extract

Component	%
Maltose	52.2
Hexoses (glucose and fructose)	19.1
Sucrose	1.8
Dextrin	15.0
Other carbohydrates	3.8
Proteins	4.6
Ash	1.5

9.2.4 Sulphite Waste Liquor

Sulphite waste liquor is a waste product in the production of pulp from wood by the sulphite method. It is a brown-yellow liquid with a density of 5 to 9° Be′ at 20°C. Its specific weight is 1.045 to 1.060. It contains 8 to 14% dry weight which consists of 18 to 20% minerals and 80 to 90% organic substances.

The inorganic part is composed chiefly of sulphur dioxide, sulphite and calcium sulphate. A part of the calcium is bound to lignin as a calcium lignosulphonate salt which is water-soluble in acidic media. Other inorganic compounds present in the liquor are salts of heavy metals such as copper, arsenic and lead. Though their amount is minute, they affect unfavourably the growth of microorganisms. The organic compounds found in the liquor, i.e. lignin and carbohydrates, are released from wood during the treatment with the sulphite cooking acid; their amount depends on the type of wood and on the mode of cooking. A short cooking at low temperatures yields a solid pulp containing a higher level of hemicelluloses and lignin; the resulting liquor is therefore poor in these substances. On the other hand, a long and intensive cooking at higher temperatures produces soft pulp containing only trace amounts of lignin and low levels of hemicelluloses; the liquor is therefore rich in carbohydrates. The liquor further contains small amounts of volatile organic acids, chiefly acetic and formic acid. Further fermentation of the sulphite liquors depends chiefly on their content of carbohydrates which are released from hemicellulose and solubilised by the action of cooking acid during pulping. The amount of carbohydrates varies according to the species of the processed wood and the technology of pulping. The composition differs in liquors from conifers and from deciduous trees. The total carbohydrates in the coniferous wood liquor usually include 10 to 40% mannose, 30 to 35% glucose, approximately 10% galactose, about 15% xylose and 5% arabinose. The ratio of hexoses to pentoses is the reverse in deciduous wood liquors which are composed of approximately 50% xylose, 15 to 20% arabinose, about 10% unidentified methylpentoses, 10% mannose and about 10% uronic acids. In addition they contain higher levels of furfural, which originates from pentoses during cooking at higher temperatures and pressures. The presence of furfural in the liquors negatively affects microbial processes.

Sulphite waste liquors cannot be used directly for fermentation in the form obtained from the cooker since they contain a considerable amount of sulphur dioxide which may be either bound (stable or labile) or free, and must be removed by a variety of procedures. The liquors must be neutralised as their pH value fluctuates between 1.5 and 3.2. The most suitable neutralisation agents are calcium hydroxide or calcium carbonate, which form insoluble salts with most of the acids present in the liquor. Other alkaline agents cannot be recommended due to higher price and operating difficulties connected with their application. Neutralised liquors contain only insignificant levels of assimilable nitrogen and phosphorus and these nutrients must therefore be added.

Sulphite liquors are used especially for the the production of ethanol (fermentation of coniferous wood liquors) and production of fodder yeast (deciduous wood liquors). Attempts were made to use sulphite waste liquor as a substrate for the production of acetone-butanol, glycerol, lactic and acetic acids, citric acid and even antibiotics, but these experiments did not extend beyond the pilot-plant scale.

9.2.5 Acid Wood Hydrolysates

The production of pulp by the sulphite or sulphate method from wood of annual plants (straw, etc.) involves the removal, from the raw material, of the majority of substances accompanying cellulose. However, the resulting pulp still contains a considerable amount of carbohydrates other than cellulose. These are removed by low concentrations of acids which hydrolyse hemicelluloses to soluble sugars; this process is called prehydrolysis. In contrast to total hydrolysis which is carried out at high concentrations of acid, high temperatures and pressures and results in the cleavage of all polysaccharides including cellulose, prehydrolysis removes only hemicelluloses, leaving cellulose intact. The composition of typical prehydrolysates obtained by treatment with sulphuric acid is shown in Table : 9.4. The conventional raw materials for prehydrolysis are deciduous wood, cereals, rape or rice straw and reed grass. The composition of prehydrolysates differs according to the raw material used, the acid used for the treatment, its concentration, and the temperature and pressure at which the treatment is performed. The prehydrolysis used in the production of viscous cellulose removes 15-20% of wood and 20-25% of straw. The straw prehydrolysate ashes contain a considerable amount of nutrients arising from industrial fertilizers. Because of their high content of pentoses, they are used for fodder yeast production.

The shortage of carbohydrates led to the introduction of total wood hydrolysis. In this process the wood pulp is treated so as to solubilise not only hemicellulose, but also the cellulose itself, which is hydrolysed by acids at elevated temperatures and pressures to glucose.

Table 9.4 : Composition of prehydrolysates of different origin obtained by sulphuric acid treatment

Component	Pine $g\ l^{-1}$	Straw $g\ l^{-1}$	Beech $g\ l^{-1}$
Reducing substances	21.65	21.76	38.0
Pentoses	6.34	21.76	30.5
Furfural	0.12	0.43	0.6
Dry weight	23.40	32.50	48.5
Ash in dry weight (%)	4.70	4.43	4.0
Total nitrogen	0.09	0.25	—
Inorganic nitrogen	0.06	0.19	—
Organic nitrogen	0.03	0.06	—
Volatile acids as acetic acid	0.89	—	1.9
Formic acid	—	1.00	—
Acetic acid	—	0.90	—
P_2O_5	0.067	0.285	—
Fe	0.196	0.033	—
Ca	0.224	0.296	—
Mg	0.194	0.097	—
K	0.038	0.70	—
SiO_2	—	0.57	—
pH	2.0	1.8	2.0

9.2.6 Vegetable Oils and Animal Fats

Vegetable oils are used as carbon sources in some microbial processes, especially in the biosynthesis of antibiotics. Animal fats are usually in a solid state and their liquefaction requires high temperatures. For this reason they are not often used. On the other hand, vegetable oils are liquids, at normal temperatures and are thus more easily manipulated. They are often used as antifoam agents. The majority of oils are readily metabolised by microorganisms. The energy value of oils, evaluated on a weight/weight basis, is two and half times higher than that of carbohydrates. The ratio of saturated to unsaturated fatty acids significantly affects the properties of oils. Table : 9.5 shows the results of analyses of several typical commercially available oils. The production of antibiotics has been examined using some animal fats including lard oil and tallow.

Table 9.5 : Composition of some typical vegetable oils

Oil	Saponification number	Iodine number	Per cent saturation	Fatty acids (% w/w)		
				Oleic	Linoleic	Linolenic
Olive	189-195	80-85	9-20	65-84	4-9	-
Peanut	189-196	85-98	18	56-65	17-21	-
Maize	188-193	117-130	12	45-47	40-42	-
Sunflower	186-194	127-136	7-10	30-35	55-65	-
Cotton-seed	191-196	103-111	25	25-30	45-50	-
Linseed	189-196	170-185	10-15	15-25	15-20	45-55
Soybean	190-193	124-133	12--13	25-36	50-55	5-8

9.2.7 Alcohols

Synthetic alcohols, especially ethanol and methanol, have considerable importance as potential raw materials for microbial processes. Synthetic ethanol is available in sufficient quantities, has a reasonable price, standard quality and homogeneity. Its physico-chemical properties place it among surface active substances; hence it increases the foaming of nutrient media. However, the foam is thin and facilitates a better distribution of air into the liquid. Synthetic alcohol is easy to produce in a pure form and is highly water-soluble. Its disadvantage is the presence of certain substances which inhibit the growth of microorganisms (croton-aldehyde). The complete utilisation of ethanol in microbial processes is not always possible because of losses caused by evaporation which may reach 10% of the total volume.

Methanol is soluble in water and some of its properties are similar to those of ethanol. In comparison with ethanol it has a higher volatility and toxicity. On the other hand, the spectrum of sources used for its production is much broader.

9.2.8 Hydrocarbons

Microorganisms are capable of utilising, to a larger or smaller extent, all hydrocarbons. While some bacteria assimilate almost all hydrocarbons, a limited number of yeasts assimilate only some of them. The best carbon sources among hydrocarbons are *n*-alkanes. According to the number of carbon atoms in the molecule, *n*-alkanes can be divided into several groups:

a) C_1–C_9 *n*-alkanes

The most widely used alkane from this group is methane since it is readily available and does not require, in contrast to other raw materials, refining. Methane is assimilated by a number of bacteria but not by yeasts. An amount of 1 kg methane yields 0.8 kg of cell dry matter.

Due to their yield and the narrow range of possible sources, C_2— C_9 *n*-alkanes are not suitable for microbial industry.

b) C_{10}–C_{20} *n*-alkanes

This group of alkanes represents the most suitable carbon sources for the growth of microorganisms. These alkanes are assimilated by bacteria, yeasts, actinomycetes and moulds. The mode of assimilation, however, varies widely, especially for *n*-alkanes with an even number of carbon atoms. Microorganisms grow better on a mixture of alkanes than on a single compound. The past data indicate that the yield of cell mass per consumed alkane is 90 to 100%.

c) C_{20} and higher *n*-alkanes

Assimilation of these substances is slower than with the preceding group. This is evidently caused by the fact that these *n*-alkanes do not occur in a liquid state at normal culture temperatures.

Table : 9.6 highlights some fermentation products using hydrocarbons as substrates by various fermenting microorganisms.

Table 9.6 : Products obtained from hydrocarbons using microorganism

Product	Culture	Substrate
Glutamic acid	*Corynebacterium hydrocarboclastus*	*n*-alkanes
Cytochrome C	*Candida albicans*	*n*-hexadecane
Carotense	*Mycobacterium smegmatis*	*n*-alkanes
Coenzyme Q	*Candida tropicalis*	*n*-alkanes
Salicylic acid	*Pseudomonas aeruginosa*	naphthalene
Lactic acid	*Arthrobacter globiformis*	1,2-propandiol
L-tryptophan	*Bacterium* sp.	hydrocarbons
L-citrulline	*Corynebacterium* sp.	*n*-alkanes
Vitamins (B group)	*Candida* sp.	*n*-alkanes
Pyrrolnitrin	*Pseudomonas cepacia*	toluene
Cunic acid	*Pseudomonas acidovorans*	*p*-cymene
Toluic acid	*Pseudomonas aeruginosa*	*p*-xylene
Trans-4-ethyl cyclohexanol	*Alcaligenes faecalis*	ethyl cyclohexane

9.2.9 Factors Determining the Choice of Carbon Source

1. The main product of a fermentation process will often determine the choice of carbon source, particularly if the product results from the direct dissimilation of it.
2. In fermentations such as ethanol or single-cell protein production where raw materials are 60 to 77% of the production cost, the selling price of the product will be determined largely by the cost of the carbon source.
3. The purity of the carbon source may also affect the choice of substrate. For example, metallic ions must be removed from the carbohydrate sources used in citric acid production.
4. The method of media preparation, particularly sterilization, may affect the suitability of carbohydrates for individual fermentation processes. It is often best to sterilize sugars separately because they may react with ammonium ions and amino acids to form black nitrogen containing compounds which will partially inhibit the growth of many microorganisms. Starch suffers from the handicap that when heated in the sterilization process it gelatinizes, giving rise to very viscous liquids.
5. The choice of carbon substrate may also be influenced by government legislation. The government's policy of import and export influences the availability and pricing of particular carbon substrate. Legislation for recognition of new products is time consuming and manufacturers may be uncertain as to whether they would benefit from carbon substrate refunds.
6. Local laws may also dictate the substrates which may be used to make a number of beverages.

9.3 NITROGEN SOURCES

Inorganic and organic substances containing nitrogen in various forms may serve as sources of nitrogen for microbial growth. The range of microbial mechanisms of nitrogen assimilation is much wider than carbon assimilation. Many microorganisms, especially lower fungi, utilize nitrogen from inorganic ammonium salts. Ammonium sulphate, the cheapest of these compounds, is most often used together with ammonium hydroxide and gaseous ammonia for the continuous regulation of pH. Another very good source of nitrogen is urea; since it decomposes at high temperatures it should be sterilised by filtration. The quality of nitrogen sources used in microbial processes cannot match the quality of pure laboratory chemicals for economic reasons.

Although many microorganisms assimilate inorganic nitrogen, they often grow faster in the presence of an organic nitrogen source. At present many suitable nitrogen sources of plant and animal origin are available as waste products.

9.3.1 Inorganic and Synthetic Organic Nitrogen Sources

(i) Ammonium sulphate is readily soluble in water; the purest form used in microbial industry is technical ammonium sulphate which contains a minimum of 20.7% nitrogen. The so called coke and gas sulphates are considerably contaminated with tar substances and are used only exceptionally for these purposes.

(ii) Diammonium hydrogen phosphate, which is used especially in yeast production, is a source of both nitrogen and phosphorus. It is usually contaminated by ammonium sulphate and contains about 20% nitrogen and 47 to 50% phosphorus pentoxide. When dissolved in water, this salt forms a voluminous sludge which does not sediment easily and contains calcium and ferrous phosphates.

(iii) Ammonia is used either in the form of a 25% aqueous solution or in gaseous form. The storage of the aqueous solution should be done with care as ammonia is readily released in the gaseous form. Equipment coming into contact with ammonia should be gasproof and made of non-corrosive material.

(iv) Urea is used in microbial industry only exceptionally, e.g. in the production of fine and high-price products duo to its relatively high price.

9.3.2 Natural Sources of Nitrogen

(i) Corn-steep liquor. One of the basic raw materials in microbial industry is corn-steep liquor which is obtained as a by-product in the production of maize starch. It contains biologically active substances which differ qualitatively and quantitatively according to the quality of the corn used and its technological processing. These substances are released into the corn-steep liquor during corn steeping and some microbial processes, especially lactic acid fermentation, participating in the release, occur during the steeping process. The composition of corn-steep liquor often varies, causing undesirable fluctuations in the microbial processes in which the liquor is used as a raw material. Corn-steep liquor was first used in penicillin production in 1946 when Moyer and Coghil found that its addition to the medium considerably enhanced the yields of the antibiotic.

Corn-steep liquor contains considerable quantities of nitrogen substances, lactic acid, reducing substances and ashes with many trace elements. Table : 9.7 shows the average chemical composition of corn-steep liquor according to Liggett and Koffler (1948) while Table : 9.8 gives the contents of amino acids according to Cardinal and Hedrick (1948); amino acids represent more than 83% of the total nitrogen.

Table 9.7 : Average chemical composition of corn-steep liquor (Liggett and Koffler, 1948)

Component	%
Water	45.0-55.0
Total nitrogen	2.7-4.5
Amino nitrogen	1.0- 1.8
Reducing substances	0.1 - 11.0
Lactic acid	5.0-15.0
Ash	9.0-10.0
Volatile acids	0.1 -0.3
$SØ_2$	0.009-0.015
Ca	0.5-1.5
Cu	0-0.001
Fe	0.01 -0.05

contd. ...

... contd.

Component	%
Mg	0.05-1.0
Mn	0.004
P	2.0-3.0
K	1.0-2.0
S	0.34
Zn	0.05

Table 9.8 : Amino acid composition of corn-steep liquor (Cardinal and Hedrick, 1948)

Amino acid	% of nitrogen	
	After hydrolysis	Before hydrolysis
Leucine	5.86	4.80
Isoleucine	3.42	2.80
Valine	3.88	2.85
Glutamic acid	7.97	3.0
Threonine	5.44	—
Lysine	3.97	—
Arginine	8.23	—
Histidine	6.78	—
Proline	4.75	—
Phenylalanine	2.05	—
Methionine	1.07	—
Aspartic acid	1.68	—
Cystine	1.22	—
Alanine	27.70	—
Tyrosine	0.74	—

Fluctuations in the biological quality of corn-steep are caused chiefly by the following factors:

a) species and quality of corn,
b) species or genus of the lactobacilli used,
c) course of steeping (extraction of soluble substances from the corn depending on sulphur dioxide content),
d) intensity of microbial processes during steeping,
e) variable content and relative proportions of amino acids,
f) temperature during corn-steep liquor condensation.

An important criterion of the quality of corn-steep liquor is the ratio of amino nitrogen to total nitrogen, which should be about 0.4, i.e. approximately 2% of amino nitrogen in dry matter. This ratio indicates the intensity of lactic fermentation during steeping. Another important criterion is the content of amino acids which should be higher than 20%.

The heating of corn-steep liquor above 100°C and its subsequent cooling gives rise to a gray-white precipitate which may interfere with the measurement of microbial growth. The precipitate can be removed by the following procedure: 1 part of corn-steep liquor (60% dry weight content) is dissolved in 2.5 parts of water and the pH is adjusted to 7.0 using saturated sodium hydroxide solution. The resulting solution is boiled for 10 minute and, after cooling, filtered through diatomaccous earth. The precipitate is then washed with water to a final filtrate volume of 5 parts; 1 ml of filtrate it then equivalent to 0.2 g of corn-steep liquor. The precipitate consists chiefly of magnesium and calcium phosphates. This treatment of corn-steep liquor is important especially in biotransformation processes which require a relatively pure microbial suspension. It was also found that nutrient media prepared in this way increase the production of antibiotics by as much as 30%.

(ii) Potato liquor and fermented extract from bran and oil seeds. Corn-steep liquor can sometimes be replaced by liquors from potato starch production and by the fermented extract from bran and oil seeds. Potato liquors are obtained by the following procedure: the liquid formed by pressing potato pulp in starch production is subjected to sulphuration and lactic fermentation by a culture of *Lactobacillus delbruckii*, and carefully dried. The composition of potato liquor (Table : 9.9) resembles, that of corn-steep liquor. The potato liquor contains a higher level of reducing substances while the following amino acids are absent: arginine, glutamic acid, histidine, leucine, isoleucine, lysine, methionine, phenylalanine, proline, threonine and valine. Potato liquor has also been tested as a possible substrate in the biosynthesis of penicillin.

Table 9.9 : Composition of potato liquor

Component	Without lactic fermentation	After lactic fermentation
	%	
Dry weight	72.7	52.5
Total nitrogen	5.1	6.0
Amino nitrogen	1.9	2.4
Reducing substances	27.5	12.5
Lactic acid	5.0	9.7
Ash	20.8	22.2

The fermented extract from bran and oil seeds is prepared in the following way: wheat bran and ground tobacco seeds are mixed with water and a small amount of lactic acid and superphosphate is added. After a 6 day lactic fermentation at 50°C the liquid is filtered and evaporated in vacuum. The composition of the extract is given in Table : 9.10.

Table 9.10 : Composition of fermented bran-oil extract (Zelinka, 1960)

Component	Weight/Volume (%)
Dry weight	70.0
Total nitrogen	4.6
Amino nitrogen (van Slyke)	2.0

contd. ...

... contd.

Component	Weight/Volume (%)
Reducing substances	19.2
Nonreducing carbohydrates	4.2
Total carbohydrates	23.4

(iii) Yeast extract. Yeast extract is available either dry or in the form of paste. The extract is prepared by autolysis or plasmolysis of baker's or brewer's yeast. To remove bitter hop substances, brewer's yeast is washed at pH 6.5 to 7.0 prior to treatment. Yeast cells must be kept in a viable state during washing and filtration in order to prevent losses of cell components. Baker's yeast need not be washed.

The autolysis is performed in large vessels under continuous stirring. The temperature is gradually increased to 48-50°C and then kept at this level for several hours. After autolysis, enzymes are inactivated by increasing the temperature above 75°C. The yeast is then transferred into a vessel containing sodium chloride solution which induces cell plasmolysis The extracts contain 30 to 40 % (w/w) of sodium chloride. After autolysis or plasmolysis the cell walls and membranes are separated from the solution by filtration or centrifugation. To prevent contamination and degradation of thermolabile substances the liquid must be concentrated rapidly at a temperature below 37°C. Yeast extract in a powder form can be obtained by drying either in an atomiser drier or spontaneous]y in vacuum. Yeast extract is a mixture of amino acids and peptides, water soluble vitamins and carbohydrates.

A typical composition of yeast extract is given in Table : 5.11.

Table 9.11 : Composition of yeast extract

Component	mg g^{-1}
Total nitrogen	75-105
Amino nitrogen	34-48
Chlorides as NaCl	0.7-13
Dry weight	300
Phosphates as P_2O_5	38
Carbohydrates	82
Sodium	56
Potassium	30
Calcium	0.1
Iron	0.05
Magnesium	2
Copper	0.05
Zinc	0.05
Manganese	0.005
Cobalt	0.005

(iv) Soybean and peanut flour. Soybean flour is an important source of nitrogen. It contains 44 to 47% Protein, 0.5 to 1.2% fat, 5.5 to 6.6% cellulose, 31 to 32% dissolved nitrogen-free substances, 5.5 to 6.0% ash, 0.3% calcium, 0.62 to 0.65% phosphorus, 2.0 to 2.5% phosphatides. The most important constituents are protein, fat and phosphorus. Soybean flour, which is obtained by extraction of oil from soybeans, has a variable quality depending on the soybeans source. Since the variable composition can affect microbial biosyntheses, biological tests are performed to determine the quality of soybean flour.

Similar to soybean flour is peanut flour obtained by oil extraction from peanuts. The most important components are nitrogen, fat and phosphorus. The composition of ash and protein is variable and depends again on the source of raw materials. The composition of peanut and soybean flour is similar but their use in microbial processes has to be carefully examined. Peanut flour is suitable for the synthesis of penicillin and lysine and may partially replace soybean flour in the biosynthesis of chlortetracycline. On the other hand, it is not suitable for the biosynthesis of tetracycline.

9.3.3 Factors Determining the Choice of Nitrogen Source

1. Preferability and suitability of fermenting organism towards nitrogen source:

Control mechanisms exist by which nitrate reductase, an enzyme involved in the conversion of nitrate to ammonium ion, is repressed in the presence of ammonia. For this reason, ammonia or ammonium ion is the preferred nitrogen source.

In fungi that have been investigated, ammonium ion represses uptake of amino acids by general and specific amino acid permeases. In *Aspergillus nidulans*, ammonia also regulates the production of alkaline and neutral proteases. So, in mixture of nitrogen sources, individual nitrogen components may influence metabolic regulation so that there is preferential assimilation of one component until its concentration has diminished.

2. The nitrogen source has been shown to influence the fermentation pattern.

e.g. (i) Antibiotic production may be inhibited by a rapidly utilized nitrogen source (Ammonium ion, nitrate, certain amino acids). The antibiotic production only begins to increase in the culture broth after most of the nitrogen source has been consumed. For example, in the production of polyene antibiotics soybean meal is considered a good nitrogen source.

(ii) In gibberellin production, the nitrogen source has been shown to have an influence on directing the production of different gibberellins and the relative proportions of each type.

3. In shake flask media experiments, salts of weak acids (e.g. ammonium succinate) may be used to serve as a nitrogen source and eradicate the source of a strong acid pH change due to chloride or sulphate ions which would be present if ammonium chloride or sulphate were used as nitrogen source. This procedure makes it possible to use lower concentrations of phosphate to buffer the medium. High phosphate concentrations inhibit production of secondary metabolites.

4. Other pre-determined aspects of the process can also influence the choice of nitrogen source. Rhodes has shown that the optimum concentration of available nitrogen for griseofulvin production showed some variation depending on the form of inoculum and type of fermenter being used. Obviously, these factors must be borne in mind in the interpretation of results in media-development programmes.

5. Some of the complex nitrogenous material may not be utilized by a microorganism and create problems in downstream processing and effluent treatment. This can be an important factor in the final choice of nitrogen substrate.

Some examples of the best nitrogen sources for production of some secondary metabolites are mentioned in Table : 9.12.

Table 9.12 : Best nitrogen sources for some secondary metabolites

Secondary metabolite product		Main nitrogen sources
1.	Bacitracin	Peanut granules
2.	Butirosin	Dried beef blood + $(NH_4)_2$ SO_4
3.	Gibbrellins	Ammonium salt + natural plant nitrogen source
4.	Novobiocin	Distillers' solubles
5.	Penicillin	Corn-steep liquor
6.	Polyenes	Soybean meal
7.	Rifomycin	Pharmamedia
8.	Riboflavin	Pancreatic digest of gelatine

9.4 GROWTH FACTORS

These are those essential metabolites which certain microorganisms are incapable of synthesizing for themselves. Some microorganisms require no growth factors whatever; for example, a number of species of bacteria grow satisfactorily on simple glucose ammonium-salts mixtures. When growth factors are needed, these are usually B-group vitamins or related compounds, certain amino acids or, sometimes, fatty acids. When a particular growth factor is required only in low concentration the microorganism is said to be auxotrophic for that factor. The growth of auxotrophs is strictly proportional to the concentration of the factor, within a particular range making it possible to use them for the biological assay of compounds such as vitamin B_{12}. Such assay can be very sensitive; for example, suitable strains of the lactic acid bacterium *Lactobacillus arabinosus* respond to the presence of 10^{-7} g of biotin/l.

Many of these growth factors are involved in enzyme reactions because they form apart of the structure of various coenzymes (Table : 9.13).

Table 9.13 : Relations between vitamins of the B group (growth factors) and certain coenzymes

Vitamin	Coenzymes	Enzyme activity
Nicotinic acid	NAD & NADP	Dehydrogenations
Riboflavin (Vitamin B_2)	Flavin nucleotides	Dehydrogenations and electron transport
Thiamine (Vitamin B_1)	Cocarboxylase	Decarboxylations
Pyridoxine	Pyridoxal phosphate	Metabolism of amino acids

contd. ...

... contd.

Vitamin	Coenzymes	Enzyme activity
(Vitamin B_6)		(trans and deamination, decarboxylation)
Pantothenic acid	Coenzyme A	Oxidation of keto acids, fatty acid metabolism
Folic acid	Tetrahydrofolic acid	Transfer of C_1 compounds
Biotin	Enzymes with biotin as prosthetic group	CO_2 fixation and transfer of the carboxyl group
Cobamide (Vitamin B_{12})	Cobamide coenzymes	Molecular rearrangements

Growth factors are usually present in the materials used as a source of carbon and of nitrogen in the formulation of a medium. It may sometimes be necessary to supplement a medium with a specific source of growth factors such as yeast extract, liver extract or distillation residues to increase yields (e.g. in the production of riboflavin by fermentation).

It is possible to have too high a concentration of a growth factor. One example has already been discussed during the study of auxotrophic mutants used for amino acid production. Thus, to obtain lysine by fermentation, the quantity of threonine or methoionine (which are growth factors for the lysine producing culture) should not exceed 200 mg/l. Again, some molasses cannot be used as a carbon source for glutamic acid production because it contains too high a concentration of biotin. The yield of glutamic acid is a function of the concentration of biotin in the medium, the optimum being about 2-5 mg/l (corresponding to about 0.5 mg biotin per gram of cells).

Vitamin contents of some industrially important raw materials are shown in Table : 9.14.

Table 9.14 : Content of vitamins in some industrially important raw materials

Vitamin ($\mu g\ g^{-1}$)	Maize flour	Barley	Soybean flour	Beet molasses	Cane molasses	Yeast extract	Stillage
Thiamine	4.5	6.5	13.5	0.8	0.8	10	3.5
Riboflavin	0.9	1.2	3.5	-	-	20	11.9
Nicotinic acid	23.0	115.0	25.2	35.0	15.0	400	75.8
Pantothenic acid	4.6	4.4	26.1	50.0	20.0	50	10.8
Pyridoxine	6.9	11.5	8.5	-	-	25	1.0
Biotin	0.1	-	0.7	0.1	1.5	1	0.3
Inositol	-	-	3850	5000	2000	1500	7170
Choline	-	1100	2880	—	—	1500	3080

9.5 INORGANIC MINERAL SALTS

Functions of some of the mesonutrient and mioronutrient mineral elements have already been given in Table : 8.1. It is difficult to determine the exact mineral salts requirement of a microörganism but they do affect the yields of some products. There are two general forms of approach:

(i) Analytical

The concentration of mineral elements in the ash, which represents about 5% of the microbial biomass, is measured and a medium is formulated to contain these elements in similar relative proportions. Generally, some 60% of this ash is in the form of phosphorus, 20% as sulphur; the rest is made up of a variety of metals many of which play an important catalytic role and may have to be provided in crucial concentrations. A defect of the analytical approach is that, because microbial assimilation is not very selective, the composition of the ash tends to vary according to the composition of the culture medium.

(ii) Synthetic

By trial and error, a medium which will give good growth and which is composed of chemically defined constituents is devised (a synthetic medium). Various mineral elements are then successively omitted to determine those which are essential. Difficulties arise from the presence of impurities in even analytical grade reagents. Thus an iron salt, though appearing to be spectroscopically pure, may still contain up to 0.1% of cobalt; the concentration of cobalt for optimum growth is, in fact, generally of the order of 10^{-12} g/ml. It is possible to add measured amounts of chelating agents (e.g. ethylenediaminetetracetic acid, EDTA), which bind very strongly to metal ions making them unavailable, to a medium create a total deficiency; the minimum quantity of a metal ion required to abolish this deficiency is then determined.

Phosphorus and sulphur requirements are generally satisfied by providing phosphates and sulphates in relatively large quantities, of the order of 0.5 g/l. Other inorganic elements are required in much smaller quantities, and are referred to as oligonutrients, but a precise concentration range is sometimes essential. **Iron,** for example, is primarily involved in the constitution of transport molecules based on haem, such as the cytochromes. It is required at a concentration between 0-.03-0.3 g/l in media for aerobic organisms, but the precise iron concentration influences the properties of the microorganisms grown and this may sometimes very important in industrial fermentations. *Streptomyces griseus* requires five times more iron to produce streptomycin than it requires for growth. *Corynebacterium diphtheriae* produces its toxin only if the concentration of iron in the medium is less than 0.8 ppm; *Candida guilliermondii* synthesizes riboflavin only in media containing less than 0.01 mg/l; the mould *Aspergillus niger* produces citric acid from glucose (or sucrose) only if the medium is deficient in iron (in this last example, it has been suggested that citric acid accumulates-only if the organism lacks one of the constituents required for activating the enzymes *cis*-aconitase and isocitrate dehydrogenase of the tricarboxylic acid cycle; when the carbon source is molasses, it is necessary to precipitate iron by adding ferrocyanide to obtain a high citric acid yield).

It is necessary to consider, also, possible interactions between ions. **Cobalt,** for example, is generally required in concentrations of about 5 ppm because it is an essential component of active vitamin B_{12}. A concentration of 10 ppm may exert a toxic effect, but this can be eliminated by the presence of magnesium. Specific requirements for oligonutrients depend both on the organisms employed and the product which is sought. Optimum growth of *Lactobacillus plantarum* requires 1 mg/l of manganese and only 0.1 mg/l is required by

Streptococcus faecalis. To obtain high yields of inosine monophosphate (IMP) by fermentation using *Brevibacterium ammoniagenes*, the concentration of manganese must be less than 10 mg/l.

A generally useful balanced mineral medium might have the following composition per litre of twice distilled water:

NH_4Cl	1 .0 g
K_2HPO_4	0.5 g
$MgSO_4.7H_20$	0.5 g
$CaCl_2$	10 mg
$FeSO_4.7H_2O$	1 mg

These salts, even when analytically pure, contain sufficient impurities to obviate the need for deliberate additions of such elements as Cu, Zn, Mn, Mo, Co. For example, when a hypothetical salt of molecular weight 100, containing 1 in 10,000 impurities, is added to a medium at the rate of 1g/l, the number of molecules of the salt will be

$$6 \times 10^{23}/10^{2}=6 \times 10^{21}/1$$

A bacterial culture containing 10^{12} organisms/litre would thus have available

$$6 \times 10^{21}/6 \times 10^{12} \times 10^{4} = 6\times 10^{5}$$

molecules of impurities available for each cell.

Some of the primary and secondary microbial products whose yields are affected by concentration of trace metals greater than those required for maximum growth are given in Table : 9.15

Table 9.15 : Trace elements influencing primary and secondary metabolites

Microbial Product	Trace element (s)
Actinomycin	Fe, Zn
Bacitracin	Mn
Citric acid	Fe, Zn, Cu
Chloramphenicol	Fe, Zn
Candicidin	Fe, Zn
Griseofulvin	Zn
Gentamicin	Co
Mitomycin	Fe
Neomycin	Fe, Zn
Patulin	Fe, Zn
Protease	Mn
Riboflavin	Fe, Co
Streptomycin	Fe, Zn
Penicillin	Fe, Zn, Cu

Chlorine does not appear to play a nutritional role in the metabolism of fungi. However, chlorine is required by some of the halophilic bacteria. Obviously, in those fermentations where a chlorine-containing metabolite is to be produced the synthesis will have to be directed to ensure that the non-chloro-derivative is not formed. The most important compounds are chlortetracycline and griseofulvin. In griseofulvin production, adequate available chloride is provided by the inclusion of 0.1% KCl in fermentation medium.

9.6 BUFFERS

Fermentation media contain buffers to retard gross changes in pH values during microbial growth. These buffers can be specifically added for their buffering capacity, or they may be normal constituents of a medium, serving a dual function as both buffer and nutritive source. During microbial growth, pH changes can occur for one of several reasons. Obviously, an acidic or alkaline fermentation product can alter the pH picture. Also, an inorganic salt component of the medium can cause pH changes; for example, ammonium sulphate supplying nitrogen as the ammonium ion for growth will leave behind the sulphate ion as sulphuric acid, thereby making the medium more acidic. Decarboxylation of organic acids or deamination of strongly basic organic amines from among the medium constituents will raise or lower the pH respectively. Also, deamination and utilization of proteins, peptides, or amino acids can raise the pH value. Calcium carbonate often is incorporated in fermentation media to provide neutralization of acidic fermentation products, although it is a relatively poor buffer; its poor solubility in water allows only reaction with acidic products. In fact, the pH value of the medium may decrease to 4 or 5 before the calcium carbonate has much effect. Media containing considerable quantities of proteins, peptides, and amino acids possess good buffer capacity in the pH range near neutrality. Additional buffering capacity in this pH range also is provided by phosphates such as the system of mono- and dihydrogen potassium or sodium phosphates. At lower pH values, acetic acid and other organic acids provide buffering capacity. In contrast to the foregoing discussion, at times it may not be desirable to employ a medium strongly buffered at the pH values initially provided at medium make-up.

9.7 PRECURSORS

Precursors required in certain industrial fermentations are provided either through crude nutritive constituents, such as cornsteep liquor, or by direct addition of the more pure compounds. **Precursors** are defined (Lee, 1951) as "Substances added prior to or simultaneously with the fermentation, which are incorporated without any major change into the molecule of the fermentation product, and which generally serve to increase the yield or improve the quality of the product." Thus, the principal criterion is that the precursor becomes incorporated into the fermentation product. Specific examples of the use of precursors are the addition to the medium of phenylacetic acid as a precursor for penicillin G, and the addition of inorganic cobalt for production of vitamin B_{12}. It should be pointed out that the precursor can also dictate as to which product from among several possibilities will be produced in greatest amount during a fermentation. Thus, with cornsteep liquor in the penicillin fermentation up to six different penicillin, are possible products but, on addition of phenylacetic acid, the major product becomes penicillin G. Precursor-requiring fermentation products are enlisted in Table : 9.16

Table 9.16 : Precursor-requiring fermentation products

Product	Precursor	Fermenting microorganisms
1. Carotenoids	Beta-Iononones	*Phycomyces blakesleeanus*
2. Chlortetracycline	Chloride	*Streptomyces aureofaciens*
3. Cyclosporin A	DL-α-Amino butyric acid	*Tolypocladium inflatum*
4. Cyclosporin C	L-Threonine	*Tolypocladium inflatum*
5. Dihydronovobiocin	Dihydronovobionic acid	*Streptomyces spp.*
6. Griseofulvin	Chloride	*Penicillium griseofulvum*
7. L-Isoleucine	α-aminobutyric acid	*Bacillus subtilis*
8. L-Isoleucine	D-Threonine	*Serratia marcescens*
9. L-Tryptophan	Anthranilic acid	*Hansenula anomala*
10. Nikkomycins	Nucleosides and bases	*Streptomyces tendae*
11. Organomycins	Para-Hydroxycinnamate	*Streptomyces organonensis*
12. Penicillin-G	Phenyl acetic acid	*Penicillium chrysogenum*
13. Penicillin-V	Phenoxy acetic acid	*Penicillium chrysogenum*
14. Riboflavin	Propionate	*Lactobacillus bulgaricus*
15. Vitamin B_{12}	Cobalt	*Propionibacterium spp.*

9.8 INHIBITORS

Inhibitors are those chemical compounds when added to certain fermentations, more of a specific product may be produced, or a metabolic intermediate which is normally metabolised is accumulated.

One of the oldest examples is the use of sodium bisulfite as an inhibitor in microbial production of glycerol. Glycerol production depends on modifying the ethanol fermentation by removing acetaldehyde. The addition of sodium bisulfite to the broth leads to the formation of the acetaldehyde bisulfite addition compound (sodium hydroxy ethyl sulfite). Since acetaldehyde is no longer available for re-oxidation of $NADH_2$, its place as hydrogen acceptor is taken by dihydroacetone phosphate, produced during glycolysis. The product of this reaction is glycerol-3-phosphate, which is converted to glycerol.

Inhibitors have also been used to affect cell-wall structure and increase the permeability for release of metabolites. The best example is the use of penicillin and surfactants in glutamic acid production.

A list of inhibitors used for various fermentation product formation is mentioned in Table : 9.17. In most fermentation cases, the inhibitor is effective in increasing the yield of the desired product and reducing the yield of undesirable related products.

Table 9.17 : Inhibitors used in fermentations

Inhibitor	Product	Microorganism	Main effect
1. Alkali metal/ phosphate	Citric acid	*Aspergillus niger*	Oxalic acid repressed

contd. ...

... contd.

	Inhibitor	Product	Microorganism	Main effect
2.	Penicillin	Glutamic acid	*Micrococcus glutamicus*	Cell wall permeability
3.	Sodium bisulfite	Glycerol	*Saccharomyces cerevisiae*	Acetaldehyde production repressed
4.	Bromide	Tetracycline	*Streptomyces aureofaciens*	Chlortetracycline formation repressed
5.	Di-ethyl barbiturate	Rifamycin B	*Nocardia mediterranei*	Other rifamycins inhibited
6.	Ethionine	7-chloro-6-demethyltetra cycline	*Streptomyces aureofaciens*	Affects one-carbon transfer reactions

9.9 INDUCERS

Inducers are those specific chemical compounds included in media which induce or provoke formation of desired industrial fermentation products. Majority of industrially important enzymes are inducible. Induced enzymes are synthesized only in response to the presence in the environment of an inducer.

Inducers are often substrates such as starch or dextrins for amylase, maltose for pullulanase and pectin for pectinases. Some inducers are very potent such as isovaleronitrile inducing nitralase. Substrate analogues that are not attacked by the enzyme may also serve as enzyme inducers. Most inducers which are included in microbial enzyme media are substrates or substrate analogues (Table : 9.18).

Table 9.18 : A partial list of industrially important enzymes and their inducers

	Enzyme	Inducer	Producer microorganisms
1.	Alpha-amylase	Starch	*Aspergillus spp.*
		Maltose	*Bacillus subtilis*
2.	Alpha-Mannosidase	Yeast mannans	*Streptomyces griseus*
3.	Pullulanase	Maltose	*Aerobacter aerogenes*
4.	Cellulases	Cellulose	*Trichoderma viridae*
5.	Proteases	Various proteins	*Bacillus subtilis*
6.	Pectinases	Pectin	*Aspergillus spp.*
7.	Penicillin acylase	Phenylacetic acid	*Escherichia coli*
8.	Nitralase	Isovaleronitrile	*Rhodococcus rhodochrous*

9.10 ANTIFOAM AGENTS

Foaming of media often accompanies microbial processes, especially in media rich in natural substances such as corn-steep liquor, flours, etc. Uncontrolled foaming causes considerable difficulties. The outflow of a large amount of foam from the fermentor may strongly reduce the volume of the culture and may lead to its contamination. Apart from various mechanical antifoam devices some natural or synthetic chemical agents can be used for the purpose. The effect of these substances is based on their ability to reduce the surface tension of liquids. Evans and Hall (1971) differentiated between antifoam and "defoam", the former being added before the foaming actually occurs and the latter being added afterwards to knock down the foam.

The natural antifoam agents used are soya, rape, coconut, sun-flower and mustard oils. Antifoam agents of animal origin are usually tallow and deodorised fish fat. Mineral oils may also be used. During the culture some of these substances are consumed by the organisms and have thus to be added continuously to the culture.

Antifoam agents should meet the following criteria: high antifoam effect, pleasant smell, light colour, purity. Another group of substances used are alcohols. The most common compound of this group is octa-decanol either pure or in combination with lard oil or mineral oil. The former combination was used in the biosynthesis of penicillin, the latter in the biosynthesis of citric acid. Alcohols and plant oils are traditional antifoam agents whose effectiveness was found purely empirically.

Specially designed antifoam agents are also available; among them silicon oils are the most important. They are usually used in water-based emulsions containing 10% silicon oil. They are more efficient when used in bacterial and yeast cultures than in processes involving filamentous microorganisms. The silicon oil emulsions are very efficient since they are not assimilated by microorganisms, and a single dose at the beginning of the process is usually sufficient to prevent the formation of foam. A similar effect is exhibited by polyalcohols with molecular weight of 2000 and alkylated glycols, e.g. Texofor D 10.

Antifoam agents are elaborately discussed in chapter-18.

9.11 WATER

Water is indispensable in fermentation processes. Water is the larger component of fermentation medium because microorganisms live in an aqueous environment in which most of their nutritive requirements are dissolved. Media usually contain more than 70% and often greater than 90% water. The water supplies trace elements, minerals such as Co, Mo, Zn, B and other substances that may be required in minute quantities.

Water is also needed in many of the ancillary services such as heating, steam production, cooling and cleaning of fermenter. A large quantity of water is also required during product recovery. Therefore clean water of consistent composition is required in large quantities from reliable permanent sources. When assessing the suitability of a water supply, it is important to consider pH, dissolved salts and effluent contamination.

High concentrations of iron salt in the water cause a discolouration of microorganisms such as yeasts; inhibit the biosynthesis of some substances, e.g. antibiotics, and may cause corrosion. Water containing less than 20-100 germs per ml (as determined by a plate count) can be regarded as high-quality water. Fluctuations in the hardness of the water and in the levels of individual cations present, especially of calcium and magnesium, may lead to instability in product yields. In an ideal case we would use deionised and demineralised water to which all the necessary substances could be added in appropriate proportions. However, such treatment could be adopted only on a laboratory scale. Generally speaking, the water quality in the environment is constantly deteriorating owing to an increasing use of chemicals in agriculture. Increasing industrial demands may exceed the available supply of water of the desired quality so that even the utilisation of sea water is being considered.

9.12 OXIDATION-REDUCTION POTENTIALS

Certain components of fermentation media tend to "poise" or hold the medium within specific ranges of oxidation-reduction potential. These agents usually present little problem in highly aerated fermentations, but they must be considered in anaerobic fermentations. Amino acids such as L-cysteine, peptides such as glutathione, and organic acids such as thioglycolic acid will poise the medium at relatively low oxidation-reduction potentials for growth of anaerobic or microaerophilic microorganisms. In addition, the medium is heated and cooled just before inoculation to drive out dissolved oxygen. Certain media, such as cooked corn mash for the anaerobic clostridia further reduce oxygen penetration by their gel-like properties.

9.13 RESTRICTED NUTRITIVE LEVELS

Media conducive to rapid microbial growth and high cell yields are not always the best media for production of fermentation products. Sometimes, media allowing only restricted growth of microorganisms provide high product yields, because relatively greater amounts of carbon and other nutritive components of the medium are shunted into fermentation products instead of growth. In the production of organic acids by fungi, the yields for some acids are greater if the growth, because of low pH or for other reasons, is relatively poor. Also, the control of the combined nitrogen level of a medium to retard microbial growth is beneficial in fermentations such as that for fat production by the yeast *Rhodotorula gracilis*. Replacement culture with fungal mycelium is another good example of the use of a restricted growth medium. A fungus, such as *Aspergillus niger* for gluconic acid production, is first grown on a medium that provides good growth as well as product formation, then the mycelium is separated from its medium and places in fresh medium high in carbon substrate (sugar), but lacking combined nitrogen so that additional growth cannot occur. Further fermentation occurs enzymatically, in this case, by the glucose oxidase of the mycelium.

9.14 SCREENING FOR FERMENTATION MEDIA

All of the previous points in relation to fermentation media must be considered when we attempt to compound a high-yielding medium for an existing or a new fermentation. Various media of differing types are tested, along with variations on these media known to provide growth of the particular microorganism under study and elaboration of its fermentation product. The cell and product yields are determined for each medium, as are the rate of product formation and the fluctuations in pH values during the fermentation. High-yielding media are then studied individually to determine what effects varying the levels of individual components will have on these fermentation parameters. These studies should yield one or two good media. However, alternate carbon and nitrogen sources also should be tested for these media as a hedge against changes in the availability and economic pictures for these substrates. The media selected are then studied from several further points of view. The initial pH value and the buffer capacity of the medium are varied so as to determine the effects of pH on cell and product yields and on the rate of product formation. Residual medium components are quantitatively assayed during the fermentation to determine whether critical carbon or nitrogen nutrients become exhausted too quickly or are present in great excess. This information can indicate either that higher levels of these components should be added at medium make-up or during the fermentation, or that certain components should be decreased in concentration. This information also can reveal whether, during the fermentation, a diauxie phenomenon is occurring in which one carbon source of the medium becomes exhausted before a second carbon source requiring enzymatic adaptation is utilized.

If a fermentation product has a chemical structure that is known to the investigator, it may be possible by study of this structure to predict the possibility for adding a portion of the molecule to the fermentation medium to serve as a precursor. Often, however, determination of the chemical structure of a fermentation product postdates the development of the fermentation. Also, the vast majority of fermentation products are not amenable to employment of precursors.

Most of the above studies are carried out without the addition of antifoam, or, if antifoam is required, it is supplied as the inert silicone type. Further studies utilize crude antifoams to determine a possible interference with or stimulation of product formation by the antifoam agent.

The fermentation is next carried out in small fermenters to detect interactions of the medium and microorganism with aeration and agitation conditions. Also, other parameters of the fermentation, as described previously, are rechecked for changes that may occur under conditions of tank fermentation. Completed fermentation broths containing high yields of fermentation product are examined for quantitative recovery and purification of the product. If recovery yields are low and with little promise for increase, alternative media for the fermentation must be tested. Finally, cost determinations are made for the economic status of the fermentation based on the expense of medium components, duration of the fermentation, recovery yields, and overhead. If these determinations present an adverse economic picture, the fermentation will need reevaluation, particularly from the standpoints of the choice of medium and microorganism.

❑❑❑

10

Introduction to Industrial Sterilization

10.1 INTRODUCTION

Physical methods for controlling microorganisms were adopted early in the work of Louis Pasteur and Robert Koch. Pasteur, for instance, utilized boiled broths to collect airborne organisms and Koch was careful to use sterile vitreous humour in his studies on anthrax bacilli. The technology for controlling microorganisms grew with the development of microbiology as physicians came to understand that sterile instruments and apparatus were necessary prerequisites to the prevention of disease. Laboratory workers also utilized physical methods of control to ensure the accuracy of their work.

Sterilization of media, air and bioreactor (fermenter) is of supreme importance in a number of microbial fermentation industries. Microbiological productions on an industrial scale require the sterilization of large volumes of production media and air passing through relatively complex devices and economic factors play an important role. Sterilization of media and air on a laboratory scale usually presents no major problems, but on an industrial scale the mechanical and technological problems of the process assume a major importance and require special consideration.

For better understanding of industrial sterilization, fundamentals of microbial control and principles of thermal kill are introduced in this chapter. Industrial sterilization of fermentation media and fermenter are discussed in chapter 11, while air sterilization is elaborately explained in chapter 12.

10.2 NEED FOR STERILIZATION (ASEPSIS) IN FERMENTATION INDUSTRY

If the fermentation is invaded by a foreign microorganism then the following consequences may occur:

(i) The medium would have to support the growth of both the production organism and the contaminant, resulting in a loss of productivity.

(ii) If the fermentation is a continuous one then the contaminant may 'outgrow' the production organism and displace it from the fermentation.

(iii) The foreign organism may contaminate the final product. e.g. single-cell protein where the cells, separated from the broth, constitute the product.

(iv) The contaminant may produce compounds which make subsequent extraction of the final product difficult.

(v) The contaminant may degrade the desired product; this is common in bacterial contamination of antibiotic fermentations where the contaminant would have to be resistant to the normal inhibitory effects of the antibiotic and degradation of the antibiotic is a common resistance mechanism, e.g. the degradation of β-lactam antibiotics by β -lactamase-producing bacteria.

(vi) Contamination of a bacterial fermentation with phage could result in the lysis of the culture.

Avoidance of contamination may be achieved by:

(i) Using a pure inoculum to start the fermentation, as discussed in Chapter 13.

(ii) Sterilizing the medium to be employed, as discussed in chapter 11.

(iii) Sterilizing the fermenter vessel. (Chapter 11)

(iv) Sterilizing all materials to be added to the fermentation during the process. (Chapter 11).

(v) Maintaining aseptic conditions during the fermentation.

(vi) Sterilizing air supply, if fermenting organism is aerobic (Chapter 12).

10.3 AN INTRODUCTION TO MICROBIAL CONTROL

Microorganisms can be removed, inhibited, or killed by several chemical agents, physical agents or physical processes, such as dry heat, moist heat, filtration radiation and chemical sterilization: they act in many different ways, and each has its own limitations.

The choice of particular method or agent depends on the kind of material to be treated (living or nonliving), the kind of microorganism to be controlled e.g. (bacteria or virus) and the environmental conditions at the time of its use (e.g. pH, temperature, microbial content etc).

Controlling agents act in many ways to kill or inhibit unwanted microorganisms. The mode of action include, the oxidation of cell components, coagulation of proteins, denaturation of proteins and nucleic acids, cell wall or cell membrane damage and interference with metabolic pathways.

Two distinct terms sterilization and disinfection are used to describe the process for the killing or removal of microorganisms.

Sterilization is process used to achieve sterility by freeing of an article from all living organisms including viable spores and **disinfection** is used to describe a process which reduces the number of contaminating microorganisms, particularly those liable to cause infection, to a level which is deemed no longer harmful to health.

One should understand two more terms in reference to microbial control.

Microbicidal agent is any agent designed to kill microorganisms. The term is used to describe many different agents such as bactericides (killing bacteria), sporicides (killing spores), fungicides (killing fungi), Virucides (killing Viruses) etc. The suffix "-cide" means "Killer" or "Killing". The terms "Microbicide" and "germicide" may be used as synonyms.

Microbiostatic agent is any agent able to inhibit or halt the action of microorganisms without necessarily killing the microorganisms. This is a general term used to describe specific agents that are able to induce bacteriostasis (inhibiting the growth of bacteria), fungistasis, or virustasis. The suffix "stasis" means "to inhibit" or "to arrest".

Fundamentals of control

To determine the most efficient way of dealing with contaminating microbes, scientists have studied the behaviour of microbial population exposed to various killing agents. Some physical and chemical control agents are effective over a wide range of microorganisms, while others are more restricted in their scope. Regardless of the nature of the control agent, its effectiveness is influenced by the conditions under which it is applied. The following discussion covers some of the basic principles of microbial control.

10.3.1 The Pattern of Microbial Death

A microbial population is not killed instantly when exposed to a lethal agent. Population death, like population growth, is generally exponential or logarithmic; that is the population will be reduced by the same fraction at constant intervals (Table : 10.1).

Table 10.1 : A theoretical microbial killing experiment

Assume that the initial sample contains 10^6 vegetative microorganisms per ml and that 90% of the organisms are killed during each minute of exposure.

Minute	Microbial Number at start of minute	Microorganisms killed in 1 minute (90% of total)	Microorganisms at end of 1 minute.	Log_{10} of survivors
1	10^6	9×10^5	10^5	5
2	10^5	9×10^4	10^4	4
3	10^4	9×10^3	10^3	3
4	10^3	9×10^2	10^2	2
5	10^2	9×10^1	10	1
6	10^1	9	1	0
7	1	0.9	0.1	-1

If the logarithm of the population number remaining is plotted against the time of exposure of microorganism to the agent, a straight line plot will result [Fig. 10.1(b)].

When the population has been greatly reduced, the rate of killing may slow due to the survival of a more resistant strain of the microorganism.

In order to study the effectiveness of a lethal agent, one must be able to decide when microorganisms are dead, a task by no means as easy as with microorganisms. It is hardly possible to take a bacterium's pulse. A bacterium is defined as dead it does not grow when inoculated into culture medium that would normally support its growth. In like manner, an inactive virus cannot infect a suitable host. In short, the term death is used in microbiology to express the loss of viability.

10.3.2 Conditions Influencing the Effectiveness of Antimicrobial Agent

Destruction of microorganisms and inhibition of microbial growth are not simple matters because the efficiency of an antimicrobial agent (an agent that kills microorganisms or inhibits their growth) is affected by following factors:

(1) **Population size**—Because an equal fraction of a microbial population is killed during each interval (Table : 10.1), a larger population requires a longer time to die than a smaller one.

(2) **Population composition**—The effectiveness of an agent varies greatly with the nature of the organisms being treated because microorganisms differ markedly in susceptibility. Bacterial endospores are much more resistant to most antimicrobial agents than are vegetative forms, and younger cells are usually more readily destroyed than mature organisms. Some species are able to withstand adverse conditions better than others. ***Mycobacterium tuberculosis***, which causes tuberculosis, is much more resistant to most treatments than the majority of bacteria.

(3) **Concentration or intensity of an antimicrobial agent**—Often, but not always, the more concentrated a chemical agent or intense a physical agent, the more rapidly microorganisms are destroyed. Dependence of effectiveness on concentration or intensity is generally not linear. Over a short range, a small increase in concentration leads to an exponential rise in effectiveness; beyond a certain point, further increases may not rise the killing rate much at all. Sometimes, an agent is more effective at lower concentrations. For example, 70% ethanol is more effective than 95% ethanol.

(4) **Duration of exposure**—The longer a population is exposed to a microbial agent, the more organisms are killed. To achieve sterilization, an exposure duration sufficient to reduce the probability of survival to 10^{-6} or less should be used.

(5) **Temperature**—An increase in the temperature at which a chemical acts often enhances its activity. Frequently, a lower concentration of disinfectant or sterilizing agent can be used at a higher temperature.

(6) **Hydration**—The presence of water in an environment affects that rate at which a particular agent kills a microbial population. For example, a high water content expedites the coagulation or jelling of cellular proteins when heat is the killing agent. A 50% solution of egg albumin in water is coagulated when the solution is heated to 55° celsius. Hot, moist air (moist heat) kills more effectively than hot dry air, because the moist air has a greater heat content

than does dry air. In addition, many chemicals must be in solution to be effective as control agents. Water affects the activity of most chemical control agents by promoting chemical reactions and the ionisation of the agents.

(7) **Local environment**—The microbial population to be controlled is not isolated, but surrounded by environmental factors that may either offer protection or aid in its destruction. For example,

(I) pH of the environment

Because heat kills more readily at an acid pH, acid foods and beverages such as fruits and tomatoes are much easier to pasteurise than more neutral foods like milk.

(II) Presence of organic matter

A Second important environmental factor is organic matter that can protect microorganisms against heating and chemical disinfectants. It may be necessary to clean an object before sterilizing or disinfecting it. Syringes and medical or dental equipment certainly need to be cleaned before being sterilized because the presence of too much organic matter could protect pathogens and increase the risk of infection. The same care must be taken when pathogens are destroyed during the supply of drinking water to the people.

10.3.3 Modes of Action of Antimicrobial Agents

Antimicrobial agents act in various ways to inhibit or kill unwanted microorganisms. Knowing the mode of action of a particular agent can help predict the conditions under which it will function most effectively. It may also reveal which species of microbes will be most susceptible to that agent.

The modes in which antimicrobial agents inhibit or kill the microorganisms include one of the following kinds of actions—

(1) Damage to the cell wall
(2) Inhibition of cell wall synthesis
(3) Alteration of permeability of the cytoplasmic membrane
(4) Alteration of the physical or chemical state of proteins and nucleic acids.
(5) Inhibition of protein synthesis
(6) Inhibition of nucleic acid synthesis
(7) Interference with metabolic pathways.

10.4. PRINCIPLES OF THERMAL KILL

Thermal procedures are usually simple, reliable, and relatively inexpensive. Much of the knowledge concerning the heat destruction ("Thermal kill") of microorganisms comes from studies by the food processing industry.

10.4.1 Terms Concerning the Thermal Kill

Although this chapter does not compare various procedures in depth, some important terms concerning the thermal kill of microorganisms should be noted and defined:

(1) Thermal death point—That temperature at which a suspension of organisms is sterilized after a 10-minute exposure.

(2) Thermal death time—The length of time required for a particular temperature to sterilize a suspension of organisms.

(3) Decimal reduction time or D Value—The time that will be required to kill 90% of the organisms in a suspension at a specified temperature. The temperature is generally expressed as a subscript as in $D_{100°C}$ or $D_{59°F}$.

The thermal death point (TDP) and the thermal death time (TDT) take into consideration the interaction of time and temperature without any particular guidelines. The "D" value is the actual number of minutes required to reduce the viable count of microbes, according to certain established guidelines.

"D" values are used to estimate the relative resistance of a microorganism to different temperatures through calculation of the **"Z" value,** the increase in temperature required to reduce D to 1/10 its value or to reduce it by one log cycle- when log D is plotted against temperature.

These **"D"** and **"Z"** values are extremely important in food-processing industries to prepare canned food. Excessive heat exposure may deteriorate the quality of food. Therefore, it is important to know the lowest temperature and shortest time that will effectively treat the product.

10.4.2 How Heat Kills Microorganisms?

Of all the control agents, high temperature (heat) is the most efficient and cost-effective sterilant. If a material is not likely to be damaged at high temperature, its sterilization with heat is recommended. Generally, when psychrophilic and mesophilic microorganisms are exposed to temperature above 55°C, their proteins denature irreversibly and they die. As the temperature is increased they die at a faster rate.

(I) Moist heat

The loss of viability in a population exposed to high heat (temperature) correlates well with the denaturation of proteins essential for microbial activities, particularly when the heat is applied in moist environments (moist heat). It also appears, however that membranes become much more fluid and extremely permeable at high temperatures. The increased permeability results in the leakage of intracellular materials and the loss of membrane potentials. This blocks the transport of nutrients into the cell and in some cases, the generation of ATP. Thus denaturation of cellular proteins and disruption of membrane integrity, whether operating alone or together contribute to efficient sterilization.

Moist heat has better and effective penetrating power.

(II) Dry heat

Heat affects microorganisms in dry environments differently than those in moist environments. Dry heat tends to dehydrate the cell, causing reversible protein denaturation. Desiccation tends to "preserve" same cells in an inactive state consequently : higher temperatures are required to kill cells in dry environments than in moist ones. The physiological state of desiccated cells is such that their cellular contents become much more resistant to irreversible damage. In general, it takes considerably more time to kill endospores using dry heat at 180°C than it does using moist heat at 121°C. Bacteria and other microorganisms eventually are killed in dry heat environments when their proteins denature and they become charred (oxidized).

10.4.3 The Sensitivity of Microorganisms to Heat at a Constant Temperature

The destruction of microorganisms by heat can be demonstrated by maintaining a suspension of bacteria or spores at a constant temperature and counting the survivors over a period of time. If the time of heating is plotted on the abscissa, and the logarithm of the number of survivors on the ordinate, then a straight line relationship is obtained (Figure 10.1). Such survival curves at constant temperature may be referred to as the Thermal Death Time Curves. From the straight line relationship it is possible to deduce that the rate of thermal destruction is directly proportional to the number of organisms *N* present at time *t*.

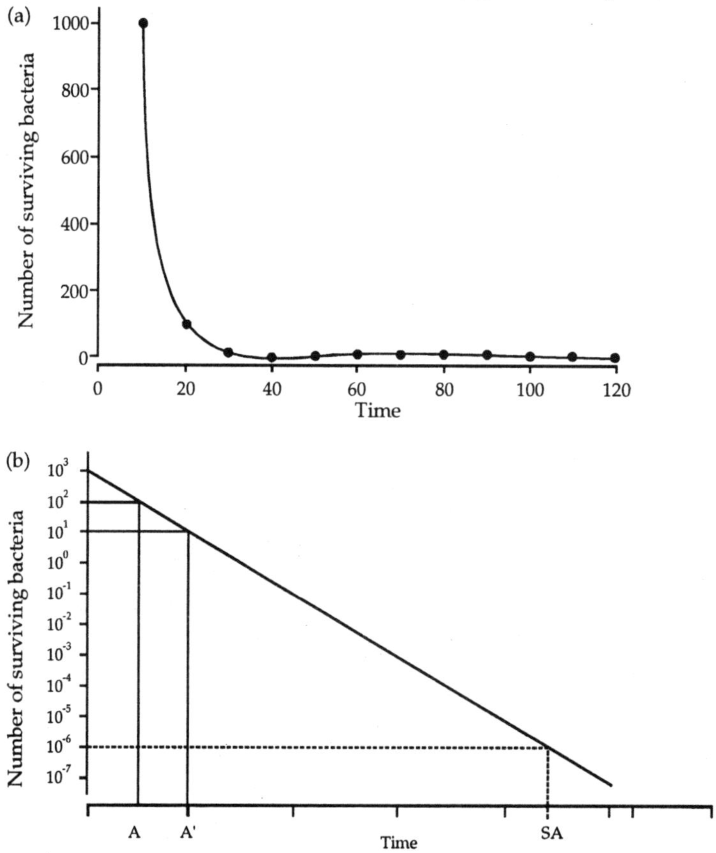

Figure 10.1 : The kinetics of cell death—Thermal death time cure. (a) During heat sterilisation, the number of living cells decreases by the same proportion per unit time, giving an exponential curve. (b) When plotted on a logarithmic scale, the decrease in numbers is seen as a straight line, whose slope is a reflection of the rate of killing. The time period between A and A′ is the decimal reduction time (D): the time taken to reduce the population to one-tenth of its size. The total period until the point SA is the sterility assurance value, when there is only a one in a million probability of any cells having survived.

$$\frac{dN}{dt} = kN \text{ or } \frac{dN}{N} = -kdt \tag{10.1}$$

$$\text{In}\frac{N_2}{N_1} = k(t_2 - t_1) \tag{10.2}$$

$$\text{In}\frac{N_1}{N_2} = k(t_2 - t_1) \tag{10.3}$$

$$\log_{10}\frac{N_1}{N_2} = \frac{k}{2.3} - (t_2 - t_1) = \frac{(t_2 - t_1)}{2.3/k} \tag{10.4}$$

The constant $2.3/k$ is called D (the decimal reduction time) and if N_0 represents the number of viable cells at the start and N the numer of survivors at time t, then the equation may be written simply as:

$$\log_{10}\frac{N_0}{N} = \frac{t}{D} \tag{10.5}$$

For the time $t = D$, $\log_{10} N_0/N = 1$ and $N_0/N = 10$ from which it can be seen that D represents the time required to reduce the population size by a factor of ten (Figure 10.1).

The values of both k and D have been determined for a number of bacterial spores at a temperature of 250° F = 121°C (Table 10.2). In fact, it is the heat resistant spores of bacteria which give particular problems during sterilization, the vegetative cells and the spores of fungi being fairly sensitive to heat treatment.

Table 10.2 : Value of *k* and *D* for various bacterial spores at 121°C

Culture	k (min^{-1})	D (min)
Bacillus subtills FS 5230	2.6-3.8	0.6-0.9
Bacillus Stearothermophilus FS 1518	0.77	3
Clostridium sporogenes PA 3679	1.8	1.8

In practice, what is required is a curve for the heat resistance of a contaminant microbial population derived from media constituents, such as corn-steep liquor, and from the environment; value of k and D are thus determined experimentally, for any particular situation rather than extrapolating from values obtained with pure cultures.

10.4.4 Sensitivity of Microorganisms to Heat as a Function of Temperature

From Figure 10.1 above showing survival curves for the same suspension of spores at two temperatures T_1 and T_2 ($T_2 > T_1$), it can be seen that increasing temperature brings about a reduction in the constant D.

It can be shown experimentally that the rate of change of D with temperature is proportional to D.

$$\frac{dD}{dT} = -K'D \text{ or } \frac{dD}{D} = K'dT \qquad (10.6)$$

$$In\frac{D_2}{D_1} = k'(T_2 - T_1) \text{ or, In } \frac{D_1}{D_2} = k'(T_2 - T_1) \qquad (10.7)$$

$$\log_{10}\frac{D_1}{D_2} = \frac{k'}{2.3}(T_2 - T_1) = \frac{(T_2 - T_1)}{2.3/k'} \qquad (10.8)$$

If the constant 2-3/k' = Z then the equation may be written simply as :

$$\log_{10}\left(\frac{D_1}{D_2}\right) = \frac{(T_2 - T_1)}{Z} \qquad (10.9)$$

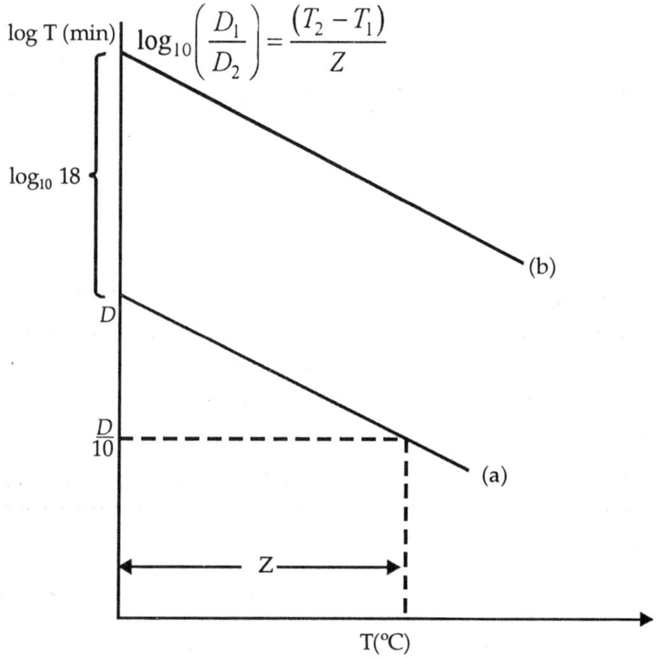

Fig. 10.2 : (a) The relation between decimal reduction time and temperature (b) The relationship between t, the time required to reduce the population by 10^{18} and temperature T.

When Z = (T_2 - T_1) then $\log_{10}$ (D_1/D_2) = 1 and D_2 = 1/10 D_1, thus Z represents the increase in temperature required to reduce the decimal reduction time D to one tenth of its value.

If the decimal reduction time D is plotted as a function of temperature on semi-logarithmic coordinates, a straight line is obtained with a gradient 1/Z [Figure 10.2(a)].

The heat treatment needed for a batch of medium is dependent on the maximum probability of residual contamination which can be tolerated. Thus for example, a bacterial spore surviving in a volume 1000 times greater than that of the fermenter used would give a probability of one batch in a thousand runs being contaminated. Yeast spores may be slightly more resistant than the vegetative cells of many bacteria but, in fact, neither these nor the spores of filamentous fungi and actinomycetes have very great heat resistance (Table : 10.3).

Table 10.3 : Heat resistance of a variety of microorganisms

Microorganism	Heating temp (°C)	D (min)	Z (°C)	Suspension medium
Bacterial spores				
Clostridium thermo saccharolyticum	132	4.4	6.7	Water
	132	3.3	10	Molasses
Bacillus stearothermophilus	115	22.6	71	Water
Bacillus megaterium	100	1	8.8	Water
Vegetative cells of bacteria				
Salmonella typhimurium	55.5	0.5	4.2	Ringer's Solution
Actinomycete spores				
Thermoactinomyces vulgaris	100	11	-	0.001 M phosphate buffer pH 7-0
Yeast spores				
Saccharomyces cerevisiae	55	0.9	-	Phosphate buffer pH 7-0
Spores of filamentous fungi				
Aspergillus chevalieri	65	50	12.8	Sucrose/malt agar
Virus				
Adenovirus	55	17	2.5	Ice cream

Example of a Calculation

For penicillin production, an industrial fermentation medium containing corn-steep liquor, a product which frequently has a high spore population. The crude corn-steep liquor is found to contain 20×10^6 viable spores/ml: thus the entire fermenter will contain

$$20 \times 10^6 \times 50 \times 10^3 \times 10^3 = 10^{15} \text{ spores}$$

An acceptable risk of contamination is considered to be 99.9% that is to say there will be a statistical expectation of only one contaminated run per 1000 fermentations. Thus the number of spores has to be reduced from 10^{15} to 10^{-3}. It will thus be necessary to heat for a period t corresponding to $(15 + 3)D = 180D$. If t (in minutes) is plotted against the temperature T on a semi-logarithmic scale the curve obtained is the same as that relating the decimal reduction time D against temperature but shifted up the ordinate by $\log_{10} 18$ (Figure 10.2b). This is because, if t = 18D, then

$$\log_{10} T = \log_{10} 18 + \log_{10} D$$

If F is the value of t for T= 121°C, and Z is the increase in temperature required to reduce the period of heating by 90% then;

$$\frac{\log_{10} \tau_T - \log_{10} F}{121 - T} = \frac{\log_{10} \dfrac{\tau_T}{F}}{121 - T} = \frac{1}{Z} \qquad (10.10)$$

This equation may be rewritten in two equivalent forms when Z and T are in °F:

$$\tau_T = F \times 10^{(250-T)Z} \quad \text{or} \quad F = \tau_T \times 10^{(T-250)Z} \qquad (10.11)$$

10.4.5 A Simple Example of the Construction of a Sterilization Chart

An acid medium containing thiamine is to be subjected to continuous sterilization in a plate heat exchanger. The treatment is required to assure the acceptable destruction of the spores of *Bacillus coagulans* the resistance characteristics of which have been previously established:

F = 0.7 : Z = 18

To simplify the calculation, it is considered that the time taken to raise the temperature to T°F, and to subsequently recool, does not significantly contribute to sterilization. There are tables giving the expression $10^{(T-250)Z}$ as a function of T and Z allowing one to calculate *t* knowing F and Z:

Thus at 245°F (118°C) r = 80 seconds

at 266°F (130°C) r = 5.4 seconds

If is required to preserve, for example, 90% of the thiamine in the medium then the curves corresponding to the destruction of *Bacillus coagulans* and to a rate of destruction of 10% thiamine are plotted on the same graph. The graph is then examined for the presence of a region fulfilling the two requirements–the desired level of destruction of spores and the destruction of less than 10% thiamine (Figure 10.3) : this region is adopted as a time/temperature chart for satisfactory heat treatment, and the necessary holding time for a given temperature can be read off.

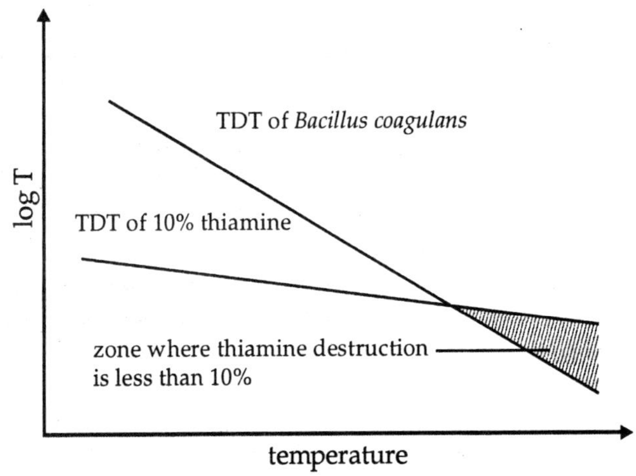

Fig. 10.3 : Determination of a zone of heat treatment reducing numbers of *Bacillus coagulans* by 10^{-18} but destroying less than 10% of the thiamine in a culture medium.

Experiments of this kind demonstrate the virtual necessity for sterilizing at relatively high temperatures for very short times (i e., continuous sterilization) if the medium contains heat labile constituents. When the heating and cooling phases of heat treatment are sufficiently long to contribute to sterilization then it is no longer possible to use calculations based on a constant temperature (Figure 10.4).

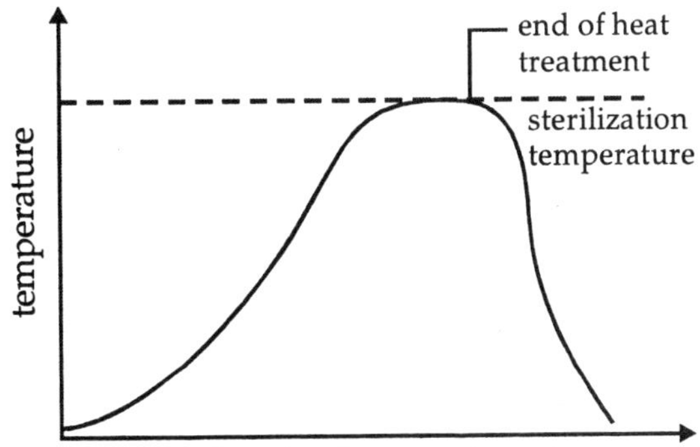

Fig. 10.4 : Change of temperature in a culture medium during the sterilization of a fermenter.

In a large industrial fermenter the duration of the periods during which the temperature is increasing and decreasing may have a more significant effect on the medium than the period at the highest temperature. For this reason the graphical method described below may be used.

10.4.6 Experimental Determination of the Heat Resistance of Microorganisms

It is required to determine the values of F and Z which will then allow the construction of a graph relating time of heat treatment and temperature, for F represents one point on that line and 1/Z represents the slope of the line (Figure 10.2). Although it may be desirable to determine the thermal resistance of unknown communities in the medium to be used because this is where problems will arise in practice, it is usual to measure the resistance of a known organism in a defined medium. The thermophile *Bacillus stearothermophilus* FS 1518 is frequently taken as a reference organism because its spores are amongst the most resistant to heat treatment.

(I) Preparation of Spore Suspension

The culture is grown on an agar medium favouring spore formation. After incubation, the sporing culture is suspended in a little sterile water and the suspension treated with lysozyme to destroy vegetative cells and the sporangia. The presence of spores is confirmed microscopically and the suspension transferred to a sterile flask containing glass beads. After shaking to break up any aggregates it is filtered through two layers of sterile cheese cloth. The number of viable spores is counted after treatment of a suitable dilution at 100°C for 5 minutes, to destroy any surviving vegetative cells and to activate germination of the spores.

(II) Tube Assay Method

Tubes, closed at one end, and about 12 cm in length, are prepared from glass tubing of 1 mm thickness and 9 mm outside diameter. To each is added 2 ml of medium (using a syringe) and suspension of a known number of spores. The tubes are then sealed by flaming at about 5 cm above the surface of the contents, and then immersed in a water bath, or a bath of oil or ethylene glycol, stirred and maintained at the required temperature.

At known periods of time, tubes are removed, plunged into cold water opened aseptically, and the contents used to inoculate tube of broth in order to demonstrate survival or destruction (The result is scored + or -). Plotting temperature along the abscissa and the logarithm of time along the ordinate two curves can be obtained which should be fairly close to one another (Figure 10.5).

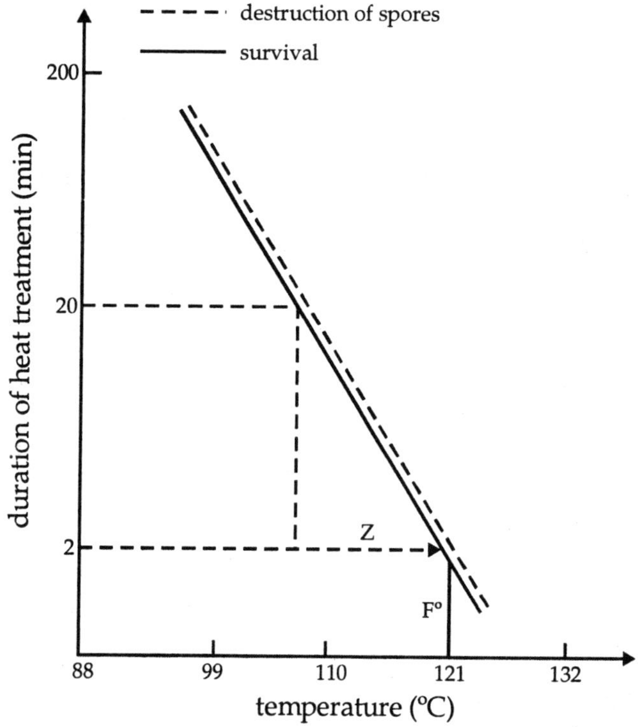

Fig. 10.5 : Curves for survival and destruction of spores.

(III) Example

The curve obtained gave the following results:

$$F = 2 \text{ minutes}$$

$$Z = 17°C$$

This result is only relevant to the destruction of the number of spores used in the experiment (174/ml). The value of F corresponds to an 18-fold decimal reduction in the number of the same spores, therefore

$$F = \frac{18}{\log_{10} 174} \times 2 = 16 \text{ minutes}$$

This method is usable only if the experimental points obtained lie reasonably close to a straight line. Otherwise it is necessary, for each temperature, to obtain the proportion of surviving spores as a function of time from which the decimal reduction time D can be calculated and the value of Z deduced from a study of the relationship between D and temperature. It is necessary then to carry out a full viable count at a range of times at each temperature (Table : 10.4).

Table 10.4 : Survival of *Clostridium sporogenes* PA 3679 at 115°C (10,000 spores to tube)

Time of heating (min) (absolute number)	Number of surviving spores per tube	
10	90	1.95
14	12	1.08
18	2	0.30
22	0.3	1.52
28	0	-
36		-

Let a be the number of spores initially (e.g.. 10,000) and b the number surviving after heating at 115°C for 10 minutes (e.g. 90).

$$D = \frac{10}{\log_{10} a - \log_{10} b} = \frac{10}{\log_{10} 10{,}000 - \log_{10} 90} = 4.88$$

By repeating the experiment at a number of temperatures the corresponding values of D can be calculated (Table : 10.5). Finally, by plotting D (on a logarithmic scale) against temperature Z can be determined from the following relationship.

$$\frac{\log_{10} D_1 - \log_{10} D_2}{(T_2 - T_1)} = \frac{1}{Z}$$

Knowing D it is easy to calculate F and with a knowledge of both F and Z for a given culture a working sterilization table can be established.

Table 10.5 : Value of D as a function of temperature for *Clostridium sporogenes* PA 3679

Temperature (°C.)	D (min)
110	19.65
113	10.46
115	4.92
118	2.56
121	1.23

10.5 CONCEPT OF 'PROTECTED' FERMENTATION

The extent to which sterilization procedures are adopted is determined by the likely probability of contamination and the nature of its consequences. Some fermentations are described as **'protected'**—that is, the medium may be utilized by only a very limited range of microorganisms, or the growth of the process organism may result in the development of selective growth conditions, such as a reduction in pH. The brewing of beer falls into this category; hop resins tend to inhibit the growth of many microorganisms and the growth of brewing yeasts tends to decrease the pH of the medium. Thus, brewing yeasts are boiled, but not necessarily sterilized, and the fermenters are thoroughly cleaned with disinfectant solution but are not necessarily sterile. Also, the precautions used in the development of inoculum for brewing are far less stringent than, for example, in an antibiotic fermentation.

❑❑❑

11

STERILIZATION OF FERMENTATION MEDIA AND FERMENTER

11.1 INTRODUCTION

The purpose of media sterilization is to prevent contamination of the end-product and to ensure that all of the nutrients are available for growth to the desired fermenting organism. Absolute or virtual sterility is easily obtained on a laboratory scale by autoclaving for short periods.

On an industrial scale, the difficulties arise from the large quantities of media involved, a single surviving spore in a batch of medium being theoretically capable of causing contamination and from the damage to medium constituents caused by lengthy heat treatment. The fermentation industry has developed some practical solutions also. Selective growth conditions can also hinder contamination, for example—the medium may be acidic (as in the yeast culture fermentation), or a thermophilic bacterium may be used; in antibiotic production the antibiotic itself confers selective protection once its formation begins.

Because of the very large volumes of liquid media to be sterilized, ionizing sterilization or ultrasonic treatment are impracticable for sterilization. High pressure steam is almost universally used for media sterilization because it is more practical, less expensive and less hazardous. Steam is used both for sterilizing media in batch cookers and for continuous sterilizing media employing either steam injection and flash cooling or plate heat exchangers.

11.2 STEAM STERILIZATION OF MEDIA

11.2.1 Principle

This is the most useful and universally employed method for the sterilization of fermentation media as well as for fermenter. A number of factors influence the success of steam sterilization:

(i) the number and type of microorganisms present,

(ii) the composition of the fermentation medium,

(iii) the pH value, and

(iv) the size of the suspended particles.

Vegetative cells are rapidly eliminated at relatively low temperatures, but for destruction of spores, temperatures of 121°C ar needed. Spores of *Bacillus stearothermophilus* are the most heat resistant. Therefore they are used as assay organisms to evaluate various steam-sterilization procedures. Table : 11.1 provides data on the relationship between temperature, k-value, and design criterion [also known as Del factor, Nabla factor and sterilization criterion represented by symbol ∇] for *B. stearothermophilus*.

A list of sterilization times and temperatures for various organisms is given in Table : 11.2.

Table 11.1 : Relationship of temperature, k-value and the design criterion in *Bacillus stearothermophilus*

T°C	k(min^{-1})	∇
100	0.019	-
115	0.666	3.154
118	1.307	6.341
121	2.538	12.549
130	17.524	90.591
140	135.9	
150	956.1	

Table 11.2 : Sterilization time and sterilization temperature of various groups of organisms

Cells	Sterilization time minute	Sterilization temperature (°C)
Vegetative cells	5-10	60
Fungus spores/yeast spores	15	80
Streptomyces spores	5-10	60-80
Bacterial spores, general	5	121
Spores of *Bacillus stearothermophilus*	15	121

11.2.2 Del Factor [Design criterion/Nabla factor/Sterilization criterion/∇]

Deindoerfer and Humphrey (1959) used the term In N_0/N_1 as a design criterion for sterilization, which has been variously called the Del factor, Nabla factor and sterilization criterion represented by the symbol ∇. Thus, the Del factor is a measure of the fractional reduction in viable organism count produced by a certain heat and time regime. Therefore:

$$\nabla = \text{In}\ (N_0/N_1)$$

but $\text{In}\ (N_0/N_1) = kt$

and $kt = A\ .\ t\ .\ e^{-(E/RT)}$

thus $\nabla = A..\ t\ .\ e^{-(E/RT)}.$ (11.1)

On rearranging, equation (11.1) becomes:

$$\text{In}\ t = E/RT + \text{in}\ (\nabla/A). \quad (11.2)$$

Thus, a plot of the natural logarithm of the time required to achieve a certain ∇ value against the reciprocal of the absolute temperature will yield a straight line, the slope of which is dependent on the activation energy, as shown in Fig. 11.1. From Fig. 11.1 it is clear that the same degree of sterilization (∇) may be obtained over a wide range of time and temperature regimes; that is, the same degree of sterilization may result from treatment at a high temperature for a short time as from a low temperature for a long time.

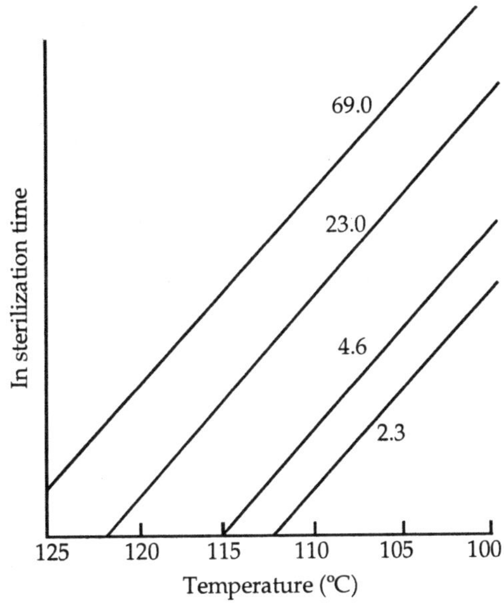

Fig. 11.1 : The effect of sterilization and temperature on the Del factor achieved in the process. The figures on the graph indicate the Del factors for each straight line (modified after Richards, 1966).

The kinetic description of bacterial death enables the design of procedures (giving certain ∇ factors) for the sterilization of fermentation broths. By choosing a value for N_t, procedures may be designed having a certain probability of achieving sterility, based upon the degree of risk that is considered acceptable. According to Deindoerfer and Humphrey (1959), Richards (1968), Banks (1979) and Corbett (1985) a risk factor of one batch in a thousand being contaminated is frequently used in the fermentation industry—that is, the final microbial count in the medium after sterilization should be 10^{-3} viable cells. However, to apply these kinetics it is necessary to know the thermal death characteristics of all the taxa contaminating the fermenter and unsterile medium. This is an impossibility and, therefore, the assumption may be made that the only microbial contaminants present are spores of *Bacillus stearothermophilus*—that is, one of the most heat-resistant microbial types known. Thus, by adopting *B. stearothermophilus* as the design organism a considerable safety factor should be built into the calculations. It should be remembered that *B. stearothermophilus* is not always adopted as the design organism. If the most heat-resistant organism contaminating the medium ingredients is known, then it may be advantageous to base the sterilization process on this organism. Deindoerfer and Humphrey (1959) determined the thermal death characteristics of *B. stearothermophilus* spores as:

$$\text{Activation energy} = 67.7 \text{ kcal mole}^{-1}$$

$$\text{Arrhenius constant} = 1 \times 10^{36.2} \text{ second}^{-1}$$

However, it should be remembered that these kinetic values will vary according to the medium in which the spores are suspended, and this is particularly relevant when considering the sterilization of fats and oils (which are common fermentation substrates) where the relative humidity may be quite low. Bader et al. (1984) demonstrated that spores of *Bacillus macerans* suspended in oil were ten times more resistant to sterilization if they were dry than if they were wet.

A regime of time and temperature may now be determined to achieve the desired Del factor. However, a fermentation medium is not an inert mixture of components, and deleterious reactions may occur in the medium during the sterilization process, resulting in a loss of nutritive quality. Thus, the choice of regime is dictated by the requirement to achieve the desired reduction in microbial content with the least detrimental effect on the medium. Figure 11.2 illustrates the deleterious effect of increasing medium sterilization time on the yield of product of subsequent fermentations. The initial rise in yield is due to some components of the medium being made more available to the process microorganism by the 'cooking effect' of a brief sterilization period (Richards, 1966).

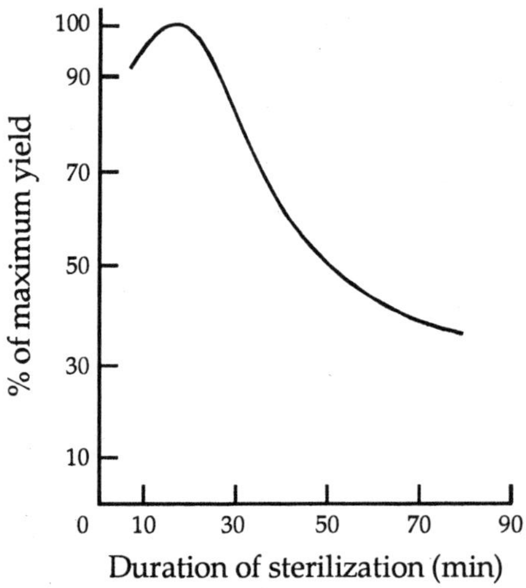

Fig. 11.2 : The effect of the time of sterilization on the yield of a subsequent fermentation (Richards, 1966).

Two types of reaction contribute to the loss of nutrient quality during sterilization:

(i) *Interactions between nutrient components of the medium.* A common occurrence during sterilization is the Maillard-type browning reaction which results in discoloration of the medium as well as loss of nutrient quality. These reactions are normally caused by the reaction of carbonyl groups, usually from reducing sugars, with the amino groups of amino acids and proteins. An example of the effect of sterilization time on the availability of glucose in a corn-steep liquor medium is shown in Table : 11.3 (Corbett, 1985). Problems of this type are normally resolved by sterilizing the sugar separately from the rest of the medium and recombining the two after cooling.

(ii) *Degradation of heat labile components.* Certain vitamins, amino acids and proteins may be degraded during a steam sterilization regime. In extreme cases, such as the preparation of media for animal-cell culture, filtration may be used and this aspect will be discussed later in the chapter. However, for the vast majority of fermentations these problems may be resolved by the judicious choice of steam sterilization regime.

Table 11.3 : The effect of sterilization time on glucose concentration and product accretion rate in an antibiotic fermentation (Corbett, 1985)

Time at 121° (min)	Amount of added glucose remaining (%)	Relative accretion rate
60	35	90
40	46	92
30	64	100

The thermal destruction of essential media components conforms approximately with first order reaction kinetics and, therefore, may be described by equations similar to those derived for the destruction of bacteria:

$$x_t/x_0 = c^{-kt} \tag{11.3}$$

where x_t is the concentration of nutrient after a heat treatment period, t,
x_0 is the original concentration of nutrient at the onset of sterilization,
k is the reaction rate constant.

It is important to appreciate that we are considering the decline in the concentration of the nutrient component, whereas we consider the decline in the number of contaminants. The effect of temperature on the reaction rate constant may be expressed by the Arrhenius equation:

$$\text{in } k = \text{In } A - E/RT. \tag{11.4}$$

Therefore, a plot of the natural logarithm of the reaction rate against $1/T$ will give a straight line, slope $-(E/R)$. As the value of R, the gas constant is fixed the slope of the graph is determined by the value of the activation energy (E). The activation energy for the thermal destruction of *B. stearothermophilus* spores has been cited as 67.7 kcal mole^{-1}, whereas that for thermal destruction of nutrients is 10 to 30 kcal mole^{-1} (Richards, 1968). Figure 11.3 is an Arrhenius plot for two reactions—one with a lower activation energy than the other. From this plot it may be seen that as temperature is increased, the reaction rate rises more rapidly for the reaction with the higher activation energy. Thus, considering the difference between activation energies for spore destruction and nutrient degradation, an increase in temperature would accelerate spore destruction more than medium denaturation.

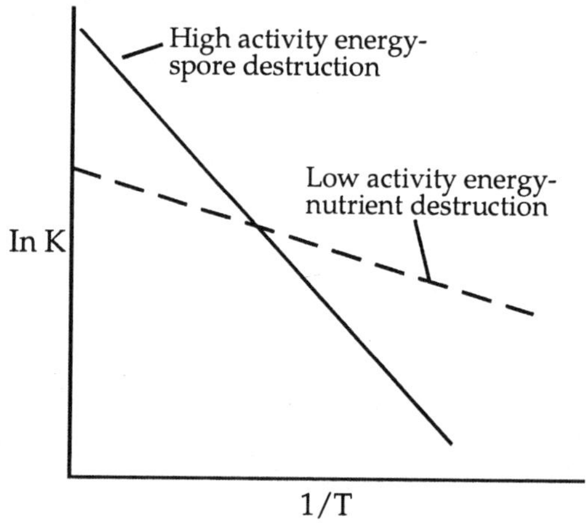

Fig. 11.3 : The effect of activation energy on spore and nutrient destruction.

In the consideration of Del factors it was evident that the same Del factor could be achieved over a range of temperature/time regimes. Thus, it would appear to be advantageous to employ a high temperature for a short time to achieve the desired probability of sterility, yet causing minimum nutrient degradation. Thus, the ideal technique would be to heat the fermentation medium to a high temperature, at which it is held for a short period, before being cooled rapidly to the fermentation temperature. However, it is obviously impossible to heat a batch of many thousands of litres of broth in a tank to a high temperature, hold for a short period and cool without the heating and cooling periods contributing considerably to the total sterilization time. The only practical method of materializing the objective of a short-time, high-temperature treatment is to sterilize the medium in a continuous stream. In the past the fermentation industry was reluctant to adopt continuous sterilization due to a number of disadvantages outweighing the advantage of nutrient quality.

11.3 CLASSICAL TECHNIQUES OF STEAM-STERILIZATION

The classical technique of making the medium sterile by the use of steam may be carried out in two ways:

(I) Batch sterilization (in fermenter or cooker or autoclave)

(II) Continuous sterilization

The relative merits of batch and continuous sterilization may be summarized as follows:

Advantages of batch sterilization over continuous sterilization

(i) Lower capital equipment costs.

(ii) Lower risk of contamination — continuous processes require the aseptic transfer of the sterile broth to the sterile vessel.

(iii) Easier manual control.

(iv) Easier to use with media containing a high proportion of solid matter.

Advantages of continuous sterilization over batch sterilization

(i) Superior maintenance of medium quality.

(ii) Ease of scale-up.

(iii) Easier automatic control.

(iv) The reduction of surge capacity for steam.

(v) The reduction of sterilization cycle time.

(vi) Under certain circumstances, the reduction of fermenter corrosion.

11.3.1 Batch Sterilization of Media by Steam

The batch sterilization of fermentation medium may be done either in fermenter (bioreactor) at 121°C or in a separate cooker.

11.3.1.1 Batch Sterilization in Fermenter or Bioreactor

Most nutrient media are presently sterilized in batch volumes in the bioreactor at 121°C. Approximate sterilization times can be calculated from the nature of the medium and the size of the fermenter. Not only the nutrient media, but also the fittings, valves and electrodes of the fermenter itself must be sterilized. Therefore, actual sterilization times are significantly longer than calculated ones and must be empirically determined for the specific nutrient solutions in the fermenter. One method of sterilization is to inject steam into the fermenter mantle or interior coils (indirect sterilization). Another method is to inject steam into the nutrient solution itself (direct procedure), in which case pure steam (free of chemical additives) is a prerequisite. Many industrial steam supplies contain potentially toxic chemicals derived from anti-corrosive additives used in the steam-manufacturing process. With direct steam injection, condensate accumulates within the fermenter and the volume of liquid thus increases during the sterilization process.

Drawbacks

The drawbacks of the heat-sterilization process are observed for a typical sterilization of a 3000 1 batch fermenter. It takes 2-3 hours to reach the sterilization temperature of 121°C, depending on the steam conduction and fermenter size. Once the proper temperature has been reached, another 20-60 minutes are required for the actual killing process, followed by cooling for about one hour. The energy required for heating must subsequently be removed in order to cool the fermenter and if the hot water obtained during the cooling cannot be put to some use, batch sterilization becomes very costly.

11.3.1.2 Batch Sterilization in Cooker or Separate Vessel

Richards, an industrial microbiologist, considered the relative merits of *in situ* medium sterilization and the use of a special vessel. The major advantages of a separate medium sterilization vessel may be summarized as:

(i) One cooker may be used to serve several fermenters and the medium may be sterilized as the fermenters are being cleaned and prepared for the next fermentation, thus saving time between fermentations.

(ii) The medium may be sterilized in a cooker in a more concentrated form then would be used in the fermentation and then diluted in the fermenter with sterile water prior to inoculation. This would allow the construction of smaller cookers.

(iii) In some fermentations, the medium is at its most viscous during sterilization and the power requirement for agitation is not alleviated by aeration as it would be during the fermentation proper. Thus, if the requirement for agitation during *in situ* sterilization were removed, the fermenter could be equipped with a less powerful motor. Obviously, the sterilization kettle would have to be equipped with a powerful motor, but this would provide sterile medium for several fermenters.

(iv) The fermenter would be spared the corrosion which may occur with medium at high temperature.

The major disadvantages of a separate medium sterilization vessel may be summarized as:

(i) The cost of constructing a batch medium sterilizer is much the same as that for the fermenter.

(ii) If a cooker serves a large number of fermenters complex pipework would be necessary to transport the sterile medium, with the inherent dangers of contamination.

(iii) Mechanical failure in a cooker supplying medium to several fermenters would render all the fermenters temporarily redundant. The provision of contingency equipment may be prohibitively costly.

Overall, the pressure to decrease the 'down time' between fermentations has tended to outweigh the perceived disadvantages of using separate sterilization vessels. Thus, sterilization in dedicated vessels is the method of choice for batch sterilization. However, as pointed out by Corbett (1985), the fact that separate batch sterilizers are used lends further weight to the argument that continuous sterilization should be adopted in preference to batch. The capital cost of a separate batch sterilizer is similar to that of a continuous one and the problems of transfer of sterile media are then the same for both batch and continuous sterilization. Thus, two of the major objections to continuous systems (capital cost and aseptic transfer) may be considered as no longer relevant.

11.3.2 Continuous Sterilization of Media by Steam

The continuous sterilization offers more flexibility in the choice of time-temperature conditions to which the fermentation medium is exposed. This process gives advantages both in the saving of time and plant space and in improved quality of the medium. For example in the production of vitamin B_{12} an increase of 160% in the yields has been obtained. The continuous sterilization of media is especially important for continuous fermentation processes.

The continuous sterilization device includes **three basic units:**

(1) a **heating unit** in which the medium is heated to the required temperature in the shortest possible time interval,

(2) a **holding vessel** in which the heated medium is kept at a constant temperature for a certain period necessary for complete sterilization,

(3) a **cooling unit** in which the medium is cooled to the culture temperature.

11.3.2.1 Working Design of Continuous Sterilization Processes

The design of continuous sterilization cycles may be approached in exactly the same way as for batch sterilization systems. The continuous system includes a time period during which the medium is heated to the sterilization temperature, a holding time at the desired temperature and a cooling period to restore the medium to the fermentation temperature. The temperature of the medium is elevated in a continuous heat exchanger and is then maintained in an insulated serpentine holding coil for the holding period. The length of the holding period is dictated by the length of the coil and the flow rate of the medium. The hot medium is then cooled to the fermentation temperature using two sequential heat exchangers: the first utilizing the incoming medium as the cooling source (thus conserving heat by heating-up the incoming medium) and the second using cooling water. The major advantage of the continuous process is that a much higher temperature may be utilized, thus reducing the holding time and reducing the degree of nutrient degradation. The required Del factor may be achieved by the combination of temperature and holding time which gives an acceptably small degree of nutrient decay.

11.3.2.2 Types of Continuous Sterilizers

There are two types of continuous sterilizer which may be used for the sterilization of fermenter media:

(I) The indirect heat exchangers, and

(II) The direct heat exchangers (steam injectors)

(I) The Indirect heat exchangers—In such type of sterilizers, indirect heating is performed using some kind of a heat exchanger (plate, tube, spiral etc.), with steam or oil heating. The holding unit is either a tube or stirred vessel. The cooling unit may be of various types but usually involves a heat exchanger, cooling with cold medium which is thus preheated, improving the economy of sterilization. The process is characterized by very short heating and cooling periods.

(a) Spiral heat exchangers—The plate and frame type heat exchangers are available. But, the most suitable indirect heat exchangers are of Rosenblad's "double spiral" type which consists of two sheets of high-grade stainless steel which have been curved around a central axis to form a double spiral, as shown in figure 11.4. The ends of the spiral are sealed by covers.

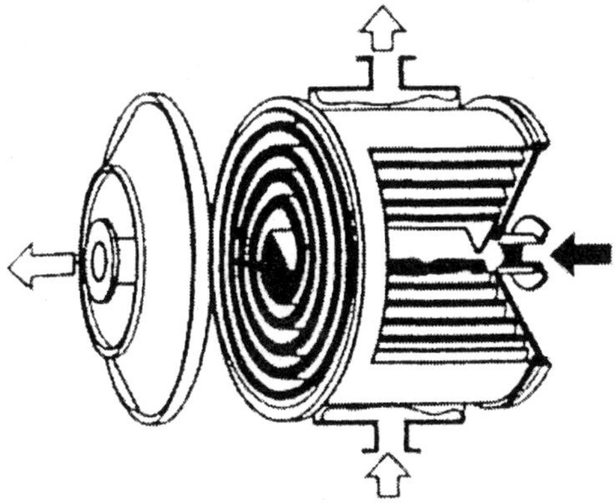

Fig. 11.4 : A schematic representation of a spiral heat exchanger (Alfa-Laval Engineering Ltd., Brentford, Middlesex).

To achieve sterilization temperature, steam is passed through one spiral and medium through the other in countercurrent streams. Spiral heat exchangers are also used to cool the medium after passing through the holding coil. Incoming unsterile medium is used as the cooling agent in the first cooler so that the incoming medium is partially heated before it reaches the sterilizer and, thus, heat is conserved.

The most widely used continuous sterilization system is that based on the spiral heat exchangers and a typical layout is shown in figure 11.5.

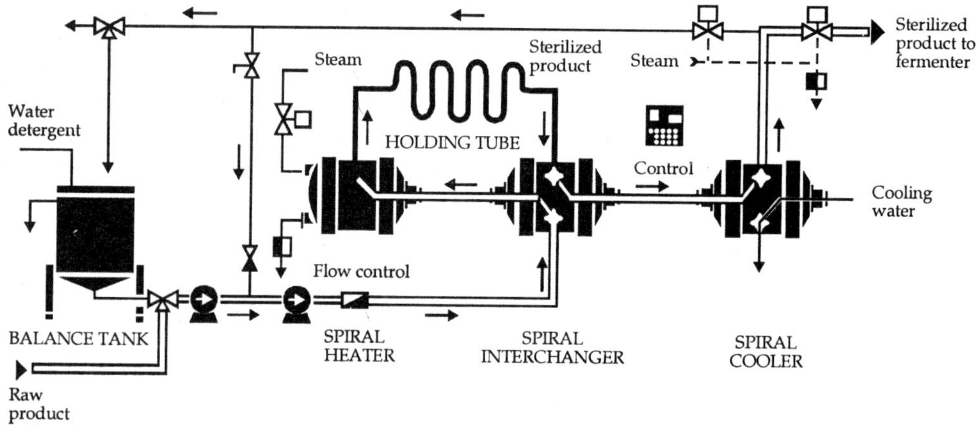

Fig. 11.5 : Flow diagram of a typical continuous sterilization system employing spiral heat exchangers (Alfa-Laval Engineering Ltd., Brentford, Middlesex).

As shown in Fig. 11.5, the plant is sterilized prior to sterilization of the medium by circulating hot water through the plant in a closed circuit. At the same time, the fermenter and the pipework between the fermenter and the sterilizer are steam sterilized. Heat conservation is achieved by cooling the sterile medium against cold, incoming unsterile medium which will then be partially heated before it reaches the sterilizer.

The major advantages of the spiral heat exchanger are:

(i) The two streams of medium and cooling liquid, or medium and steam, are separated by a continuous stainless steel barrier with gasket seals being confined to the joints with the end plates. This makes cross contamination between the two streams unlikely.

(ii) The spiral route traversed by the medium allows sufficient clearances to be incorporated for the system to cope with suspended solids. The exchanger tends to be self-cleaning which reduces the risk of sedimentation, fouling and 'burning-on'.

(b) Plate heat exchangers

Indirect plate heat exchangers consist of alternating plates through which the countercurrent streams are circulated. The plates are separated by gaskets and failure of these gaskets can cause cross-contamination between the two streams. Also, the clearances between the plates are such that suspended solids in the medium may block the exchanger and, thus, the system is only useful in sterilizing completely soluble media. However, the plate exchanger is more adaptable than the spiral system in that extra plates may be added to increase its capacity.

The working layout of plate-heat exchanger is shown in figure 11.6.

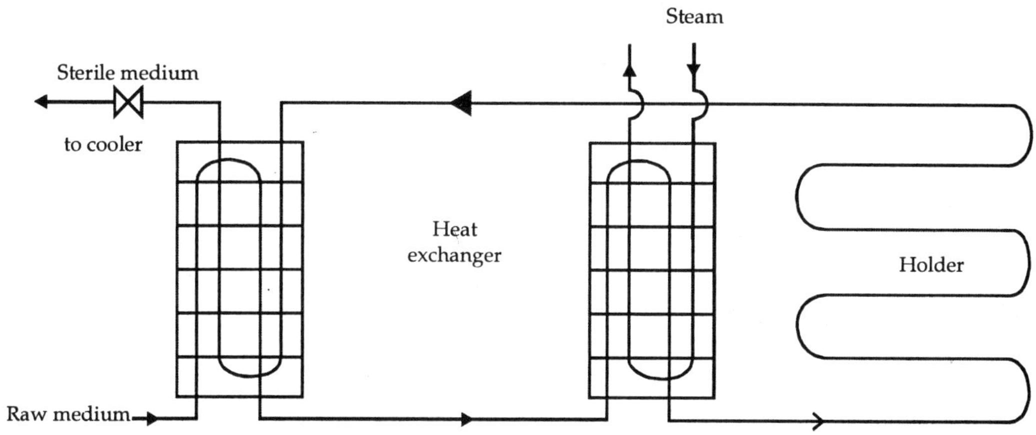

Fig. 11.6 : Working layout of continuous sterilization using plate heat exchanger.

As shown in figure 11.6, this method involves passing of production medium through a heat-exchanger, a holding coil and a cooler. The temperature of the medium undergoing sterilization is raised to the desired level in the heat-exchanger. The medium then passes on to a holding coil, where it is maintained at the sterilizing temperature for a predetermined time period. Finally, the medium is rapidly cooled by counter circulating it in the exchanger against the cool input medium, and then against cold water. The possibility of short holding time allows the use of sterilization temperatures higher than 120°C, without any adverse effects on the nutritional qualities of the culture medium.

Drawbacks:

The early continuous sterilizers were constructed as plate heat exchangers and these were unsuitable on two accounts:

(i) Failure of the gaskets between the plates resulted in the mixing of sterile and unsterile streams.

(ii) Particulate components in the media would block the heat exchangers.

(II) The Direct heat exchangers

In such type exchangers, the medium in the sterilizer is heated to the required temperature by steam directly via an injector.

(a) Continuous sterilization by steam injection : Flash cooling

A slightly modified system involves the injection of high temperature steam into the medium. Here, the required temperature (< 140°C.) is reached in microseconds. The cooling of the medium is carried out in the flash cooler in a vacuum chamber (Fig. 11.7).

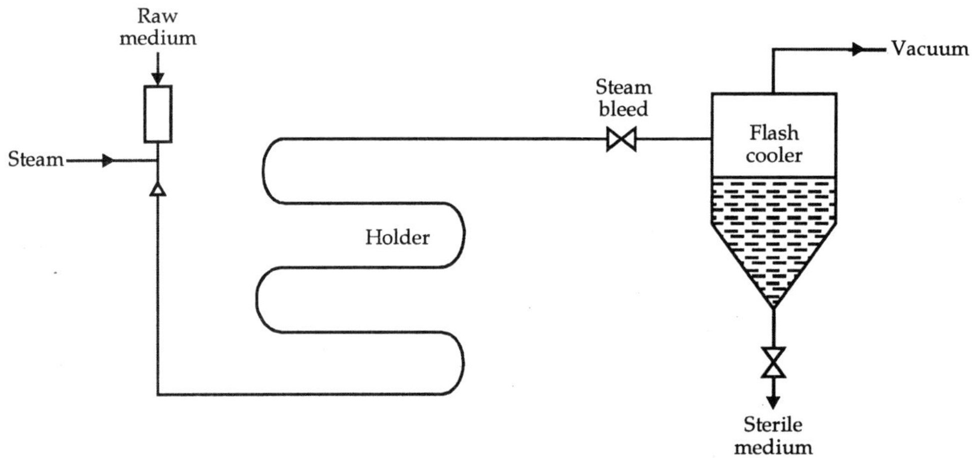

Fig. 11.7 : A working layout showing continuous sterilization by steam injection : Flash cooling.

(b) Modified continuous injector-flash cooler sterilizer

In some cases the injection system is combined with flash cooling, where the sterilized medium is cooled by passing it through an expansion valve into a vacuum chamber. Cooling then occurs virtually instantly. A flow chart of a continuous sterilization system using direct steam injection is shown in Fig. 11.8.

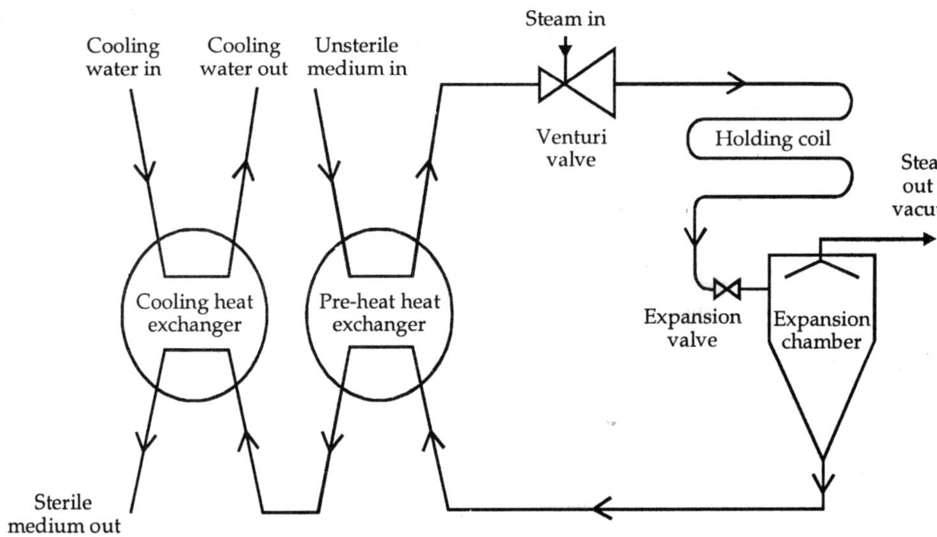

Fig. 11.8 : Flow diagram of a typical continuous injector-flash cooler sterilizer.

Advantages

(i) Very short (almost instantaneous) heating up times.

(ii) It may be used for media containing suspended solids.

(iii) Low capital cost.

(iv) Easy cleaning and maintenance.

(v) High steam utilization efficiency.

Disadvantages

(i) Foaming may occur during heating.

(ii) Direct contact of the medium with steam requires that allowance be made for condense dilution and requires 'clean' steam, free from anticorrosion additives.

In some cases a combination of direct and indirect heat exchangers may be used (Svensson, 1988). This is especially true for starch-containing broths when steam injection is used for the pre-heating step. By raising the temperature virtually instantaneously the critical gelatinization temperature of the starch is passed through very quickly and the increase in viscosity normally associated with heated starch colloids can be reduced.

Parameters involved in continuous sterilization by steam injection (direct) and heat-exchangers (indirect) are mentioned in Table : 11.4.

Table 11.4 : Parameters involved in continuous sterilization

Details	Steam injection	Heat-exchanger
Holding time	1-5 minutes	10 minutes
Holding temperature	140°C	120°C
Heating up/Cooling down period	Negligible	1% contribution

11.4 FILTER STERILIZATION OF MEDIA

Filter sterilization is often used for all components of nutrient solutions which are heat sensitive and would thus be denatured through the steam sterilization process normally used in industrial fermentation. Vitamins, antibiotics or blood components are examples of heat-labile compounds which must be sterilized by filtration. Such materials must be completely dissolved before filtration, otherwise they would be filtered out of the mixture with the microorganisms. If the solution is clear, the most suitable method is filtration.

Because of its simplicity, it is sometimes preferred over steam sterilization even in solutions containing no heat-labile components. The possibility to sterilize media by filtration is limited since most fermentation media contain colloidal or solid particles. Sterilization by filtration is thus used mostly on laboratory or pilot-plant scale. But, media for animal-cell culture are sterilized by filtration only.

11.4.1 Filters used for Media Sterilization

Media sterilization by filtration can be performed using

(i) Seitz filters, or

(ii) Candle filters, or

(iii) Sintered-glass filters, or

(iv) Membrane filters.

Each of these types of filters has its merits and shortcomings. Seitz filters, porcelin candle filters and sintered-glass filters are relatively cheap and easily sterilized but they clog rapidly and are difficult to clean.

The most suitable filters for sterilization of clear solutions are membrane filters-thin membranes with millions of pores per square centimetre, manufactured from various materials, most often cellulose acetate or cellulose esters.

The filters used for filtration of clear solutions usually have a pore diameter of 0.2 micrometer since none of the known bacteria is smaller than 0.2 micrometer and all bacteria can thus be trapped on the filter under all conditions. The choice of filter holder depends on the volume of the solution to be filtered, its viscosity and turbidity.

The filter is sterilized in an autoclave at 121°C for 10 to 45 minutes depending on the filter holder size; after cooling, the solution is passed through a filter under pressure.

The rate of filtration of clear solutions through a filter of 290 mm diameter, pore size of 0.45 micrometer and at 140 kPa pressure is 50 liters per minute; the total volume of solution that can be filtered is 100 hl.

11.4.2 Disadvantages

The disadvantages of filtration are: 1) certain components of the nutrient solution may be adsorbed on the filter material, and 2) high pressures must be used (up to 5 bar), which are undesirable in industrial practice.

One approach which is cost-effective is the filtration of just the water which is to be used in the preparation of the culture medium. For instance, in steroid bioconversion processes, a concentrated nutrient solution is sterilized by heat in the fermenter and is then diluted to the normal concentration with water which has been filter-sterilized.

11.5 MEDIA STERILISATION BY CHEMICAL AGENTS

The main criteria for evaluation of a suitable chemical sterilising agent are the following:

1. Appreciable biocide effect even at low concentrations.
2. No adverse effect on the medium.
3. Residual concentration of the agent must be readily removed or must have no effect on the growth of microorganisms and on product synthesis.
4. Effective at normal pressures and temperatures.
5. Easy application, low price, commercial availability and rapid action.
6. The agent must not be inflammable, explosive or toxic.

Sterilisation by chemical agents is influenced by a number of factors such as time, temperature, concentration of the agent in the medium, pH, surface tension. The appropriate chemical agent must then be chosen allowing for these factors.

11.5.1 Examples of Chemical Sterilants

Although a number of chemical sterilising agents are known, only oxirane (ethylene oxide) and 2-oxetanone (β-propiolactone) are suitable for the sterilisation of media. The properties of these substances are summarised in Table : 11.5.

Table 11.5 : Physical constants of 2-oxetanone and oxirane

Formula	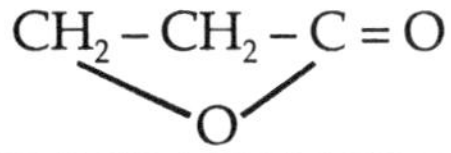	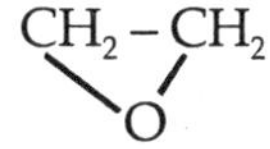
Molecular weight	72.06	44.05
Boiling point	155°C	10.7°C
Melting point	–33.4°C	–112°C
Refractive index	20_D 1.4131	4_D 1.3614
Density ($d^{20}4$)	1.149	0.891
Flammability point	74°C	–10°C and less
Solubility	soluble in most organic solvents; in water 37 % (v/v) at 25°C	completely soluble in most organic solvents and in water

The purity of commercially available 2-oxetanone is 96% or 99% (specially purified for biological use). The chemical reactions of 2-oxetanone are determined by the four-memberd ring which is readily cleaved. It reacts rapidly with amino, hydroxyl, carboxyl, sulphydryl and phenol groups contained in proteins. The agent has a high biocidal activity but is readily inactivated by rapid hydrolysis; the hydrolytic product, β-hydroxypropionic acid, has no biological effects. The elimination of pathogenic fungi requires a concentration of 0.25 to 0.50 %, bacteria require 0.1 to 0.2 % and spore suspensions 0.5 to 0.75 % of the agent and a treatment performed at 37°C for 2 h. The agent does not degrade proteins.

11.5.2 Device

The device for the continuous sterilisation of media by 2-oxetanone is shown in Fig. 11.9. It consists of a stock vessel with 2-oxetanone (A), pump (B), hydrolyser (C) made of a polyethylene tube of a total volume of 1 litre placed in a frame in a water bath of constant temperature, thermometer (D), relay for temperature regulation (E) and an immersed heater (F). The medium passes from the hydrolyser to a cooler (G) and to a neutralisation vessel, a 750 ml Erlenmeyer flask fitted with a magnetic stirrer (I) and a side inlet for glass electrode (M), overflow tube (H) and an opening for the addition of neutralising solution. The neutralising solution, 5N NaOH, is added from a storage flask (J) via an electromagnetic valve (L) connected with a pH-titrator (K) which regulates the addition of the solution. The neutralised and sterile solution is accumulated in a collecting vessel (N).

At normal temperature and pressure, oxirane is a colourless gas, below its boiling point, i.e. 11°C, and at atmospheric pressure it is a colourless, free flowing fluid. As shown in Table : 11.5, oxirane is highly inflammable and its vapour forms an explosive mixture with air in all proportions from 3 to 100%. The explosion may be prevented by mixing the agent with an inert gas like CO_2 or halogenated hydrocarbons. The inexplosive mixture is manufactured under the trade mark "Cryoxide". During sterilisation the vessel is filled with 0.5 to 1% oxirane and the medium is exposed to the gas for 1 h at 4°C. The kinetics of sterilisation by oxirane have been described by Ernst (1974).

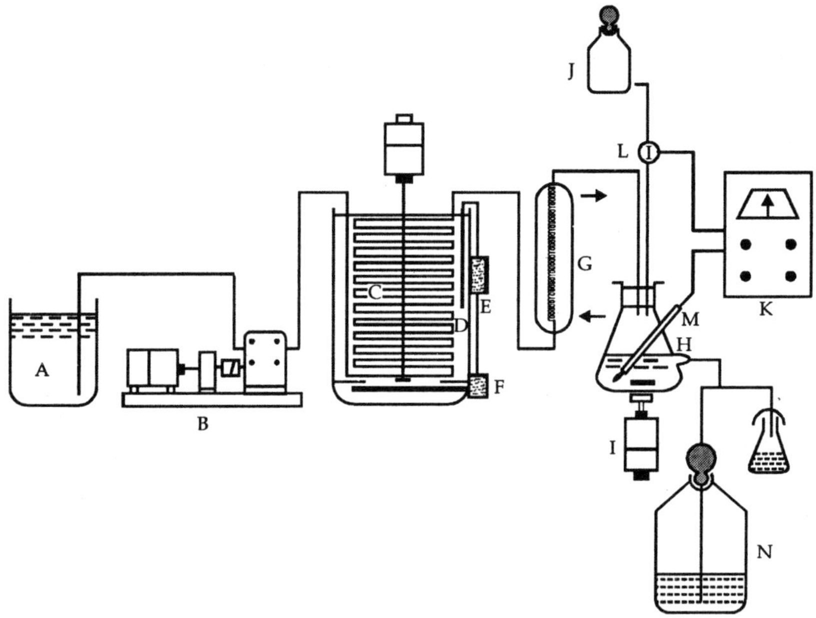

Fig. 11.9 : Scheme of continuous medium sterilisation device using 2-oxetanone (Toplin. 1962). Explanation of symbols in the text

11.6 MEDIA STERILIZATION BY RADIATION

Radiation sterilisation of liquids is mostly performed using the electromagnetic X- and γ-rays and the particulate cathode rays. An important characteristic of the sterilisation effect of cathode rays is their penetration which is directly proportional to the energy of the electron beam and inversely proportional to material density.

$$\text{Penetration} \propto \frac{\text{Eelectron density}}{\text{Material density}}$$

In view of the nature of the reaction mechanisms involved, the optimum sterilisation effect does not occur on the surface of the treated material; with γ-rays it is achieved close beneath the surface while with cathode rays it occurs several centimetres deeper.

Table 11.6 : Lethal doses of different kinds of radiation

Microorganism	Lethal, dose, kGy		
	Cathode rays (a)	γ-rays (b)	X-rays (c)
Vegetative, non-pathogenic	1 -2.5		
Vegetative, pathogenic	4.5-5.5	1.5-2.5	0.3-5
Bacterial spores	5-21	about 15	5-20
Fungi	2.5-11.5	2-3	2.5-10
Yeasts	5-10	3	2.5-15

(a) from Van de Graaf generators (b) from ^{60}Co (c) from a 3 MeV source

The radiation resistance of microorganisms varies according to species and strain of microorganism, as seen in Table : 11.6, as with other treatments, the highest resistance against radiation is exhibited by bacterial spores. The efficiency of the radiation treatment is adversely influenced by the presence of particulate matter in the medium and is appreciably affected by other factors such as the "oxygen" effect, presence of reducing substances and sensitising agents, freezing and drying. A lowering of the oxygen concentration in the medium below 10 mg/l reduces the lethal effect of radiation by 3 to 4 times. When irradiated in the frozen state, a suspension of *Escherichia coli* displays a lowered sensitivity to the treatment owing to retardation of diffusion of free radicals through the medium. On the other hand, *Bacillus subtilis* exhibits a higher sensitivity in the frozen state; this may be attributed to the rupture of the spore walls which makes the spores more sensitive to the treatment. *Bacillus thermoacidurans* and *Clostridium sporogenes* showed no sensitivity differences between the frozen and non-frozen state. Interesting results were obtained by adding acceptors of free radicals such as sodium ascorbate to different media. In a *Bacillus subtilis* suspension irradiated by cathode rays a 1% concentration of this salt reduced the sensitivity of spores to the treatment both in physiological saline and in a complex medium whereas for a suspension of *Bacillus thermoacidurans* this concentration of the salt had a protective effect in the complex medium but not in saline.

According to Kempe, radiation can be used in two ways: radio-Pasteurisation and radio-sterilisation. Radio-Pasteurisation is carried out at low doses, about 2 kGy of γ-rays, and is suitable when the liquid contains only Gram-negative and non-sporulating bacteria. This type of treatment is used with fluids and solutions in which strong radiation would cause changes in composition. Radiation sterilisation of media was tested in the production of lactic acid. γ-rays (25 and 45 kGy) were first used to sterilise separately corn-steep liquor and malt sprouts; these were then aseptically added to the basic medium previously sterilised by steam. A complete medium with corn-steep liquor and malt sprouts was sterilised by radiation in parallel. The radiation sterilisation of the complete medium was found to result in lower lactic acid yields and in a lower fermentation rate whilst the separate sterilisation gave higher yields and higher fermentation rate.

These examples show that considerable experimental work will have to be done before radiation sterilisation can be used economically in microbial industry. The basic method of media sterilisation is thus still by heat treatment using steam.

11.7 STERILIZATION OF THE FEEDS

A variety of additives may be administered to a fermentation during the process and it is essential that these materials are sterile. The sterilization method depends on the nature of the additive, and the volume and feed rate at which it is administered. If the additive is fed in large quantities then continuous sterilization may be desirable, for example, Aunstrup *et al.* (1979) cited the use of continuous heat sterilization for the feed medium of microbial enzyme fermentations. Aunstrup *et al.* also referred to the use of filtration for the sterilization of certain feeds. Stowell (1987) described the use of a continuous sterilizer for the addition of oil feeds to industrial scale fermenters, but stressed that each oil has its own characteristics which makes it impossible to predict the performance of a sterilizer for different oil feeds. Batch sterilization of feed liquids normally involves steam injection into the material held in storage vessels. Whatever the sterilization system employed it is essential that all ancillary equipment and feed pipework associated with the additions are sterilizable.

11.8 STERILIZATION OF LIQUID WASTES

Process organisms which have been engineered to produce 'foreign' products and therefore contain heterologous genes are subject to strict containment regulations. Thus, waste biomass of such organisms must be sterilized before disposal. Sterilization may be achieved by either batch or continuous means but the whole process must be carried out under contained conditions. Batch sterilization involves the sparging of steam into holding tanks, whereas continuous processes would employ the type of heat exchangers which have been discussed in the previous section. A holding vessel for the batch sterilization of waste is shown in Fig. 11.10. Whichever method is employed the effluent must be cooled to below 60°C before it is discharged to waste. The sterilization processes have to be validated and are designed using the Del factor approach considered in the previous sections. However, the kinetic characteristics used in the calculations would be those of the process organism rather than of *B. stearothermophilus*. Also, the N_1 value used in the design calculations would be smaller than 10^{-3} which is used for medium sterilization and would depend on the assessment of the hazard involved should the organism survive the decontamination process. Thus, the sterilization regime used for destruction of the process organism will be different from that used in sterilizing the medium.

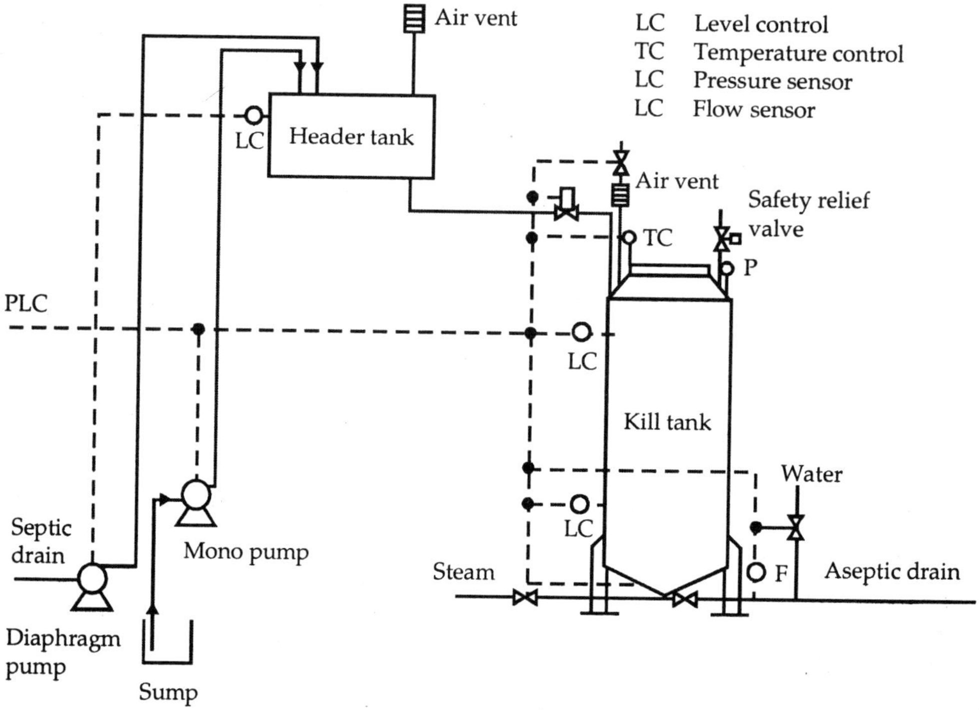

Fig. 11.10 : A vessel for the batch sterilization of liquid waste from a contained fermentation (Jansson *et al.*, 1990).

11.9 STERILIZATION OF THE FERMENTER (BIOREACTOR)

If the medium is sterilized in a separate batch cooker, or is sterilized continuously, then the fermenter has to be sterilized separately before the sterile medium is added to it. This is normally achieved by heating the jacket or coils of the fermenter with steam and sparging steam into the vessel through all entries, apart from the air outlet from which steam is allowed to exit slowly. Steam pressure is held at 15 psi in the vessel for approximately 20 minutes. It is essential that sterile air is sparged into the fermenter after the cycle is complete and a positive pressure is maintained; otherwise a vacuum may develop and unsterile air be drawn into the vessel.

Apart from sterilization and maintenance of sterile condition of the tanks (fermenter and seed tank), it is equally important to sterilize and to maintain sterility of ancillary equipment attached to the fermenters. Ancillary equipment includes many extra-connections, namely:

(i) Agitator shaft passing through the lid of the vessel—There should be an adequate seal or gland for maintaining pressure in the vessel. Moreover, this prevents the entry of microbes to the vessel from outside.

(ii) Sparger, usually beneath the agitator and a pipe for expelling exhausted air from the fermenter.

(iii) A connection between the defoamer tank and the fermenter.

(iv) A connection for introducing the inoculum.

(v) A connection for drawing samples during the fermentation.

(vi) A connection for unloading the fermenter after the completion of the fermentation.

Thus, these above mentioned extra-connections add to the complications of the sterilization of the fermenter (whether empty or with a charge of medium). It is a usual practice to supply each connection with its own pressure air-line. This is advantageous, since the ancillary equipment can be steamed out, while the fermenter is held under steam pressure at a temperature of 120°C. Such ancillary equipment is used in aerobic fermentations, requiring much more care for sterility of the system.

Certain points are to be kept in mind while sterilizing and maintaining sterile conditions in the fermentation equipment. They are:

(i) No direct connection should be made permanently between the non-sterile and the sterile parts of the system.

(ii) Welded construction should be used, if possible and convenient.

(iii) Where joints have to be used, they should be of a high quality valve, using rubber or other impervious material as the seal.

(iv) The type of valves used and service lines should be carefully selected, so that they can be easily sterilized and serviced when necessary. Moreover, they should be capable of maintaining pure culture conditions during the fermentation cycle.

(v) After sterilization, all parts of the system which are to be kept sterile should be kept under a positive pressure. This may be accomplished either with the sterile air or the sterile liquid used during the period of cooling and fermentation.

(vi) Each part of the system should be capable of independent sterilization without interfering with the operation of the rest of the plant.

(vii) Isolated danger spots which are difficult to sterilize at the start of a batch, should be provided with pressure steam connection for continuous or intermittent use.

❑❑❑

12

AIR STERILIZATION

12.1 INTRODUCTION

Most industrial aerobic fermentations are operated under vigorous aeration and the air supplied to the fermenter must be sterilized. The number of particles and microorganisms in air varies greatly depending on location of the fermentation plant, air movement, and previous treatment of the air. On the average, outdoor air has 10-100,000 particle per m^3 and 5-2,000 microorganisms/m^3, of these, 50% are fungus spores and 40% are bacteria.

Fermenters generally work with aeration rates of 0.5-1.0 vvm (air volume/liquid volume/minute). A fermenter having a working volume of 50 m^3 with an aeration rate of 1 vvm needs 3000 m^3 sterile air per hour. Therefore, sterilization of air has gained importance with the rapid development of industries which require large quantities of sterile air, especially those microbiological processes in which oxygen limitation may be a decisive factor in product formation.

The methods available for air sterilization include:

(i) Filtration,

(ii) Heat,

(iii) Ultra-violet and other radiations,

(iv) Electrostatic sedimentation or precipitation,

(v) Chemical agents or sterilants,

(vi) Ozone gas injection and scrubbing

Of these above methods, only filtration and heat are practical at an industrial level. Other methods are used only sparsely and will therefore be mentioned only briefly.

12.2 TYPES OF AIR STERILISATION

Air can be sterilised by heat. The reduction in the initial number of *Bacillus subtilis* spores by a factor of 10^7 was found to require an exposure at 220°C for 24 s. On increasing the temperature by 30°C the necessary exposure time is reduced by half. The air may be heated in electrical furnaces or by adiabatic compression carried out in a thermally insulated system. The utilisation of this procedure for sterilising large volumes of air has as yet been hindered by high costs associated with the rapid heating and cooling of large volumes (when sterilisation is performed at high temperatures) or by difficulties inherent in maintaining large volumes of air under constant conditions for long periods of time (when the process is accomplished at lower temperatures). Air can also be sterilised by exposure to UV-light or other types of radiation. At present this method of sterilisation is uneconomical for the continuous production of large volumes of sterile air; it is used rather as a complementary sterilisation of boxes, etc. Washing of air with chemicals such as sulphuric acid is also unsuitable since small particles in the air are not wetted during the passage through the washing fluid and, in addition, particles of the fluid or its fumes can be carried along in the air stream and may affect negatively the fermentation process.

Another possible method of air sterilisation, i.e. an electrostatic sedimentation of dust and microorganisms, is not sufficiently effective. It can be used for prepurification of air before sterilisation by a fibrous filter but on a large scale this process is also inconvenient. Filtration of air through fibrous or porous filters thus remains the most convenient method of air sterilisation for industrial purposes.

12.3 STERILISATION OF AIR BY FIBROUS MATERIALS

Two types of filters are currently used for the purpose:

A. The filter material is compressed in a cylinder, creating a homogeneous layer of considerable thickness:

B. Fine fibres are layered onto a firm substratum, mostly textile, so that the filtering unit is formed by a relatively thin layer.

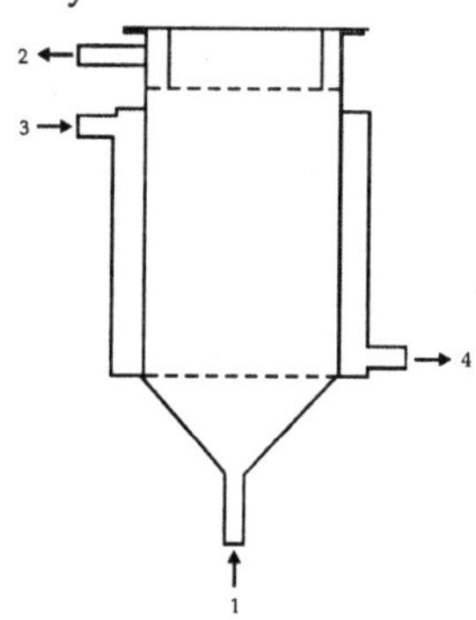

Fig. 12.1 : Scheme of filter for air sterilisation 1—pressurized air inlet, 2—sterile air outlet, 3—steam inlet into filter jacket (filter bed is steam-dried after sterilisation), 4—condensate outlet. Filter bed is enclosed in the space between the two perforated bottoms.

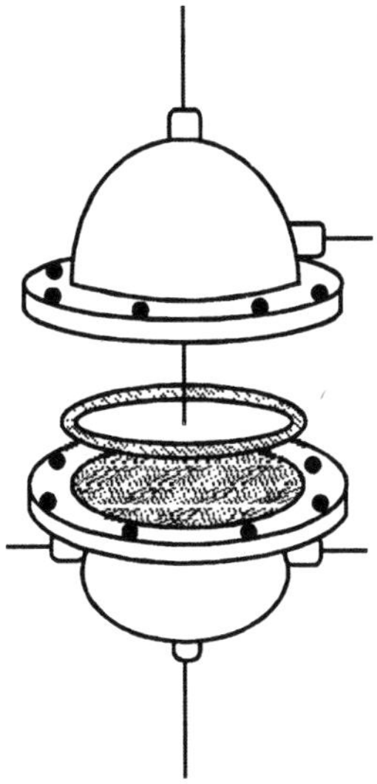

Fig. 12.2 : Air sterilisation filter (Eikoh Kasei Co., Ltd.)

A filter of type A (Fig.12.1) is a jacketed cylinder filled with the filtering material. i.e. cellulose, viscose or, most often, glass wool. The cylinder itself is metal. Porous filter materials such as powdered coal, slag or coke are used mostly as prefilters to remove dust and oil droplets arising from the compressor. Glass wool filters are sterilised by steam and dried thoroughly by a stream of hot steam passing through the jacket. The quality of the filter bed is very important since damage to the filter results in a sharp deterioration of performance. The filter material must be distributed homogeneously throughout the filter volume, preventing the formation of channels and the passage of air along the walls. If the filter is wet its performance deteriorates markedly, thus, when the air to be filtered is too humid, it has to be heated above the dew point to preclude condensation.

Filters of type B (Fig. 12.2) are most suitable and are being more and more widely used. The filter beds consist in this case mostly of plastic fibres with a diameter (less than 1 μm) much less than that of glass or cotton wool. They have a high filtering efficiency and the replacement of the filter layer is easy. The filter layer is reinforced by a firm substratum, and thus forms filter cloth. Several layers of the cloth are usually placed into filters constructed as filter presses or Seitz filters although a single layer is often sufficient. Since the filter layer offers a high resistance to the air stream the filters have a large area to prevent a pressure drop on the filter. The filters cannot usually be steam sterilised and formaldehyde fumes are therefore used for the purpose. The recently manufactured polyvinylalcohol fibre filters, however, can be steam-sterilised.

Successful air sterilisation depends largely on the choice of a suitable filter material and design, with special reference to filtering efficiency and air pressure drop across the filter although, initially, filter design was purely empirical. In laboratory devices the choice of a filter with sufficient efficiency is relatively easy since the costs of small filters are negligible and the required efficiency can be readily achieved by using filters with much larger dimensions. However, in industrial devices the filters must, for reasons of economy, not exceed certain dimensions. Filters of excessive dimensions are costly, take up considerable space and the replacement of filter bed is longer and more laborious. Both filtration efficiency and economy thus dictate the design of art optimum filter. The efficiency of large filters cannot be directly assessed since it is not feasible to determine the presence of microorganisms in the whole air volume passing through the filter at the usually high flow rates (e. g., 10000 l min^{-1}). Determination of microorganisms present in a mere fraction of the passing volume is meaningless since the values obtained cannot be extrapolated to the whole volume with anything like the required accuracy. Satisfactory results were achieved by the Bourdillon method, i. e. by forcing the air at a high flow rate through a narrow slit against the surface of a solid culture medium situated at a short distance from the slit.

The per cent efficiency of a filter, E, is defined as follows:

$$E = 100\frac{N_1 - N_2}{N_1}, \tag{12.1}$$

where N_1 — number of incoming microorganisms,

N_2 — number of microorganisms leaving the filter, which is equivalent to the per cent decrease in microbial count due to passage through the filter.

The Bourdillon method was originally developed for determining the filter bed thickness necessary for 100% filtration efficiency. The difference between 99% and 100% efficiency, which is highly important in terms of the probability of contamination of the process, is difficult to assess by this method since the low number of germs detected may arrive both from the filtered air and from other sources of contamination of the culture medium.

The theory of air filters received a strong impetus from their use in the nuclear industry, i.e. in the removal of small radioactive particles from air. The resulting data were applied to the filtration of bacterial aerosols. The development of air filtration theory in microbial industry was largely due to Gaden and Humphrey (1956), Maxon and Gaden (1956), Humphrey (1960), Aiba *et al.*, (1961, 1963). Model filters were designed by the authors to reduce complicating factors. Glass wool was pressed into thin plates and the fibres were arranged so as to avoid their mutual displacement during operation. The plates were equipped with flanges to prevent air leakage along the walls; the air was dried to a constant level of 15% relative humidity. The results obtained with model filters allow for the formulation of general mathematical relationships. The ratio of incoming to leaving microorganisms was found to be proportional to filter layer thickness L according to the formula:

$$\text{In}\ \frac{N_1}{N_2} = kL. \tag{12.2}$$

This readily understandable formula implies that each element of the filter layer retains only a limited number of microorganisms. The formula holds only with the assumption that microorganisms do not migrate inside the filter matrix: this condition is satisfied only when the filter is dry.

12.3.1 Determination of Optimum Filter Thickness

Optimum filter thickness, i.e. the required number of layers, can be determined in three ways:

1. From results achieved with a small model filter made of the same material and with the same filter bed as the large filter operating under the same conditions in an industrial device;

2. From the data given by the manufacturer of the filter material, especially in filters of a new type;

3. By calculation from appropriate theoretical relationships into which actual values (fibre diameter, particle size, etc.) have been inserted.

A useful characteristic of filter efficiency is the decimal reduction filter thickness L_{90}, i.e. filter thickness reducing the microbial count in the air by 90%. Thus:

$$\log \frac{N_1}{N_2} = \frac{k}{2.3} L. \tag{12.3}$$

According to the definition

$$1 = \frac{k}{2.3} L_{90}, \tag{12.4}$$

$$\therefore \log \frac{N_1}{N_2} = \frac{L}{L_{90}}, \tag{12.5}$$

$$L = L_{90} \log \frac{N_1}{N_2}. \tag{12.6}$$

This equation (12.6) defines the dependence of filter efficiency on filter layer thickness (in filters of type A) or on number of filter layers (filters of type B) under constant conditions, i.e. with the same filter bed material, particle size, air pressure and flow rate, etc.

The determination of filter efficiency is complicated by the fact that the decimal reduction thickness is not a constant for a given filter and given conditions (particle size, etc.) but depends on the air flow rate (Purchas, 1966). It should be noted that the air flow rate is not measured in units of $m^3\ min^{-1}$ but in $m\ min^{-1}$. The dependence of L_{90} on air flow rate reflects the dependence of the mechanism of particle trapping in the filter matrix on the flow rate. At low flow rates the particles are retained mainly by gravitational and electrostatic forces whereas at higher flow rates these forces cease to act and the L_{90} rises. Above a certain critical flow rate the particles are retained by inertial forces proportional to the flow rate. Above this critical flow rate the L_{90} decreases appreciably.

The L_{90}/flow rate relationship has important practical consequences since no device works at a completely constant air flow rate. The flow rate varies within wide limits since different processes require a different aeration intensity; variations may also be due to compressor failure, etc. A filter design which is reliable under all kinds of conditions has to be based on the maximum L_{90} value, i.e. the value determined at the critical flow rate at which the filter efficiency is lowest. This value is then used in all further calculations. When designing large filters from results obtained with small ones, it is therefore necessary to use different air flow rates in order to determine the maximum L_{90}. When the L_{90} is calculated according to equations (12.3) to (12.6), critical air flow rate values have to be inserted.

12.3.2 Factors affecting Particle Retention

The filter layer can be envisaged simply as a row of uniformly distributed fibres exposed to a laminar air stream. The following mechanisms of particle retention have to be considered:

a) Direct retention—this comes into play for particles which, during their passage through the filter, come in contact with the fibres. The phenomenon is almost independent of particle size and air flow rate.

b) Diffusion mechanisms—particles oscillate due to Brownian movement, impinging on fibre surfaces. The efficiency of diffusion mechanisms decreases with increasing particle size and air flow rate.

c) Electrostatic forces—this mechanism depends on the ratio of fibre charge to particle charge and is strongly affected by humidity; it decreases usually with flow rate.

d) Retention due to inertial forces—the flow of air plus particles around the fibres is not laminar; the particles tend to move in straight trajectories, colliding with fibres. The effect of these forces rises markedly with increasing flow rate and slightly less with increasing particle mass.

Under given conditions, the proportion of individual mechanisms varies: they do not necessarily all act simultaneously.

12.3.3 Filter Design for Air Sterilisation

With a basic theory of air filtration we can design a suitable filter for given conditions. The procedure is usually based on equation (12.3). The required efficiency is determined by the acceptable probability p that a single germ will pass through the filter throughout the whole operation period, i.e. within the time interval between individual filter sterilisations. The probability is usually put equal to 0.001, i.e. during a 1000-fold use of the filter the fermenter will probably be contaminated only once via filtered air. The mean number of contaminating microorganisms per unit volume of air can be used to calculate the number entering the filter during the operation period; this yields the required filter efficiency N_1/N_2. Let us denote the air flow F, operation period t and the number of germs per unit air volume a; then

$$N_1 = Fta,\ N_2 = p,\ \frac{N_1}{N_2} = \frac{Fta}{p}. \qquad (12.7)$$

L_{90} is determined experimentally in a device represented schematically in Fig. 12.3.

The equation $L = L_{90} \log (N_1/N_2) = L_{90} \log (Fat/p)$ yields the necessary filter thickness for required efficiency.

Another important characteristic to be defined is the filter cross-sectional area. Filters with large cross-sectional area require higher investment costs and take up more space. On the other hand, they offer lower resistance to the air stream, causing a lower air pressure drop behind the filter and thus lowering operation costs. The danger of filter rupture or settling is also considerably reduced.

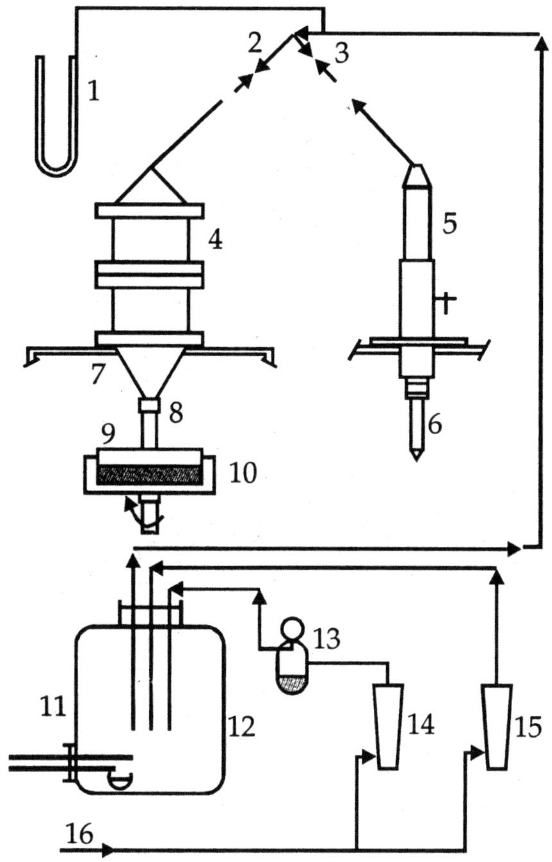

Fig. 12.3 : Device for L_{90} determination (Aiba *et al.*, 1961) 1—manometer, 2—valve B, 3—valve A, 4—test filter, 5—control sampling bypass, 6—nozzle, 7—sterile glass chamber, 8—nozzle, 9—Petri dish with agar, 10—rotating dish, 11—psychrometer, 12—air chamber, 13—atomizer, 14, 15—rotameter, 16—air compressor

12.4 CLASSIFICATION OF AIR FILTERS AND THEIR INDUSTRIAL-LEVEL APPLICATIONS

12.4.1 Depth Filters (Non-Fixed Pore Filters)

Depth filters are those filters in which the pores in the filter are larger than the particles which are to be removed. Such type of filters are composed of glass wool fibres, felts, woven yarns, asbestos pads and loosely packed fibre glass.

In older industrial systems, pure depth filters such as glass wool filters were installed, in which particles would be trapped by a combination of physical effects. As shown in figure 12.4, particle filtration involve inertial effects, blocking effects, diffusion, gravity separation, and electrostatic attraction. The last two mechanisms have a minimal effect on the removal of particles.

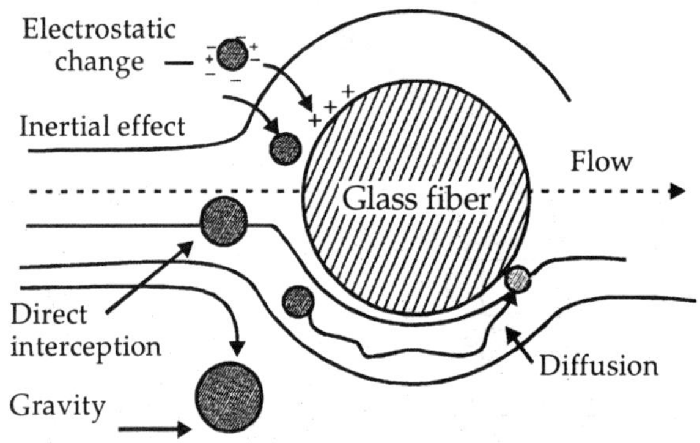

Fig. 12.4 : Mechanism by which particles are removed by a depth filter.

The design of air-filter using depth filter (slag wool material) was constructed by Terjesen and Cherry (as shown in figure 12.5). They used a preformed slab of slag wool 3 inches thick, packed to a density of at least 1716/ft.^{3}, and of mean fibre diameter less than 6 micrometer. The air velocity through the slab was kept below 1ft./second to avoid channelling through the slag wool material.

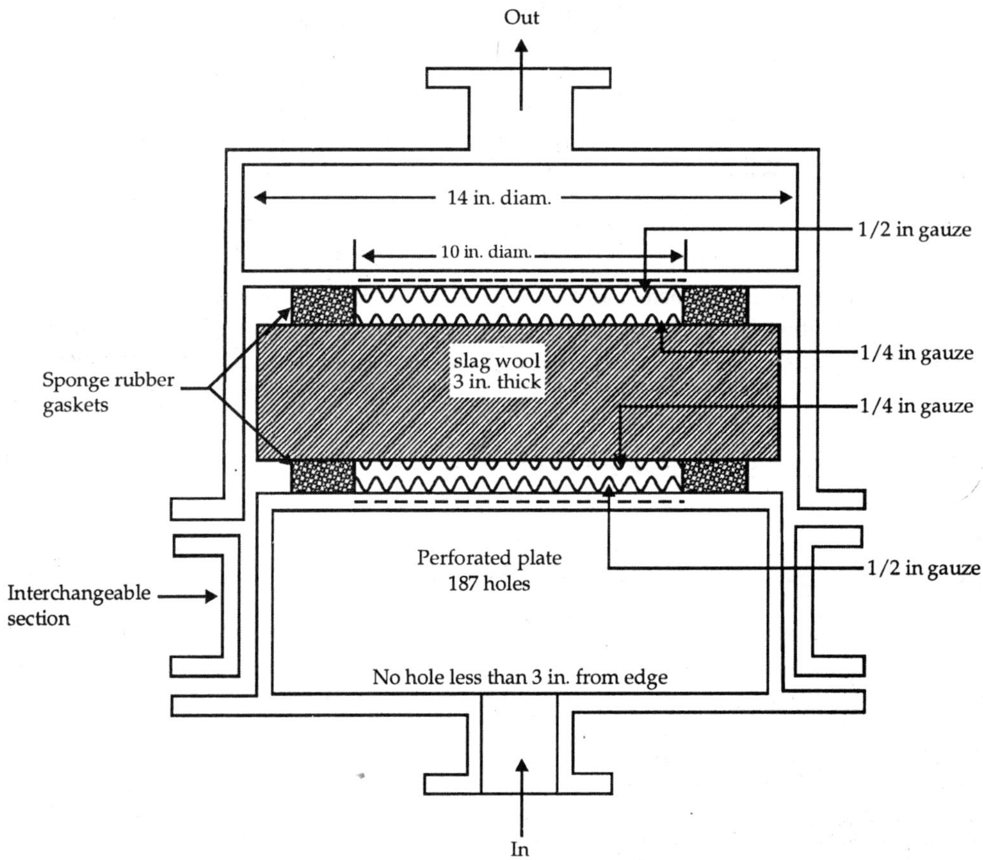

Fig. 12.5 : Design of Depth filter using slag-wool.

Theoratically, the removal of microorganisms by a depth filter (made up of fibrous materials) cannot be absolute as there is always the possibility of an organism passing through the filter, regardless of the filter's depth. Also, because the fibres are not cemented into position an increase in the pressure to which the filter is subjected may result in movement of material, producing larger channels through the filter. Increased pressure may also result in the displacement of previously trapped particles.

In reference to above discussed problems of depth filters, Terjesen and Cherry performed many experiments to achieve maximum air-sterilization efficiency from depth filters. They made the following suggestion to get best performance from depth filters:

(a) The slab to be installed in the filtration system must follow certain specifications (e.g. fibre diameter, uniformity, density of the slab, etc.).

(b) Attention should be paid to the air which may escape by passing around the edges of the filtering medium.

(c) No holes should be drilled less than 3 inches from the edges of the perforated supporting plates.

(d) A non-continuous sponge-rubber gasket should be inserted between the slab and the 3 inches blank annulus of the supporting plates.

(e) The rubber gasket around the edges of the supporting frame can be eliminated, provided the metal faces around the edges of the supporting frame are very smooth.

(f) Air velocities higher than 0.5 ft./sec. should not be used, since there is an increased pressure drop across the filtering medium and an increased danger of channelling.

(g) Sterilization of the slabs and the body of the filter should be practised by means of dry heating at 160 to 180°C for 2 hours.

(h) Prefilters should be used to remove dust particles and related suspended matter, thus increasing the working life of the sterilizing filter.

Disadvantage

The disadvantages of glass wool like fibrous filters are shrinkage and solidification during steam sterilization.

It is important to realize that filters should be steam sterilized before and after operation. Thus, the materials must be stable at high temperatures and the steam must be free of particulate matter because the filter modules are particularly vulnerable to damage at high temperatures. Thus, the steam itself is filtered through stainless steel mesh filters rated at 1 μm.

12.4.2 Absolute Filters (Fixed Pore Filters)

Absolute filters are those filters in which the pores in the filters are smaller than the particles which are to be removed. Absolute filters are claimed to be 100% efficient in removing microorganisms. The 'fixed pore filters' term is also used for an absolute filters.

Fixed pore filters are constructed so that the filtration medium will not be distorted during operation so that the flow patterns through the filter will not change due to disruption of the material. Pore size is controlled during manufacture so that an absolute rating can be quoted for the filter—i.e. the removal of particles above a certain size can be guaranteed. Thus, interception is the major mechanism by which particles are removed. Because fixed pore filters have depth they are also capable of removing particles which are smaller than the pores by the mechanisms of inertial impaction, diffusion and attraction and these mechanisms do play significant roles in the filtration of gases. Fixed pore filters are superior for most purposes in that they have absolute ratings, are less susceptible to changes in pressure and are less likely to release trapped particles.

Disadvantage

The major disadvantage associated with absolute filters was the resistance to flow they present and, hence, the large pressure drop across the filters which represents a major operating cost. However, modern absolute filter cartridges which have been developed by many filtration companies contain pleated membranes with very large surface areas, thereby minimizing the pressure drop across the filter.

12.4.3 Modern Absolute Filter Cartridge

New cartridge filter systems using pleated membranes are now widely available. The advantage of these filters is that they are substantially smaller than glass wool or activated carbon towers. Operating the systems has become much simpler; because of the removable cartridge construction, replacing used filter elements is easy. Constructed of cellulose ester, polysulfone, or nylon, these membrane filters have the same structure as depth filters, but because they have a membranous structure they have an absolute filter effect.

Such modern air filter systems consist of pleated membrane cartridges designed to be accommodated in stainless steel modules. The most common construction material used for the pleated membranes for air sterilization is PTFE, which is hydrophobic and is therefore resistant to wetting. Also, PTFE filters may be steam sterilized and are resistant to ammonia which may be injected into the air stream, prior to the filter, for pH control.

As was seen for the filter sterilization of liquids, it is essential that a prefilter is incorporated up-stream of absolute filter. The prefilter traps large particles such as dust, oil and carbon (from the compressor) and pipescale and rust (from the pipework).

The use of a coalescing prefilter also ensures the removal of water from the air; entrained water is coalesced in the filter (air flow being from the inside of the filter to the outside) and is discharged via an automatic drain. Figure 12.6, illustrates the layout of such a filtration unit showing the steam sterilization system having prefilter and barrier filter (dual hydrophobic filter system).

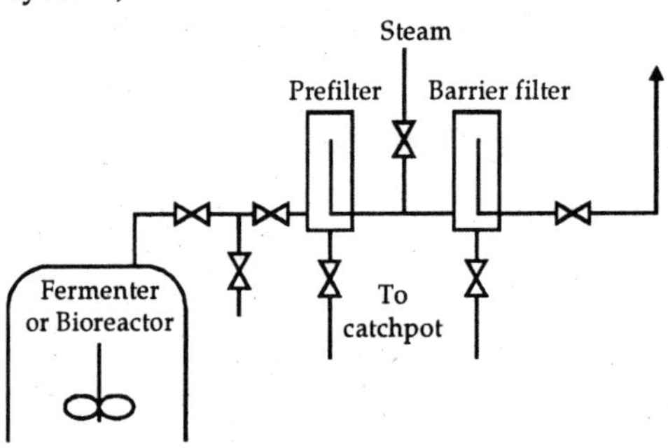

Fig. 12.6 : Dual hydrophobic filter system for sterilization of off-gas from a fermenter.

Figure 12.7 illustrates a fermenter installation with incoming and outgoing air filters made up of filter cartridges like PTFE.

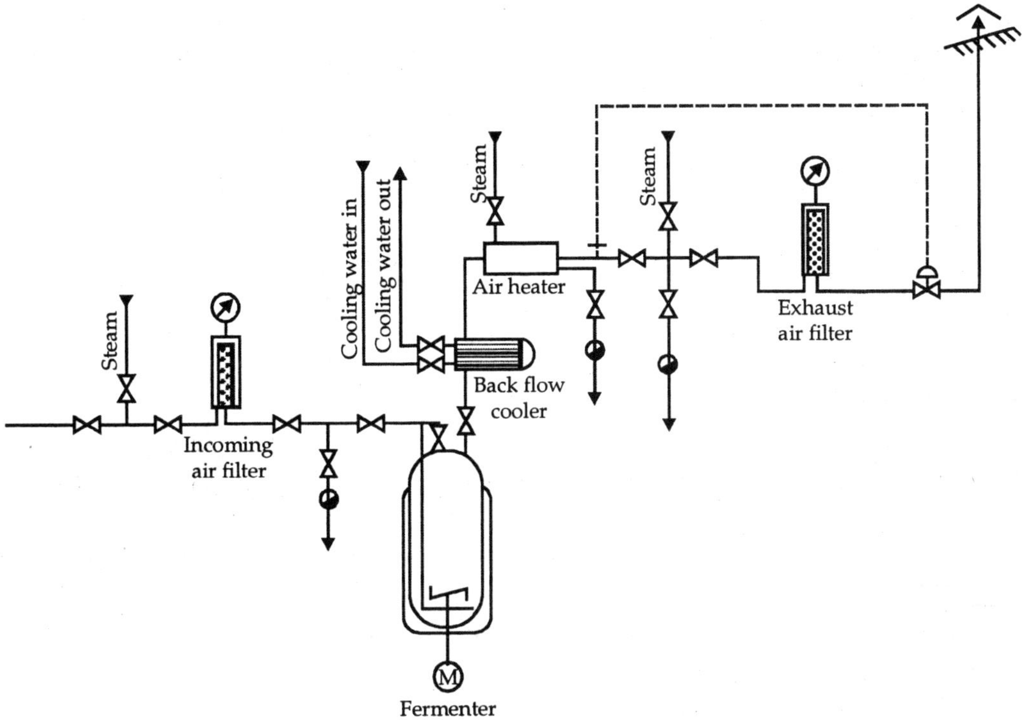

Fig. 12.7 : Installation of an air filter system (incoming and outgoing) in a fermenter. Air filters are made up of modern absolute PTFE cartridge filtering material.

Limitation

The limitation of most systems installed today is that there is not yet any **absolute filter for bacteriophages** in industrial use. Bacteriophages can cause total failure in a system: for example, in the production of glutamic acid with *Corynebacterium glutamicum* or in the production of penicillin acylase with *Escherichia coli.*

12.5 STERILIZATION OF FERMENTER EXHAUST AIR

In many traditional fermentations the exhaust gas from the fermenter was vented without sterilization or vented through relatively inefficient depth filters. With the advent of the use of recombinant organisms and a greater awareness of safety and emission levels of allergic compounds the containment of exhaust air is more common (and in the case of recombinant organisms compulsory). Fixed pore membrane modules are also used for this application but the system must be able to cope with the sterilization of water saturated air, at a relatively high temperature and carrying a large contamination level. Also, foam may overflow from the fermenter into the air exhaust line. Thus, some form of pre-treatment of the exhaust gas is necessary before it enters the absolute filter. This pretreatment may be a hydrophobic prefilter or a mechanical separator to remove water, aerosol particles and foam.

The pretreated air is then fed to a 0.2-µm hydrophobic filter. Again, it is important to appreciate that the filtration system must be steam sterilizable.

Fig. 12.8 illustrate the use of prefilter system, while figure shows the use of mechanical separator and hydrophobic filter (PTFE) system for sterilization of exhaust air or off gas from a fermenter.

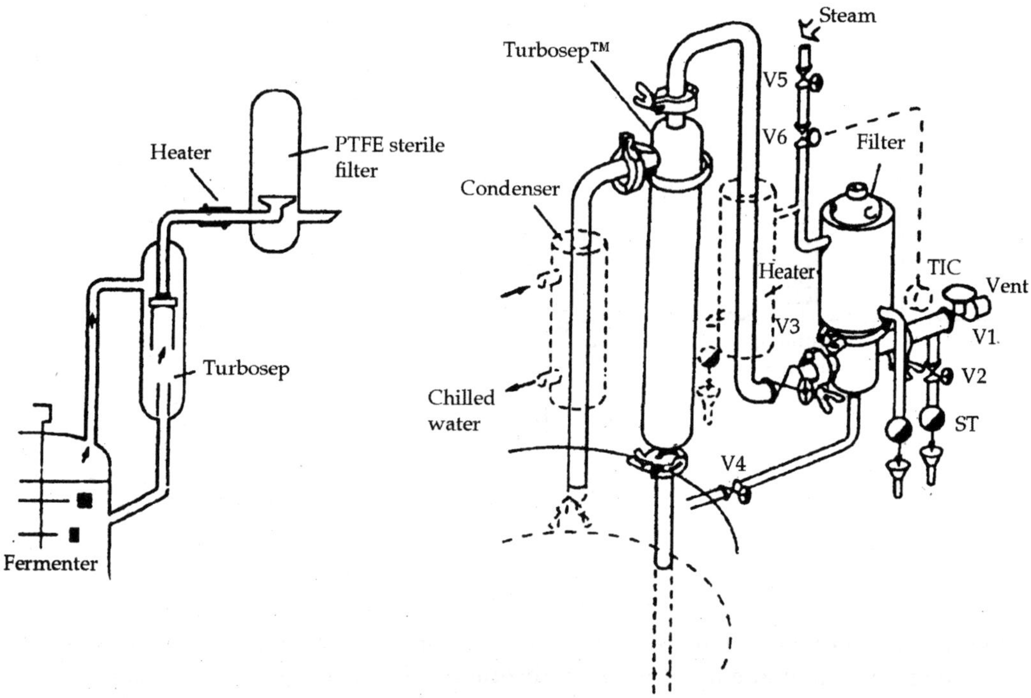

Fig. 12.8 : A mechanical separator and hydrophobic filter system for the sterilization of off-gas from a fermenter. Left. Cut-away diagram. Right. Equipment arrangement, showing steam supply. V1-V6, valves; O/ST–steam traps (Domnick Hunter Ltd., Birtley, Co., U.K.)

❑❑❑

13

Inoculum Development

13.1 INTRODUCTION

The handling of a microbial culture, from the time it is transferred from its preserved state until it is inoculated into the final location where the microbial activity of interest is expressed, is referred to as **inoculum development.**

The **primary purpose of inoculum development** is to provide microbial mass, of predictable phenotype, at a specific time, and at a reasonable cost for the productive stage of microbial activity. Until now, inoculum development has been more art than science. There remains a need, especially at the shake flask or spore-generating stages of the process, for time and "it looks good" criteria to be replaced with biochemical, physiological, or morphological markers as both descriptors of an optimum inoculum and indicators for optimum timing of inoculum transfer.

Inoculum preparation is a critical stage in an industrial fermentation process. The microbial cells or spores required to inoculate large quantities of fermentation medium in a fermenter (minimum capacity about 50,000 litres) are in high numbers, since a loopful of culture cannot meet the requirements of completing the fermentation run in the economically shortest time. Therefore, fermenting microbes in the stock culture are serially transferred to a special inoculum medium with increasing the volume in each step in the sequence.

13.2 CRITERIA FOR AN IDEAL INOCULUM

It is essential that the culture used to inoculate a fermentation process satisfies the following criteria:

(1) Inoculum must be in a healthy, active state thus minimizing the length of the lag phase in the subsequent fermentation.

(2) It must be available in sufficiently large volumes to provide an inoculum of optimum size.

(3) It must be in a suitable morphological form.

(4) It must be free of contamination.

(5) It must retain its product-forming capabilities.

So, the process adopted to produce an inoculum satisfying the above criteria is called inoculum development.

Much of the variation observed in small-scale laboratory fermenters is due to poor inoculum being used and, thus, it is essential to appreciate that the establishment of an effective inoculum development programme is equally important regardless of the scale of the fermentation.

13.3 PRINCIPLES AND VARIOUS ASPECTS OF INOCULUM DEVELOPMENT

13.3.1 Inoculum Size

The quantity of inoculum normally used is between 3 and 10% of the medium volume. A relatively large inoculum volume is used to minimize the length of the lag phase and to generate the maximum biomass in the production fermenter in as short a time as possible, thus increasing vessel productivity. Thus, starting from a stock culture, the inoculum must be built up in a number of stages to produce sufficient biomass to inoculate the production-stage fermenter This may involve two or three stages in shake flasks and one to three stages in fermenters, depending on the size of the ultimate vessel. Throughout this procedure there is a risk of contamination and strain degeneration necessitating stringent quality-control procedures. The greater the number of stages between the master culture and the production fermenter the greater the risk of contamination and strain degeneration. Therefore, a compromise may be reached regarding the size of the inoculum to be used and the risk of contamination and strain degeneration. Another factor to be considered in the determination of the inoculum volume is the economics of the process. A seed fermenter 10% of the size of the production fermenter represents a considerable financial investment and must be justified in terms of productivity.

13.3.2 Inoculum Media

Inoculum media usually differ in composition from production media. Inoculum media are compounded to quickly yield large numbers of microbial cells in their proper physiological and morphological states, but without sacrificing genetic stability of the cells. If the utilization of some component of the production medium requires that the cells be enzymatically adapted to this substrate, then this particular component should also be included in the inoculum medium. This precaution prevents deadaptation, if the cells already are adapted, or eliminates the lag period during the initial stage of production which may occur while adaptation is being accomplished.

We are not interested in the accumulation of a fermentation product other than cells in the inoculum medium, and, hence, the inoculum medium is not balanced for product formation. Thus, inoculum media are usually less nutritious overall than the production media, and they usually contain a considerably lower level of the main nutritive carbon source. For some fermentations, the cells are separated from the inoculum medium before being added to the production medium, and this fact should be considered in the design of the inoculum medium for these fermentations. For example, a carbon source in the inoculum medium which becomes exhausted before harvest of the inoculum can cause certain becateria to flocculate and settle from the medium; thus, the cells can be drained from the bottom of the tank or the supernatant fluid can be siphoned off. However, when inoculum cells are separated from the medium by sedimentation or centrifugation, any particulate components of the medium also separate with the cells. Thus, if inoculum cells devoid of adhering inoculum medium components are desired, then completely soluble components are of value.

Lincoln (1960) stated that the lag time in a fermentation is minimized by growing the culture in the 'final-type' medium. Lincoln's argument is an important one, so the inoculum development medium should be sufficiently similar to the production medium to minimize any period of adaptation of the culture to the production medium, thus reducing the lag phase and the fermentation time. Furthermore, Hockenhull (1980) pointed out the dangers of using very different media in consecutive stages. Major differences in pH, osmotic pressure and anion composition may result in very sudden changes in uptake rates which, in turn, may affect viability. Hockenhull also emphasized that for antibiotic fermentations the inoculum medium should contain sufficient carbon and nitrogen to support maximum growth until transfer, so that secondary metabolism remains repressed during growth of the inoculum. If secondary metabolism is derepressed in the seed fermentation, then selection may enrich the culture with non-producing variants having a growth advantage over high-producing types.

Examples of inoculum media and production media for some commercial fermentation products are shown in Table : 13.1, from which it may be seen that inoculum media are, generally less nutritious than production media and contain a lower level of carbon.

Table 13.1 : Inoculum development and production media for some products

Product	Inoculum development medium		Production medium	
Griseofulvin	Whey powder	to	Lactose	7%
	Lactose	give:	Corn-steep	
	Lactose	3.5%	liquor solids	
	Nitrogen	0.05%	to give:	
	Corn-steep		Nitrogen	0.2%
	liquor solids	0.38%	Limestone	0.8%
	(to give approx.		KH_2PO_4	0.4%
	0.04% N)		KCl	0.1%
	KH_2PO_4	0.4%		
	KCl	0.05%		

contd. ...

... contd.

Product	Inoculum development medium		Production medium	
Clavulanic acid	Soybean flour	1.0%	Soybean flour	1.5%
	Dextrin	2.0%	Oil	1.0%
	Pluronic L81 (Antifoam)	0.03%	KH_2PO_4	0.1%
		(g dm^{-3})		(g dm^{-3})
Vitamin B_{12}				
	Molasses	70		105
	Sucrose	—		15
	Betaine	—		3
	$NH_4H_2PO_4$	0.08		—
	$(NH_4)_2SO_4$	2		2.5
	$MgSO_4$	0.2		0.2
	$ZnSO_4$	0.02		0.08
	5-6 Dimethyl benzimidazole	0.005		0.025
Lysine	Cane molasses	5%		20%
	Corn-steep liquor	1%		—
	$CaCO_3$	1%		—
	Soybean meal hydrolysate	—		1.8%

13.3.3 Sterilization of Inoculum Medium

One of the more important variables in the scale-up of a fermentation process is the effect of sterilization on the productivity of the culture and on the production of foam.

The impact that sterilization procedures may have on an inoculum and ultimately on the productivity of a culture ideally should be determined at the shake flask level prior to pilot-scale fermentations. One problem, however, is how to correlate the data generated in a shake flask with what might occur on the fermenter scale beyond the generalization that the fermentation is or is not susceptible to damaging sterilization effects. Perhaps one can use the observation that the process of sterilization resulting in products of the Maillard reaction also results in a time-and temperature-related increase in foam formation by the resulting medium.

To prevent the loss of important medium ingredients and to prevent the formation of caramelization products during sterilization, separate sterilization of sensitive or particularly reactive components or continuous sterilization methods can be followed.

13.3.4 Seed Transfer Timing

The timing of transfer from seed fermenters into production can, if the fermenters have the appropriate instrumentation, take advantage of online measurements such as CO_2, fluorescence, dissolved oxygen, and the derived variables respiratory quotient (RQ) and specific growth rate to characterize the status of the inoculum. After optimization for productivity, these parameters can be used to direct the timing of the last inoculum transfer. For the earlier stages of a seed train and for fermentations where seed growth is performed only in shake flasks, other measures of the status of an inoculum are needed to standardize the optimum transfer time. Respiration rate, motility, shear, specific growth rate, product formation, dehydrogenase, and nucleic acid level have all been used for this purpose (Table : 13.2).

Table 13.2 : Time-independent measures for optimum inoculum transfer

Measure	Transfer when measurement is:	Organism	Product and effect
Motility	Maximum	*Clostridium acetobutylicum*	2X > BuOH
RNA	Falling	*Claviceps paspali*	Maximum alkaloid
Declorization of methylene blue	Time $\leq$ 2.2 min	*Micromonspora* spp.	Peak gentamicin
Lactic acid level	Before inhibitory concentration	*Lactococcus lactis*	Eliminate lag phase
Respiration	Maximum	*Acremoniuim chrysogenum*	Maximum cephalosporin
Specific growth rate	Maximum	*A. chrysogenum*	Maximum cephalosporin
Shear	80-120 mPa	*Streptomyces noursei*	Optimum nystatin

One of the most common studies of inoculum optimization is the determination of the effect of the age of transfer of the culture on productivity (Table : 13.3). The optimum titer of proteinase, for example, surprisingly clustered around a 6 to 7-day age of the slant used to prepare *Aspergillus oryzae* spores. Slants younger than 5 days and older than 9 days resulted in spore inocula with lower proteinase titers. The basis of this effect should be investigated because it points to an unexplained variation in spore inoculum competence. Unless all steps before that point are rigidly controlled, the age of the inoculum can be of disappointing predictive value for subsequent fermentations.

Yamamoto *et al.* proposed elimination of lactic acid produced during inoculum development to prevent the lag in production during the final fermentation. Generally, product synthesis during inoculum development results in lower final titers.

The quality of the inoculum can be determined for some organisms by measuring a biochemical characteristic of the inoculum source. Nucleic acid content and enzyme levels have been used (Table : 13.3).

Table 13.3 : Key variables affecting inoculum quality

Variable	Organism	Optimum	Effect
Inoculum age	*S. hygroscopicus*	Juvenile (<25 h)	Decrease passage loss
Medium	*S. hygroscopicus*	Reduced carbohydrate	Decrease passage loss
Inoculum level	*S. hygroscopicus*	$\geq 1\%$	Maximum productivity
No. of stages	*S. hygroscopicus*	≤ 2	Minimize loss on passage
Enzyme level	*S. antibioticus*	Higher than control	Earlier onset of oleandomycin
Krebs cycle			
Inoculum level	*A. niger*	10^2 spores/litre	Maximum polygalacturonase
Inoculum level	*A. niger*	10^8 spores/litre	Maximum pectin lyase
Inoculum age	*A. oryzae*	Spores from 6-day slant	Peak proteinase
Inoculum level	*A. oryzae*	$\geq 6 \times 10^7$ spores/g of substrate	Proteinase peak
Enzyme ratio	*P. chrysogenum*	G6PDH/ICDH near 3	Optimum penicillin
Enzyme ratio	*P. patulum*	G6PDH/ICDH near 1	Optimum griseofulvin
Inoculum level	*A. mediterranei*	40 mg/flask	Highest specific production of rifamycin
RNA level	*C. purpurea*	≥ 6 µg/mg (dry weight)	Produced alkaloid >3 mg/litre
DNA level	*C. purpurea*	Maximum	Optimum alkaloid

In recent years, probes have been developed for on-line assessment of biomass and these could be invaluable in estimating the time of inoculum transfer. Boulton *et al.* (1989) reported the use of a biomass sensor (the Bugmeter) to control the yeast pitching rate (inoculum level) in brewing. The probe measures the dielectric permittivity of viable yeast cells and is unaffected by the presence of dead cells, air bubbles or detritus, making it ideal for the routine monitoring of yeast inoculum. Using the probe, these workers developed an automatic inoculum dispenser allowing a preset viable yeast mass to be transferred from a yeast storage vessel to the brewery fermentation.

Afford *et al.* (1992) reported the use of a real-time expert computer system to predict the time of inoculum transfer for industrial-scale fermentations. The system involves the comparison of on-line fermentation data with detailed historical data of the process. A problem with the interpretation of carbon dioxide production rate figures is that the data are not available continuously because the analyser is not dedicated to any one fermenter, but is analysing process streams from a large number of vessels via a multiplexer systems. Thus, a fermentation may have passed a critical stage between monitoring times. Also, occasional false readings may be generated. The expert system enabled the verification of data points as well as prediction of the outcome of the fermentation from early information. Data from seed fermentations were analysed by the expert system and the transfer time predicted. As a result of this approach operators were able to plan their work more effectively, the need for manual sampling was reduced and early warning of contamination was provided if the seed-culture profile predicted from early readings was abnormal.

13.3.5 Cell Morphology

Yeast in pellet form are the preferred morphological form for production in a continuous-process reactor. Induction of cell aggregation in a column reactor requires a high cell density, a low flow rate, and, in some cases, the presence of ethanol in the growth medium. When these aggregated cells are used as inoculum, they retain their aggregated form unless they are dispersed by blending before inoculation. Cells dispersed with a blender become non-flocculant and sediment at rates comparable to inoculum prepared from agar slants.

Using *Aspergillus terreus*, Gbewonyo *et al.*, were able to show that one of the very significant differences between inoculum in the form of pellets and inoculum in the form of dispersed mycelia was the dispersed mycelia under production fermenter conditions are likely to experience dissolved-oxygen limitations.

Cell morphology has a dramatic impact on culture productivity by filamentous organisms. Production is usually optimal either when pellets are formed or when growth is in a dispersed mycelial form. Braun and Vecht-Lifshitz have reviewed reports on the correlation between mycelial morphology and metabolite production. Their suggestions on how to induce pellet formation are summarized in Table : 13.4. Since their review, other approaches to the induction of pellet formation have been presented; they include supplementation with $MgSO_4 . 7H_2O$ or phosphate, nitrilo-triacetic acid, or 8-hydroxyquinoline, and CO_2 supplementation of intake air. Carboxymethyl cellulose, on the other hand, enhanced the production of the dispersed mycelial morphological form.

Table 13.4 : Possible approaches for inducing mycelial pellet formation in fermentation culture

(1)	Use nitrogen-limited medium.
(2)	Do not use readily assimilated carbon sources.
(3)	Add cationic polyelectrolytes.
(4)	Add mild non-ionic detergents.
(5)	Add Ca^{+2} ions.
(6)	Use a small inoculum.
(7)	Reduce shear.
(8)	Reduce temperature.
(9)	Maintain medium at about pH 5-6.
(10)	Increase dissolved-oxygen concentration.

13.3.6 Inoculum Source

The necessity for standardization of conditions used for the preparation and harvest of inocula has been emphasized, since analysis of error pointed to variability of the test inoculum as a major factor in the lack of test reproducibility. After various inoculum regimens were tested, the original source of culture from culture collections was implicated as a significant source of variability.

Spore formation of *Streptomyces coelicolor* on agar was dependent upon the type of agar used, the inclusion of trace elements, the nitrogen source, and a C/N ratio between 40 and 100.

Spore storage, however, could be a problem, since the spores lost 72% of their viability after storage for 1 week in buffer at 4°C. Many strains isolated from nature and often strains that have been subjected to a mutation programme result in an "unstable" culture, whose productivity can be rapidly lost. For such strains, a single spore selection step or its equivalent is a necessity for maintenance of productivity.

13.3.7 Immobilization

The production of microbial inoculum for use in bioremediation, agricultural applications, and waste treatment is limited by the ability of the microorganism to compete in these environments and to be metabolically effective. One of the methods by which microbial inocula are being improved for these applications is the use of immobilization technology.

The unique characteristics of immobilization inocula include

(i) enhanced inoculum viability,

(ii) protection from stress during manufacture,

(iii) enhanced ecological competence,

(iv) increased metabolite production,

(v) ultraviolet resistance,

(vi) the opportunity to use immobilized cells as a source of continuous inoculum, and

(vii) the opportunity to introduce mixed culture inocula into a process.

Immobilization of *Phanerochaete chrysosporium* in calcium alginate for use in bioremediation applications was improved when immobilized *P. chrysosporium* in alginate was supplemented with either sawdust or corncob grits.

13.3.8 Acclimatization

A number of commercial-level microbiological processes use as the inoculum, at least in part, culture growth that has been part of a previous "production phase".

The use of an acclimatized inoculum has been reported to result in significant improvements in operational efficiencies for xylose conversion to xylitol by *Candida guilliermondii* grown on a sugar cane hemicellulosic hydrolysate.

In the brewery industry, the reuse or pitching of yeast is a common practice. The effect of serial pitching of the yeast inoculum on subsequent refermentation has not been well characterized. The condition of the yeast cell surface as measured by flocculation can be predictive of subsequent fermentation performance.

For fermentation processes involved in the degradation of waste materials, a very important variable is the extent of acclimatization of the inoculum source. The process lag period before initiation of biodegradation decreases with increased numbers of competent microorganisms. High degradation rates are obtained when acclimated sewage sludge operated in a plant with low retention times is used as the inoculum. Assessment of the ability of a waste-water plant to degrade a material could greatly be underestimated if the inoculum has not been acclimatized to the tested waste material.

13.3.9 Shear

The rheological behaviour of a culture broth reflects the growth of the culture and its physiological state. In the course of determining the optimum seed transfer time for the production of nystatin by an industrial strain of *Streptomyces noursei*, Ettler found that the optimum inoculum occurred in a very tight time window when the broth shear stress was between 80 and 120 mPa at a shear rate of 698 S^{-1}. It was also a period when the culture broth response shifted from a Newtonian to a non-Newtonian behaviour. Perhaps shear measurement can become a valuable measure and indicator for the timing of optimum seed transfer.

A procedure for measuring the shear rate of culture media in a shake flask fermentation has been reported and involves a rolling-sphere viscometer and movement of calibrated stainless steel spheres on inclined surfaces.

13.3.10 Antifoam Addition

For commercial-scale production facilities, the final stages of inoculum development often take place in fermenters that requires the addition upon a culture and upon its productivity is an important study that can be undertaken at the shake flask level early in the culture development program. It is found that glycol-, oil-, and silicone-based antifoams often have different effects upon cell productivity.

The effect of antifoam addition on a cell or fermentation can be far reaching. *N*-Hexadecane, besides its antifoam properties, has been reported to cause changes in the morphology of *Penicillium chrysogenum* and to be an oxygen transfer enhancer. Since the cell morphology of fungi including *P. chrysogenum* has been reported to be an important variable for productivity, the use of N-Hexadecane in a seed fermenter could have an effect beyond its antifoam activity.

13.3.11 Mutations

It is evident that many generations of microbial cells must occur during inoculum production between the starting point of a stock culture and the termination point of a suitable inoculum for a large production tank. We assume that, during this cellular multiplication, the microorganisms does not change physiologically or genetically. However, mutations do occur among these cells, although the occurrence of a particular mutation usually is infrequent and, when it does occur, it usually is lethal to the cell or provides no selective advantage to the cell in its growth competition with non-mutated cells. Thus, the occurrence of a mutation is not serious in inoculum build-up unless it provides the microorganism with a distinct growth advantage and, at the same time, with an altered ability to produce the fermentation product.

A serious problem can exist, however, if the organism used for the fermentation is in itself a mutant strain. Mutations are not always stable, and the frequency of occurrence of "back mutation", or loss of a particular mutation, can be rather high; thus, inoculum for production tanks can contain a considerable portion of back mutants. Obviously, fermentation and media conditions tending to select for growth of the mutant fermentation strain or against growth of the back mutant are of definite benefit in inoculum production for this type of organism. For example, an *Escherichia coli* mutant strain requiring L-lysine for growth should not be allowed to exhaust the supply of this amino acid in the medium during inoculum production. Should this occur, the mutant could not compete in growth with the nonlysine-requiring back mutant, and this back mutant would become dominant in the inoculum. Obviously, stock-culture maintenance of mutated fermentation strains should be rigorously controlled so that the least possible numbers of back mutants will find their way into inoculum development.

13.3.12 Enzymatic Adaption

Enzymatic adaption in microorganisms resembles a mutational change, but it is not an inherited characteristic. A microorganism does not continuously produce high levels of enzymes to attack all of the possible substrates for which it has the inherent capability. It produces in quantity only those enzymes that it really needs. Some of these enzymes may be "adaptive enzymes" produced only in response to the presence of a substrate that the organism wishes to utilize for energy and growth. If such a substrate is deleted from the medium or becomes exhausted, the microorganism in succeeding generations decreased its output of the respective adaptive enzyme.

Industrial fermentations often utilize the activities of the adaptive enzymes of microorganisms and, hence, these enzymes should be present in the highest possible quantities during production. Therefore, provision should be made that the "inducing" substrate be present either in all the steps of inoculum development or, at least, in the final stages. If this consideration is not met, a distinct lag in growth may be observed in the production tank immediately after inoculation while the microorganisms are adaptively forming enzymes to allow utilization of the inducing substrate.

13.3.13 Water Activity (Aw)

Water activity (Aw) is defined as the relative humidity of the gaseous atmosphere in equilibrium with the substrate. Water activity is a critical variable for inoculum prepared on solid substrates. It is also a critical determinant for the survival of dried cultures.

On-line sensors have been developed for Aw measurements in liquid and in solid substrate fermentations. Xavier and Karanth have adapted the proximity equilibration cell method for Aw measurement of solid substrates.

13.3.14 Inoculum Level

Inoculum level can have a dramatic effect upon productivity. *Aspergillus niger* produced maximum levels of polygalacturonase and maximum pectin lyase with inoculum levels of 10^2 and 10^8 spores/litre, respectively (See Table : 13.3).

13.3.15 Contamination

Contamination is an ever-present risk in inoculum production. Microbial contamination detection usually relies upon the use of differential media and conditions to encourage the growth of likely contaminants in the presence of the inoculated microbe. Detection of contamination in a mixed-culture fermentation provides the microbiologist with a different challenge.

Polymerase chain reaction (PCR) has provided a rapid, effective technique for the detection of a contaminant present at low levels in a sample. This protocol can be applied to mixed-culture fermentations either for the detection of a particular contaminant of interest or for the detection of an indicator organism, such as the detection of *Escherichia coli* as an indicator of fecal contamination.

Phage contamination is a constant threat to the productivity of any bacterial fermentation process. This problem is particularly prevalent in fermentations of dairy products. Selection of plasmids that confer phage resistance to lactic streptococci was proposed by Klaenhammer as a possible solution. Selection of phage-resistant strains for any bacterial process is an important piece of insurance, if multiple large-scale batches are contemplated. The report that alginate-immobilized streptococci were protected from attack by phages is potentially an interesting alternative approach.

Inoculum for anaerobic fermentations should be tested for contamination by culturing under both anaerobic and aerobic incubation conditions so as to detect the presence of facultative anaerobes. For some fermentations, such as acetone-butanol fermentation, there are characteristic fermentation patterns which provide a clue to the presence of contamination. During growth of this organism, there is an initial production of acetic and butyric acids accompanied by adaptive enzyme formation for the conversion of these acids to acetone and Butanol. Concurrent with the production of the latter solvents, the levels of acetic and butyric acids decrease markedly—the "acid-break", this phenomenon can be observed by titrating successive samples of inoculum or production culture with dilute alkali. Inoculum of this organism should be used only after a decisive acid-break has occurred, since contaminating microorganisms prevent the detection of a sharp acid-break.

13.4 INOCULUM PREPARATION PROCEDURES

A typical inoculum-development programme will follow any of the following procedure in fermentation industry.

13.4.1 Traditional Procedure

The master culture is reconstituted and plated on to solid medium; approximately ten colonies of typical morphology or high producers are selected and inoculated on to slopes as the sub-master cultures, each sub-master culture being used for a new production run. At this stage, shake flasks may be inoculated to check the productivity of these cultures, the results of such tests being known before the developing inoculum eventually reaches the production plant. A sub-master culture is used to inoculate a shake flask (250 or 500 cm^3 containing 50 or 100 cm^3 medium) which, in turn, is used as inoculum for a larger flask, or a laboratory fermenter, which may then be used to inoculate a pilot-scale fermenter. Culture purity checks are carried out at each stage to detect contamination as early as possible. Although the results of these tests may not be available before the culture has reached the production plant, at least it is known at which stage in the procedure contamination occurred.

13.4.2 Lincolon Procedure

Lincoln suggested a more elaborate procedure for the development of inoculum for bacterial fermentations which, with minor modifications, is applicable to any type of culture. The procedure involved the use of one sub-master culture to develop a bulk inoculum which was subdivided, stored in a frozen state and used as inocula for several months. A single colony, derived from a sub-master culture, was inoculated into liquid medium and grown to maximum log phase. This culture was then transferred into nineteen times its volume of medium and incubated again to the maximum log phase, at which point it was dispensed in 20-cm^3 volumes, plug frozen and stored at below—20°. At least 3% of the samples were tested for purity and productivity in subsequent fermentation and, provided these were suitable, the remaining samples could be used as initial inocula for subsequent fermentations. To use one of the stored samples as inoculum it was thawed and used as a 5% inoculum for a seed culture which, in turn, was used as a 5% inoculum for the next stage in the programme. This procedure ensured that a proven inoculum was used for the penultimate stage in inoculum development.

13.4.3 Recovery of Culture from the Preserved State

The following procedures have generally allowed good recovery of cultures from a preserved state.

(1) Frozen ampules, wrapped in a sterile gauze pad, are open after being submerged in disinfectant. The contents are shaken into 2 ml of a suitable recovery medium and incubated for at least 1 hour before being transferred to agar slants or liquid media.

(2) Agar plugs that have been frozen are removed and streaked onto agar slants or plates.

(3) For organisms preserved in soil, a small aliquot is removed to either broth or agar plates.

(4) Cultures stored on paper are placed upside down on agar plates and shifted after 24 hours to allow for better aeration of the growing culture.

(5) Rehydration of lyophilised cultures in broth rather than in saline or distilled water has been recommended. This minimized the growth lag and allowed the elimination of a solid-medium cell proliferation step, a step which can generally result in large variation in initial inoculum density.

13.5 THE DEVELOPMENT OF INOCULA FOR BACTERIAL PROCESSES

The main objective of inoculum development for traditional bacterial fermentations is to produce an active inoculum which will give as short a lag phase as possible in subsequent culture. A long lag phase is disadvantageous in that not only is time wasted but also medium is consumed in maintaining a viable culture prior to growth. The length of the lag phase is affected by the size of the inoculum and its physiological condition (Meyrath and Suchanek, 1972). As already stated, the inoculum size normally ranges between 3 and 10% of the culture volume. Lincoln (1960) stressed that bacterial inocula should be transferred in the logarithmic phase of growth, when the cells are still metabolically active. The age of the inoculum is particularly important in the growth of sporulating bacteria, for sporulation is induced at the end of the logarithmic phase and the use of an inoculum containing a high percentage of spores would result in a long lag phase in a subsequent fermentation.

Examples of Bacterial Fermentations

13.5.1 Bacterial Enzyme-Protease and Amylase

Keay *et al.* (1972) quote the use of a 5% inoculum of a logarithmically growing culture of a thermophilic *Bacillus* for the production of proteases. Aunstrup (1974) described a two-stage inoculum development programme for the production of proteases by *Bacillus subtilis*. Inoculum for a seed fermenter was grown for 1 to 2 days on a solid or liquid medium and then transferred to a seed vessel where the organism was allowed to grow for a further ten generations before transfer to the production stage. Priest and Sharp (1989) cited the use of a 5% inoculum, still in the exponential phase, for the commercial production of *Bacillus* α-amylase.

13.5.2 Vitamin B_{12}

The inoculum development programme for vitamin B_{12} production using *Pseudomonas denitrificans* is outlined in Figure : 13.1.

STOCK CULTURE
Lyophilised with skim milk
↓
MAINTENANCE CULTURE
Agar slope incubated 4 days at 28°
↓
SEED CULTURE - FIRST STAGE
2 dm^3 flask containing 0.6 dm^3 medium inoculated with culture from one slope; incubated with shaking for 48h at 28°
↓
SEED CULTURE - SECOND STAGE
40-80 dm^3 fermenter containing 25-50 dm^3 medium inoculated with 1- 1.2% first stage seed culture. Incubated 25-30h at 32°
↓
PRODUCTION CULTURE
500 dm^3 fermenter with 300 dm^3 medium inoculated with 5% second stage seed culture. Incubated at 32° for 140 - 160 h

Fig. 13.1 : The inoculum development programme for a vitamin B_{12} pilot scale fermentation using *Pseudomonas denitrificans*.

13.5.3 Vinegar

The necessity to use an inoculum in an active physiological state is taken to its extreme in the production of vinegar. The acetic-acid bacteria used in the vinegar process are extremely sensitive to oxygen starvation. Therefore, to avoid disturbing the system, the cells at the end of a fermentation are used as inoculum for the next batch by removing approximately 60% of the culture and restoring the original level with fresh medium (Conner and Allgeier, 1976). The advantage of a highly active inoculum apparently outweighs the disadvantages of possible strain degeneration and contaminant accumulation.

13.5.4 Lactic Acid

In the lactic-acid fermentation the producing organism may be inhibited by lactic acid. Thus, production of lactic acid in the seed fermentation may result in the generation of poor quality inoculum. Yamamoto *et al.* (1993) generated high quality inoculum of *Lactococcus lactis* IO-I on a laboratory scale using electrodialysis seed culture which reduced the lactate in the inoculum and reduced the length of the lag phase in the production fermentation.

13.5.5 Acetone-butanol

An example of the inoculum development programme for anaerobic bacterial fermentation process is explained by the clostridial acetone-butanol fermentation as outlined in Table : 13.5.

Table 13.5 : The inoculum development programme for the clostridial acetone-butanol fermenation (Spivey, 1978)

Stage	Cultural conditions	Incubation time (hours)
1.	Heat-shocked spore suspension inoculated into 150 cm^3 of potato glucose medium	12
2.	Stage 1 culture used as inoculum for 500 cm^3 molasses medium	6
3.	Stage 2 culture used as inoculum for 9 dm^2 molasses medium	9
4.	Stage 3 culture used as inoculum for 90,000 dm^3 molasses medium	

13.5.6 Recombinant Bacterial Products-Biotin

Sabatie *et al.* (1991) demonstrated that plasmid stability and productvity in an *E.coli* biotin fermentation was greatly improved if stationary, rather then exponential, phase cells were used as inoculum. They postulated that the plasmid copy number may be higher in stationary cells than in exponential ones, resulting in a lower plasmid loss in the subsequent fermentation when a stationary culture is used as inoculum. A stationary phase inoculum would result in a lag phase, but this disadvantage was more than compensated for by the considerable improvement in plasmid retention and biotin production compared with that obtained using an exponential inoculum.

13.6 THE DEVELOPMENT OF INOCULA FOR YEAST PROCESSES

Whilst the largest industrial fermentations utilizing yeasts are the brewing of beer and the production of biomass, recent processes have also been established for the production of recombinant products.

13.6.1 Brewing

It is common practice in the British brewing industry to use the yeast from the previous fermentation to inoculate a fresh batch of wort. The brewing terms used to describe this process are 'crop', referring to the harvested yeast from the previous fermentation, and 'pitch', meaning to inoculate.

The dangers inherent in this practice are the introduction of contaminants and the degeneration of the strain, the most common degenerations being a change in the degree of flocculence and attenuating abilities of the yeast. In breweries employing top fermentations in open fermenters these dangers are minimized by collecting yeast to be used for future pitching from 'middle skimmings'. During the fermentation the yeast cells flocculate and float to the surface, the first cells to do this being the most flocculent and the last cells the least flocculent. As the head of yeast develops, the surface layer (the most flocculent and highly contaminated yeasts) is removed and discarded and the underlying cells (the 'middle skimmings') are harvested and used for subsequent pitching. Therefore, the 'middle skimmings' contain cells which have the desired flocculence and which have been protected from contamination by the surface layer of the yeast head. The pitching yeast may be treated to reduce the level of contaminating bacteria and remove protein and dead yeast cells by such treatments as reducing the pH of the slurry to 2.5 to 3, washing with water, washing with ammonium persulphate and treatment with antibiotics such as polymixin, penicillin and neomycin.

The difficulties outlined above and the likelihood of strain degeneration and contamination mean that yeasts are rarely used for more than five to ten consecutive fermentations (Thorne, 1970; Reed and Nagodawithana, 1991) which necessitates the periodical production of a pure inoculum. This would involve developing sufficient biomass from a single colony to pitch a fermentation at a level of approximately 2 grams of pressed yeast per litre. Hansen (1896) pioneered the use of pure inocula and devised a yeast propagation scheme utilizing a 10% inoculum volume at each stage in the programme and employing conditions similar to those used during brewing. However, modern propagation schemes use inoculum volumes of 1% or even lower and may use conditions different from those used during brewing. Therefore, continuous aeration may be used during the propagation stage which seems to have little effect on the beer produced in the subsequent fermentation (Curtis and Clark. 1957). Yeast inoculum produced in this way would also be sterol rich, obviating the need for aerated wort.

A number of yeast propagators (which are basically closed, aerated vessels) have been described in the literature. The simplest type of propagator is a single stage system resembling an unstirred, aerated fermenter which is inoculated with a shake-flask culture developed from a single colony (Gilliland, 1971). Curtis and Clark (1957) and Thorne (1970) described two-stage systems which could be operated semi-continuously. Thorne's propagator consisted of two linked vessels, 1.5 and 150 dm^3 respectively. The smaller vessel was filled with wort, sterilized, cooled, aerated and inoculated with a flask-grown culture. After growth for 3 to 4 days the culture was forced by air pressure into the second vessel which had been filled with sterilized, cooled wort and aerated. An aliquot of 1.5 dm^3 was forced back into the first vessel after mixing. In a further 3 to 4 days the larger vessel contained sufficient biomass to pitch a 1000 dm^3 fermenter and the first vessel contained sufficient inoculum for another second stage. However, although this procedure should produce a pure inoculum there is a danger of strain degeneration occurring in such a semi-continuous system.

13.6.2 Baker's Yeast

The commercial production of bakers' yeast involves the development of an inoculum through a large number of aerobic stages. Although the production stages of the process may not be operated under strictly aseptic conditions a pure culture is used for the initial inoculum, thereby keeping contamination to a minimum in the early stages of growth. Reed and Nagodawithana (1991) discussed the development of inoculum for the production of bakers' yeast and quoted a process involving eight stages, the first three being aseptic while the remaining stages were carried out in open vessels. The yeast may be pumped from one stage to the next or the seed cultures may be centrifuged and washed before transfer, which reduces the level of contamination. The yields obtained in the first five stages are relatively low because they are not fed-batch systems, whereas the last three stages are fed-batch.

13.7 THE DEVELOPMENT OF INOCULA FOR MYCELIAL PROCESSES

The preparation of inocula for fermentations employing mycelial (filamentous) organisms is more involved than that for unicellular bacterial and yeast processes. The majority of industrially important fungi and streptomycetes are capable of asexual sporulation, so it is common practice to use a spore suspension as seed during an inoculum development programme. A major advantage of a spore inoculum is that is contains for more 'propagules' than a vegetative culture. Three basic techniques have been developed to produce a high concentration of spores for use as an inoculum.

13.7.1 Sporulation on Solidified Media

Most fungi and streptomycetes will sporulate on suitable agar media but a large surface area must be employed to produce sufficient spores.

Examples

(i) Penicillin

Parker (1950) described the 'roll-bottle' technique for the production of spores of *Penicillium chrysogenum*. Quantities of medium (300 cm^3) containing 3% agar were sterilized in 1 dm^3 cylindrical bottles, which were then cooled to 45° and rotated on a roller mill so that the agar set as a cylindrical shell inside the bottle. The bottles were inoculated with a spore suspension from a sub-master slope and incubated at 24° for 6 to 7 days. Parker claimed that although the use of the 'roll-bottle' involved some sacrifice in ease of visual examination, it provided a large surface area for cultivation of spores in a vessel of a convenient size for handling in the laboratory.

(ii) Clavulanic acid

Butterworth (1984) described the use of a Roux bottle for the production of a spore inoculum of *Streptomyces clavuligerus* for the production of clavulanic acid. The spores produced from one bottle containing 200-cm^2 agar surface could be used to inoculate a 75 dm^3

seed fermenter which, in turn, was used to inoculate a 1500 dm^3 fermenter. The clavulanic acid inoculum development programme is illustrated in Fig. 13.2.

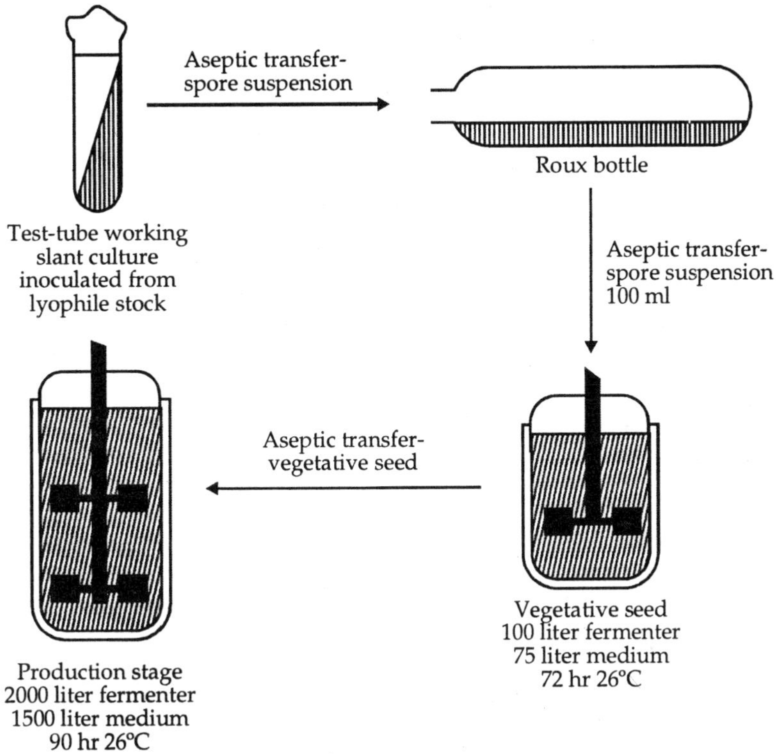

Fig. 13.2 : The inoculum development programme for the production of clavulanic acid from *Streptomyces clavuligerus.*

The composition of solidified media suitable for the sporulation of *Penicillium chrysogenum* and *Aspergillus niger* are mentioned in Table : 13.6.

Table 13.6 : Solidified media suitable for the sporulation of some representative fungi

	Fungus	Medium (g dm^{-3})
Penicillium chrysogenum	Glycerol	7.5
	Cane molasses	7.5
	Curbay BG	2.5
	$MgSO_4 . 7H_2O$	0.05
	KH_2PO_4	0.06
	Peptone	5.0
	NaCl	4.0
	Agar	20
Aspergillus niger	Molasses	300
	KH_2PO_4	0.5
	Agar	20

13.7.2 Sporulation on Solid Media

Many filamentous organisms will sporulate profusely on the surface of cereal grains from which the spores may be harvested. Substrates such as barley, hard wheat bran, ground maize and rice are all suitable for the sporulation of a wide range of fungi. The sporulation of a given fungus is particularly affected by the amount of water added to the cereal before sterilization and the relative humidity of the atmosphere, which should be as high as possible during sporulation.

Examples

(1) Singh *et al.* (1968) have described a system for the sporulation of *Aspergillus ochraceus* in which a 2.8-dm^3 Fernbach flask containing 200 grams of 'pot' barley or 100 grams of moistened wheat bran produced 5 x 10^{11} conidia after six days at 28° and 98% relative humidity. This was 5 times the number obtainable from a Roux bottle batched with Sabouraud agar and 50 times the number obtainable from such a vessel batched with Difco Nutrient Agar, incubated for the same time period. Vezina *et al.* (1968) have published a list of fungi which are capable of sporulating heavily on cereal grains.

(2) El-Sayed (1992) quoted the use of cooked rice for the production of spores of *Penicillium* and *Cephalosporium* in *penicillin* and *cephalosporin* inoculum development.

(3) Podojil *et al.* (1984) quoted the use of millet for the sporulation of *Streptomyces aureofaciens* in the development of inoculum for the *chlortetracycline* fermentation and the process is outlined in figure 13.3.

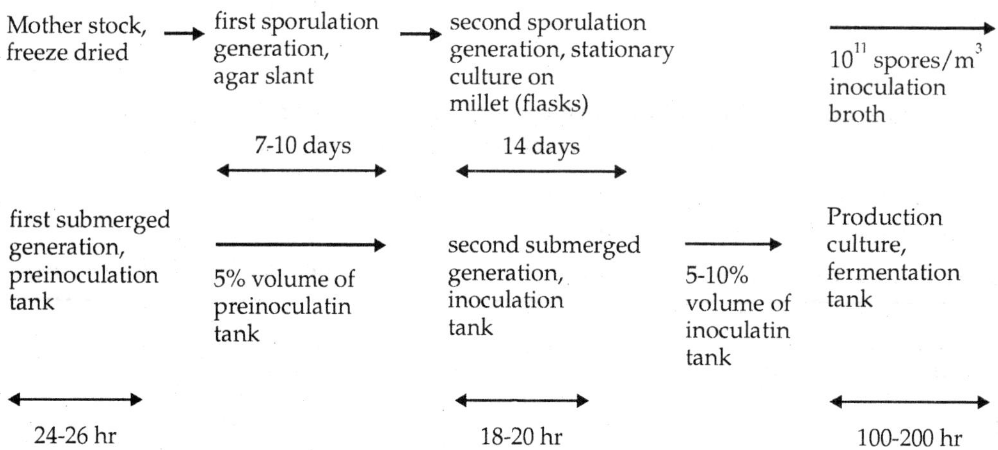

Fig. 13.3 : The inoculum development programme for the production of chlortetracycline by *Streptomyces aureofaciens* (Podojil. 1984).

13.7.3 Sporulation in Submerged Culture

Many fungi will sporulate in submerged culture provided a suitable medium is employed (Vezina *et al.*, 1965). This technique is more convenient than the use of solid or solidified media because it is easier to operate aseptically and it may be applied on a large scale.

Examples

(1) The technique was first adopted by Foster *et al.* (1945) who induced submerged sporulation in *Pencillium notatum* by including 2.5% calcium chloride in a defined nitrate—sucrose medium. An example of the use of this technique for the production of inoculum for an industrial fermentation is provided by the griseofulvin process.

(2) Rhodes *et al.* (1957) described the conditions necessary for the submerged sporulation of the griseofulvin-producing fungus, *Penicillium patulum*, and the medium utilized is given in Table : 13.7. These authors found that for prolific sporulation the nitrogen level had to be limited to between 0.05 and 0.1% w/v and that good aeration had to be maintained. Also, an interaction was demonstrated between the nitrogen level and aeration in that the lower the degree of aeration the lower the concentration of nitrogen needed to induce sporulation. Submerged sporulation was induced by inoculating 600 cm^3 of the above medium, in a 2-dm^3 shake flask, with spores from a well-sporulated. Czapek-Dox agar culture and incubating at 25° for 7 days. The resulting suspension of spores was then used as a 10% inoculum for a vegetative seed stage in a stirred fermenter, the seed culture subsequently providing a 10% inoculum for the production fermentation.

Table 13.7 : Media for the submerged sporulation of selected fungi

Rhodes *et al.* (1957): *Penicillium patulum*		
Whey powder, to give	Lactose 3.5% Nitrogen 0.05%	
KH_2PO_4		0.4%
KCl		0.05%
Corn-steep liquor solids to	give approx. 0.04% N	0.38%
Foster *et al.* (1945): *Penicillium notatum*		
Sucrose		2.0%
$NaNO_3$		0.6%
KH_2PO_4		0.15%
$MgSO_4 . 7H_2O$		0.05%
$CaCl_2$		2.5%
Vezina *et al.* (1965): *Aspergillus ochraceus*		
Glucose		2 5%
NaCl		2.5%
Corn-steep liquor		0.5%
Molasses		5.0%

Most actinomycetes do not sporulate in submerged culture (Whitaker, 1992) and, thus, solid or solidified media are used for the production of spore inocula.

13.7.4 The Large-scale use of Spore Inoculum in Fermentation Industry

The stage in an inoculum development programme at which a large-scale spore inoculum is used varies according to the process, it appears to be common practice that the penultimate stage is so inculated, but this will depend on the scale of the production fermentation.

Examples

(1) In the inoculum development programme for the early penicillin fermentation described by Parker (1950) the penultimate stage was inoculated with a spore suspension (from a 'roll-bottle') and this stage may have produced either a vegetative or a submerged spore inoculum for the final fermentation.

(2) For the griseofulvin process, Rhodes *et al.* (1957) stated that the spore suspension obtained from the submerged sporulation stage could either be used for direct inoculation of the production fermentation or it could be germinated in an inoculum development medium to yield a vegetative inoculum for the final fermentation. The latter course was preferred and an inoculum volume of 7-10% was used.

(3) From Fig. 13.2 to 13.4 it can be seen that in the clavulanic acid process the spore inoculum is used to inoculate the final seed stage, in the chlortetracyciine process a vegetative stage is interspersed between the spore inoculated batch and the production fermentation, and in the sagamicin process the spore inoculum is used at a very early stage followed by vegetative growth.

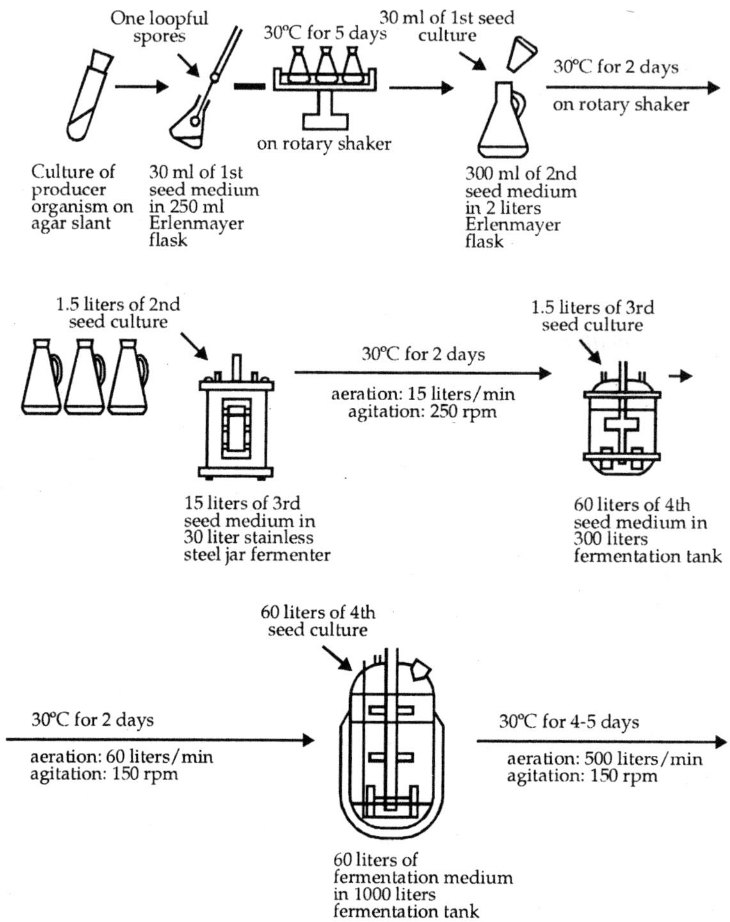

Fig. 13.4 : The inoculum development programme for the production of sagamicin by *Micromonospora sagamiensis* (Podojil, 1984)

13.7.5 Inoculum Development for Vegetative Fungi

Some fungi will not produce asexual spores and, therefore, an inoculum of vegetative mycelium must be used. *Gibberella fujikuroi* is such a fungus and is used for the commercial production of gibberellin.

Example

(i) Gibberellin

Hansen (1967) described an inoculum development programme for the gibberellin fermentation. Cultures were grown on long slants (25 × 10 mm test tubes) of potato dextrose agar for 1 week at 24°. Growth from three slants was scraped off and transferred to a 9-dm^3 carboy containing 4 dm^3 of a liquid medium composed of 2% glucose, 0.3% $MgSO_4 . 7H_20$, 0.3% NH_4Cl and 0.3% KH_2PO_4. The medium was aerated for 75 hours at 28° before transfer to a 100-dm^3 seed fermenter containing the same medium.

The major problem in using vegetative mycelium as initial seed is the difficulty of obtaining a uniform, standard inoculum. The procedure may be improved by fragmenting the mycelium in an homogenizer, such as a Waring blender, prior to use as inoculum. This method provides a large number of mycelial particles and therefore a large number of growing points.

❑❑❑

14

INDUSTRIAL FERMENTER DESIGN

14.1 INTRODUCTION

A fermenter (or bioreactor) is a system consisting of a few pieces of equipment which provide controlled environmental conditions for the growth of microbes (and/or production of specific metabolites) in liquid culture whilst preventing entry and growth of contaminating microbes from the outside environment.

Fermenters are the core of any fermentation—based or biotechnology-based production process be it for antibiotics, enzymes, amino acids, proteins, organic acids, solvents, vitamins, biopolymers, vaccines, enzymic or microbial transformations, bioremediation and biodegradation, and microbial inoculants for use as biofertilizers, biopesticides, etc. Microorganisms, biocatalysts, animal or plant cells, are produced and maintained in bioreactors.

A production facility typically has a train of fermenters (bioreactors) ranging from few litres to thousand litres. Still larger vessels are encountered in certain processes. In a great majority of processes, the fermenters are operated in a batch mode, under sterile or mono-septic conditions. The most common operational practice starts with culturing microorganisms or cells in the smallest fermenter. After a predetermined batch time the contents of this reactor is transferred to a larger, presterilized, medium-filled reactor and this process is repeated until the production fermenter, the largest reactor in the train, is reached. Most commercial processing is carried out as submerged culture with a nutrient medium in a suitable fermenter.

14.2 IMPORTANT CRITERIA FOR DESIGNING THE FERMENTER

Industrial fermenters of different types have been developed to provide the best possible of varying environment for the optimum growth and metabolism of industrial processed strains to produce quantities of commercially desired product. Provisions to allow ease of manipulations for all operations associated with the use of fermenters are designed. The design of fermenter should consider the factors such as

(i) maintenance of aerobic/anaerobic conditions to support the growth for optimum product formation;

(ii) maintenance of the process temperature at the desired level;

(iii) maintenance of the culture in pure state throughout the process;

Such factors are influenced by the design and operating conditions, and operation of the fermenter.

The various criteria are to be considered in the design and development of the fermenters, and these include,

(i) the fermenter tank should be capable of being operated aseptically for a long period and be reliable throughout the operation;

(ii) adequate aeration and agitation should be provided to meet the metabolic requirements of the microorganisms;

(iii) power consumption should be as low as possible;

(iv) a system of pH control should be provided;

(v) sampling facilities should be included;

(vi) a provision for temperature control be made;

(vii) evaporation loss from fermenter should not be excessive;

(viii) the vessel should be designed to require the minimal use of labour in operation, harvesting, cleaning and maintenance;

(ix) the vessel should be suitable for variety of reactions;

(x) the vessel should be constructed to ensure smooth internal surfaces, using welds instead of flame joints wherever possible;

(xi) the fermenter vessel should be of similar geometry to both smaller and larger vessels in the plant or pilot plant to facilitate scale-up;

(xii) the cheapest material that enable satisfactory results to be achieved should be used;

(xiii) there should be adequate service provisions for the bioreactor;

(xiv) the material from which they are fabricated must not be corroded by the fermentation product, nor contribute toxic ions to the medium.

14.3 INDUSTRIAL FERMENTERS

The most widely used industrial fermenters are stirred, baffled, aerated tanks provided with systems of temperature, pH, and foam formation control. A schematic representation of such a batch fermenter is shown in Fig 14.1.

Although, the vast majority of fermenters in commercial operation rely on mechanical agitation to achieve good mixing and oxygen transfer, some fermenters rely upon very high gas inputs without mechanical agitation.

Industrial fermenters can be grouped into three categories

(i) Mechanically agitated fermenter

(ii) Non-mechanically agitated fermenter

(iii) Non-agitated fermenter

The appropriate design of a bioreactor, for a particular product will depend on the structure and environmental requirement of the organism.

The design should be such that it is easy to maintain an aseptic condition throughout the operation period. Operating cost involved for aeration and agitation (mixing) of fermented broth has to be specially considered.

Mechanically agitated stirred tank reactors are the most commonly used bioreactors in industries. Due to the non-Newtonian nature of the fermented broth the power requirement is very high. As the agitator shaft has to rotate, microclearance between the shaft and stuffing box cannot be neglected and overtime may increase due frictions. To ensure asepticity, special arrangements to avoid contamination mean additional expenses. Another problem of mechanically agitated bioreactor is the high shearing and tearing effect of its impeller blades. Therefore, this type is not suitable for mycelial fungal strains or for tissue cultures. In very large capacity vessels, uniformity in mixing is difficult to achieve. The other type newly developed bioreactors include:

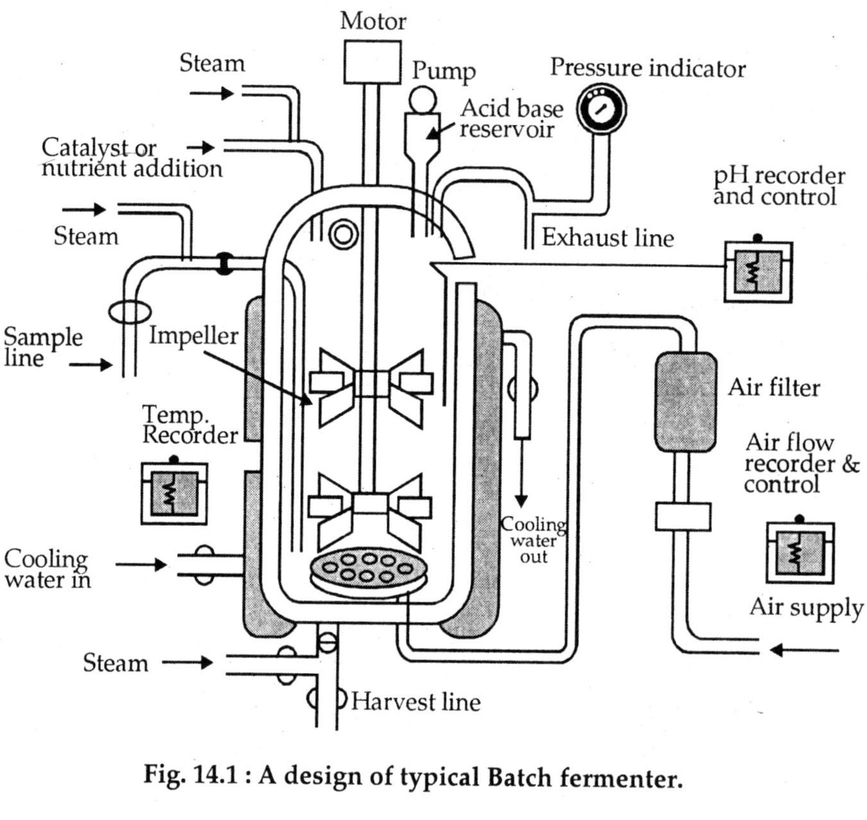

Fig. 14.1 : A design of typical Batch fermenter.

(i) Air-lift bioreactors
(ii) Cyclone type bioreactor
(iii) Air-lift fermenter with external loop
(iv) ICI Air-lift fermenter with internal loop
(v) Deep jet fermenter
(vi) Membrane recycle bioreactor
(vii) Continuous fermentation stirred tank
(viii) Modified air lift bioreactor, with both internal and external loop type (IIT Kharagpur).
(ix) Packed bed bioreactor
(x) Fluidized bed bioreactor
(xi) Hollow fibre bioreactor
(xii) Novel seesaw effect bioreactor (under patent)

14.4 FABRICATION MATERIALS FOR FERMENTERS

In fermentation with strict aseptic requirements, it is important to select materials that can withstand repeated steam-sterilization cycles.

Laboratory fermenters

On a small scale (1 to 30 dm^3), it is possible to use glass and/or stainless steel. Glass is better because it gives smooth surfaces, and is also non-toxic and corrosion proof. On a small scale, two basic types of laboratory fermenters are used:

(i) A glass vessel with a round or flat bottom and a top flaged carrying plate,
(ii) A glass cylinder with stainless-steel top and bottom plate.

Pilot and Industrial fermenters

The fermenters vessel for pilot scale and larger industrial scale is usually made of stainless-steel or copper. Most bench scale fermenters are glass vessels with stainless-steel cover plates. Wood, plastic and concrete have been used when contamination is not a problem in a fermentation process.

In fabrication of industrial fermenters, type 316 steel is usually employed, with the jacket and coverplate usually being type 304. Type 316L (low carbon) steel is sometimes used to minimize pitting, especially from chlorides. This is generally felt to be necessary for tissue culture work. Generally all wetted parts are of either 316 or 316L steel.

Addition vessels are usually made of 316 stainless-steel, but this material is not suitable for hydrochloric or sulphuric acids. If it is desired to use these acids, then other materials must be substituted, for example, Hastelloy B or C, zirconium, or an inert internal coating. These materials are all moderately expensive.

The choice of material for fittings and pipe or tubing that connect the addition vessel to the fermenter may present an additional problem. Hastelloy and zirconium tubings are available, as well as Hastelloy fittings, but these are very expensive. Fluorocarbon tubing and fittings are readily available, but these do not work well because the line must be sterilized with the fermenter, and this material has undesirable cold-flow properties.

Sight glasses are almost always made from Pyres glass. Gaskets and "O" rings are generally made from ethylene propylene (epdm) or Viton. Epdm is relatively inexpensive compared to viton and works for most fermentation applications. Epdm does, however, swell when exposed to hydrocarbon and cannot be used with hydrocarbon fermentations.

The mechanical shaft seals consist of two parts. The rotating portion is usually made of a carbon composition. The composition best suiting fermentation applications can be supplied by the major seal manufacturers. The stationary ring is usually made from tungsten carbide. In some of the smaller fermenters, it is ceramic.

Thickness of the construction material will increase with scale. At 3,00,000 to 4,00,000 dm^3 capacity, 7 m.m. plate may be used for the sides of vessel and 10 m.m. plate for top and bottom which should be hemispherical to withstand pressure.

Size—Batch fermenters are available with varying capacities.

(i) **Small laboratory (Baby) fermenters** are in the size of 1 to 2 litres with a maximum upto 12 to 15 litres.

(ii) **Pilot plant fermenters** have a total volume of 25 to 100 gallons; upto 2000 gallons of total volume.

(iii) **Industrial or production fermenters** range from 10,000 to 1,00,000 gallons are used; upto maximum 5,00,000 gallons.

Actually, the working volume in a fermenter is always less than that of the total volume. A "head space" is left at the top of the fermenter above the aqueous medium to allow aeration process. This head space usually occupies a fifth to a quarter of the total volume of the fermenter.

14.5 THE COMPONENTS OF A FERMENTER AND THEIR USES

The main subdivisions of a bench-scale fermenter are as follows:

1. Base components including drive motor, heaters, pumps, gas control, etc.;
2. Vessel and accessories;
3. Peripheral equipment such as reagent bottles;
4. Instrumentation and sensors.

These components perform the following functions—

1. Provide operation free from contamination;
2. Maintain a specific temperature;
3. Provide adequate mixing and aeration;
4. Control the pH of the culture;
5. Allow monitoring and/or control of dissolved oxygen;
6. Allow feeding of nutrient solutions and reagents;
7. Provide access points for inoculation and sampling;
8. Use fittings and geometry relevant to scale-up;
9. Minimize liquid loss from the vessel;
10. Facilitate the growth of a wide range of organisms.

14.6 TYPICAL FERMENTER VESSEL AND ITS MAJOR COMPONENTS

The vessel is a single-walled cylinder of borosilicate glass with a flat bottom or as a glass-jacketed system having a typically round bottom. The top plate is made from stainless steel and is compressed onto the vessel flange by an easily released clamping system. The vessel glass and the top plate is separated by a seal. Port fittings of various sizes are provided for insertion of probes, inlet pipes, exit gas cooler, cold fingers, sample pipes, etc. These work by compressing the sides of the probe/pipe against the seal. A special inoculation port will have a membrane seal held in place with a collar. Culture can be withdrawn into a sampling device or a reservoir bottle via a sample pipe situated in the bulk of the fermenter fluid. A gas sparger is also fixed into the top plate and this terminates in a special assembly, which ensures that incoming air is dispersed efficiently within the culture by the flat-bladed

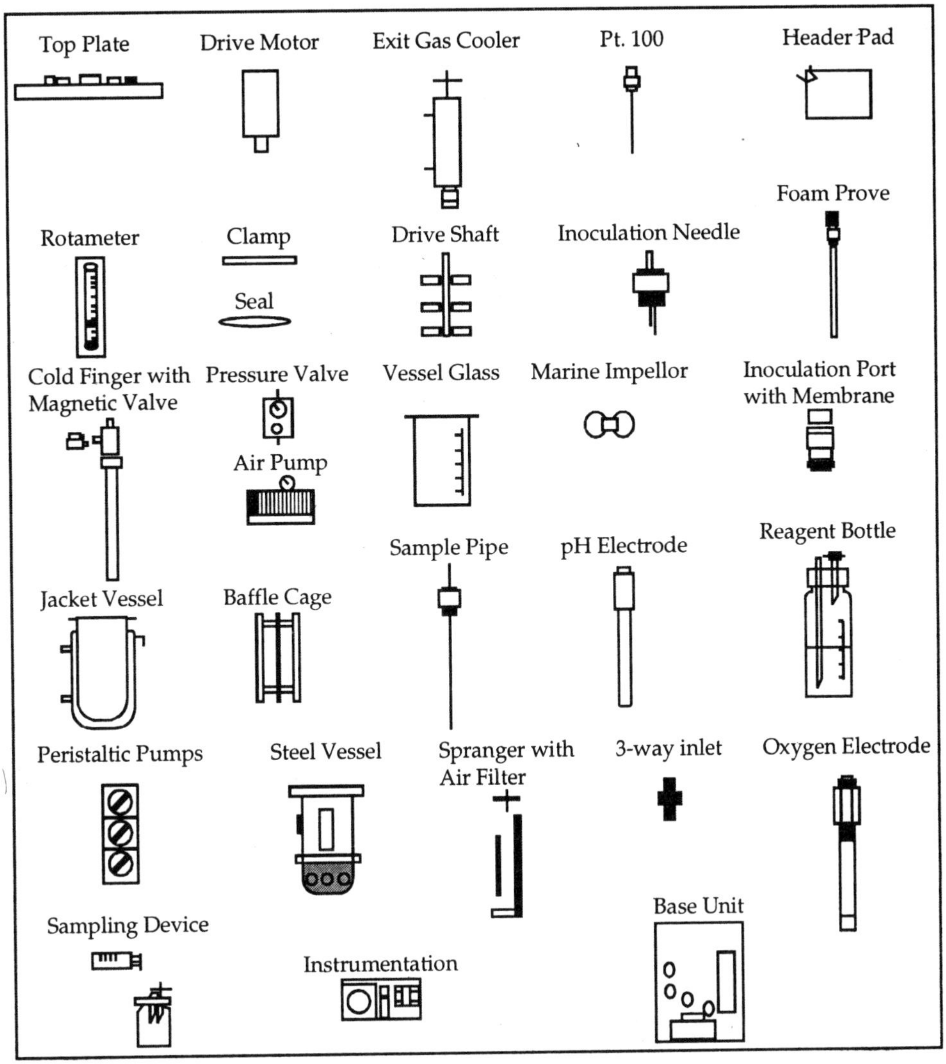

Fig. 14.2 : Illustration of major components of fermentor.

'Rushton-type' impellors fixed to the drive shaft. A drive motor is attached for the purpose of stirring to the drive shaft and is usually fitted directly to the drive hub on the vessel top plate. An exit gas cooler is also provided to the vessel, which works like a simple Liebig condenser to remove as much moisture as possible from the gas leaving the fermenter to prevent excessive liquid losses during the fermentation. It also prevents the wetting of the exit air filter.

A narrow platinum resistance temperature sensor is also attached with the vessel, which control the temperature either by direct heating with a heater pad or by circulating warm water around the vessel jacket. A cold finger is also used to cool the vessel contents, if direct heating is used.

Figure 14.2 illustrates the major components of a fermenter vessel.

The sensors which fit into the vessel do so either by direct coupling with a thread on the body of the electrode, as is the case in the gel-filled type of pH electrode, or by a special fitting on the vessel top plate, as in the case of the dissolved oxygen electrode (polarographic type). Another system involves the use of a simple compression fitting which holds the body of the electrode, as with the foam probe. In this case, the height is variable and the tip of the probe is above the culture fluid.

14.7 PERIPHERAL PARTS

14.7.1 Reagent Pumps

Pumps are used for the essential parts of the instrumentation system for measuring pH and antifoam and control. Peristaltic pumps are basically used and the flow rate is normally fixed with a timed system of control. Flow rates will depend on the bore of the tubing used and it is a good idea to move the tubing every few days so that one piece does not become worn out by the pump rollers. A dip tube goes to the bottom of the bottle. A filter in the top of the reservoir bottle is kept open during autoclaving to prevent a build-up of pressure. For extremely high accuracy of addition, the reservoir bottles are placed on analytical balances, which can be used to determine how much reagent has been pumped in a given time.

14.7.2 Medium Feed Pumps and Reservoir Bottles

Medium feed pumps are often used with variable speeds so as to give the maximum possible range of feed rates. The reservoir bottles must be changed several times during a long continuous culture experiment, so it is usual to fit an aseptic coupling in the tubing. An effluent pump is often used to remove culture fluid from the fermenter vessel into a storage reservoir bottle.

14.7.3 Rotameter/Gas Supply

A laboratory air supply is provided from a separate pump, which must be oil free. A variable area flow meter or rotameter is used to control the airflow rate into the fermenter vessel. A pressure regulator valve is also attached before the rotameter, which ensures safer operation. A sterile filter (usually 0.22 μm) is used as a bridge between the tubing from the rotameter and that connected to the air sparger of the fermenter.

14.7.4 Sampling Device

This allows culture fluid to be removed aseptically during the fermentation at intervals decided by the user. The frequency of sampling and the size of each sample are determined empirically according to the needs of the experiment.

14.8 ADDITIONAL ACCESSORIES

14.8.1 Feed Pumps

For fed-batch or continuous applications a separate feed pump is used. The type of pump depends on the size of the vessel, the required flow rates, the nature of the additive (e.g. viscous, vorrosive, etc.) and the speed at which the addition should be made.

14.8.2 Peristaltic Pump

The pumps used for medium feed are typically variable speed units, which can accept signals from an external source such as a computer or process controller to allow adjustment of feed rates based on, for example, respiratory quotient (RQ) or optical density (OD). This is the most common type of addition pump used in fermenter applications. The key advantages of peristaltic pumps are as follows:

1. The liquid being pumped is not contaminated by any part of the pump.
2. The liquid being pumped does not contaminate the pump.
3. A wide range of flow rates is possible using tubing of different bores.
4. The accuracy of delivery is good over a wide range of flow rates.
5. The pump can run safely even when the tubing is empty.

The disadvantages are:

1. The flow is pulsed and cannot 'push' against high backpressure.
2. The tubing wears over time and flattens, reducing accuracy and increasing the risk of bursting.
3. Silicone tubing is not suitable for all liquids and alternative tubing material may be too rigid to allow correct peristaltic action.

The problem with tubing wear can be partly overcome by leaving extra length in the tubing when connecting it to the fermenter vessel and moving a new length into the pump head every few days to stop wear in just one small section.

14.8.3 Syringe Pump

This type of pump is typically used in medical applications for the precise delivery of drugs. Most now allow for computer control. In this case, a syringe of various sizes from 1µl to 100 ml, is filled with the liquid to be added and secured onto the syringe pump. The advantages of syringe pumps are:

1. Very precise and accurate delivery; non-pulsile flow.
2. Very low flow rates possible.
3. Can work against backpressure.
4. Can pump very viscous fluids, e.g. oils.

The disadvantages are:

1. Must be replaced or refilled regularly.
2. Range of flow rates not as wide as with peristaltic pumps.
3. Usually more expensive.

14.8.4 Diaphragm/Positive Displacement Pump

It uses a flexible membrane and soft tubing to displace the liquid and control its flow but also employs a piston and pump heads of fixed volume. A stainless steel pump head is connected to soft tubing and autoclaved. It is then located onto the pump and held securely. Advantages of this type of pump include:

1. Accurate delivery.
2. More choice in type of tubing for corrosive liquids.

Disadvantages are:

1. Less flexibility on flow rate ranges for each pump head.
2. Stainless steel chamber and diaphragm may not be suitable for some liquids.
3. Pulsed flow even more pronounced than with a peristaltic pump.

14.8.5 Exit Gas Analysis

The exit gas may need to be conditioned (e.g. moisture removed) before going into thc analyzer, depending on the type of instrument used.

14.8.6 Infra-red Carbon Dioxide Analyzer

An infrared detector measures the amount of radiation reaching it after the gas absorbs some. This type of detector allows continuous measurement with good sensitivity, accuracy and selectivity.

14.8.7 Paramagnetic Oxygen Analyzer

A dumb-bell of gas-filled spheres is linked to a support mechanism suspended in a magnetic field created by permanent magnets. If oxygen is present in the test gas, its attraction to the magnetic field will cause the dumb-bell to be displaced. The movement is detected using a mirror in the centre of the balance system, which displaces a beam of light shown via the mirror onto a photocell. The signal form the photocell will be proportional to the amount of oxygen present in the sample. A low flow rate is needed for this system to work properly, so inlet gas is typically pumped into the detector cell with most being discarded through bypass pipe -work.

14.8.8 Mass Spectrometer

With the help of mass spectrometer the analysis is rapid, so it allows a multi-inlet system to be used, which 'shares' the analyzer between separate vessels within a bank of fermenters. This is metal plate, which generates a tiny electric current whenever a gas ion strikes it. This signal can be amplified and displayed. By re-tuning the magnetic field very quickly (milliseconds) different ions can be detected and measured in a sequential way, which is so fast that it appears to give an almost simultaneous measurement.

14.9 DIFFERENT TYPES OF FERMENTER INSTRUMENTATION

Various kinds of fermenter instrumentation are being used in fermentation technology. These can range from simple analogue control modules arranged as a 'stack', to powerful, embedded microprocessors which operate as on-board computers to directly operate the complete system.

14.9.1 Analogue Controllers—Rack System

For compact, flexible instrumentation the individual modules for the measurement and control of each parameter are made to plug into a rack system. A bus system in the racking transmits signals from one module to another to allow measured values to be passed to controllers and for one controller to influence another, e.g. oxygen levels altering speed is one of the such examples.

14.9.2 Analogue Controllers — Separate Modules in Housings

The modules may be stacked one on top of another and can be supplied with 'pass through' power connections so that only one power socket is needed. In this each measurement/control module has a separate housing and is operated in isolation. This system allows gradual building up or swapping of instrument modules between several fermenters.

14.9.3 Digital Controllers—Embedded Microprocessor

The measurement sensors link directly to the single control module and several parameters are displayed immediately on a single screen. Control is by direct action on heaters, valves, etc. (Direct Digital Control or DDC). The microprocessor is permanently embedded in the instrumentation and may even be a single chip. Operation is usually via a simple menu system.

14.9.4 Digital Controllers—Process Controllers

The controller is usually 'programmed' using simple commands to input set-point values, etc. Usually from a production control environment, a complete process controller is added to a housing containing all the signal processing and control actuators for the fermenters. It controls in the same way as an embedded controller but is essentially a 'plug-in' component.

14.9.5 Digital Controllers—Direct Computer Control

The processing power and speed of modern PCs made the process more than capable of replacing separate PLCs or embedded controllers. A special Input/Output (I/O) card is needed for the computer which allows the input values to be accessed by the measurement and control software and sends signals out to operate relays, e.g. to turn a heater on or off. In this case, there is no external instrumentation or processor between the actuators and the computer. A printed circuit board with operational amplifiers for the probe input signals is the only electronic part which may be present in the fermenter base unit. The advantage with this system is that the computer display and control software is integrated totally with the fermenter.

14.10 ALTERNATIVE VESSEL DESIGNS

There are various types of alternative vessel designs commonly found for the fermentation processes. Usually, alternative vessel designs are tried when the standard vessel configuration does not allow adequate growth of the organism. Shear forces in fermenters with turbine impellors often disrupt for example animal cells. These are also used when the scale-up criteria require a different design of bioreactor. For example production of large quantities of single-cell protein is cheaper on a large scale if airlift fermenters are used to eliminate energy costs associated with a drive system. A number of these special designs are available at the bench/pilot scale of operation to allow small-scale research into the suitability of a particular method.

14.10.1 Air-lift Bioreactor and Modified Air Lift Bioreactors

The air lift bioreactor works on the basic principle of an air lift pump and may be of either internal loop type (Fig. 14.3) or external loop type (Fig. 14.4). The volume of the reactor is determined from the capacity, kinetic data, and specific growth rate of the organism used. Air flow rate of the reactor depends on the volumetric mass transfer coefficient in the reactor system. The air-lift bioreactor used in the industries has a uniform cylindrical cross section for both riser and down-comer, while the modified air-lift type developed at IIT-Kharagapur has an internal loop or external loop riser configuration, diverging and converging. The change in the configuration improves the O_2 transfer rate *vis-à-vis* mass transfer coefficient for a particular air flow rate. Consequently, to achieve particular dissolved O_2 concentration to satisfy the O_2 uptake rate of particular microbial system, the modified bioreactor requires less air flow. This reduces the operating cost for pumping air through the bioreactor.

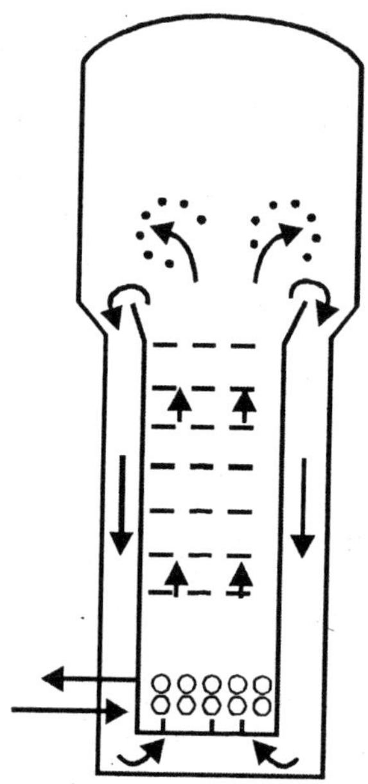

Fig. 14.3 : Air-lift fermenter with internal loop.

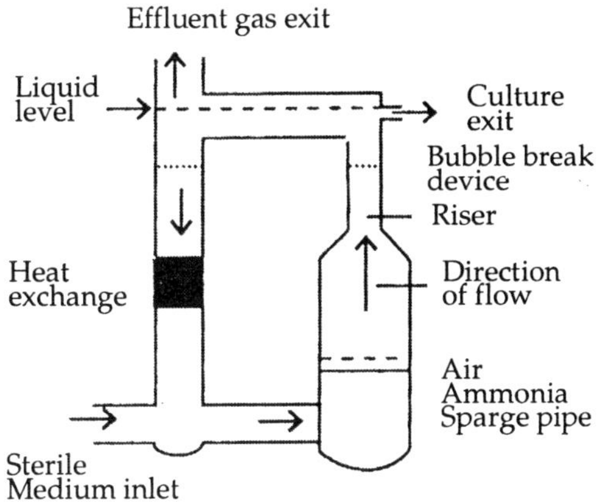

Fig. 14.4 : Airlift fermenter with external loop.

14.10.2 Bubble Columns

A bubble column bioreactor is shown in Fig. 14.5. Usually, the column is cylindrical with an aspect ratio of 4-6 (height-to-diameter). Gas is sparged at the base of the column through perforated pipes, perforated plates, or sintered glass or metal micro-porous spargers. O_2 transfer, mixing and other performance factors are influenced mainly by the gas flow rate and the rheological properties of the fluid. Internal devices such as horizontal perforated plates, vertical baffles and corrugated sheet packings may be placed in the vessel to improve mass transfer and modify the basic design. The column diameter does not affect its behaviour so long as the diameter exceeds 0.1 m. One exception is the axial mixing performance. For a given gas flow rate, the mixing improves with increasing vessel diameter. Mass and heat transfer and the prevailing shear rate increase as gas flow rate is increased. In bubble columns the maximum aeration velocity does not usually exceed 0.1 m s^{-1}. The liquid flow rate does not influence the gas—liquid mass transfer coefficient so long as the superficial liquid velocity remains below 0.1 m s^{-1}.

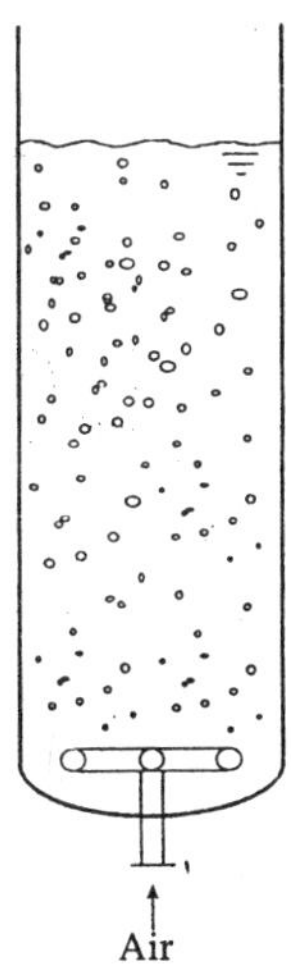

Fig. 14.5 : A bubble column bioreactor

14.10.3 Fluidised Beds Bioreactor

Fluidised bed bioreactors are suited to reactions involving a fluid-suspended particulate biocatalyst such as the immobilised enzyme and cell particles or microbial flocs. An up-flowing stream of liquid is used to suspend or 'fluidise' the solids as in Fig. 14.6. Geometrically, the reactor is similar to a bubble column except that the top section is expanded to reduce the superficial velocity of the fluidising liquid to a level below that needed to keep the solids in suspension. Consequently, the solids sediment in the expanded zone and drop back into the narrower reactor column below; hence, the solids are retained in the reactor whereas the liquid flows out.

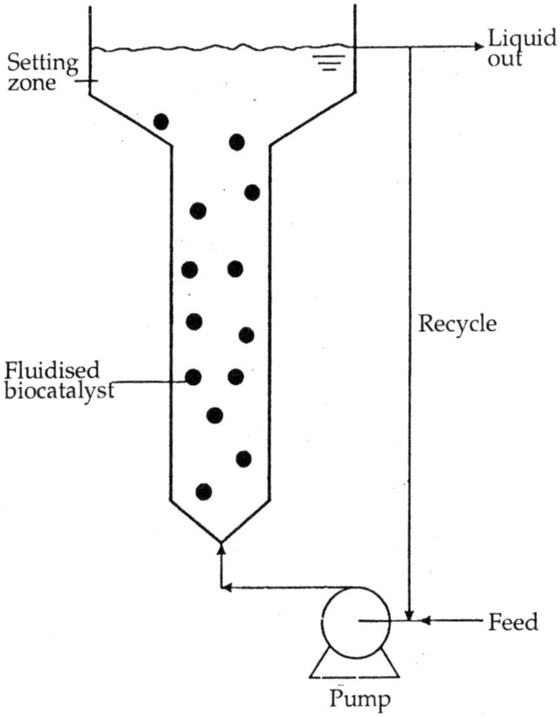

Fig. 14.6 : A fluidised bed bioreactor.

A liquid fluidised bed may be sparged with air or some other gas to produce a gas-liquid-solid fluid bed. If the solid particles are too light, they may have to be artificially weighted, for example by embedding stainless steel balls in an otherwise light solid matrix. A high density of solids improves solid-liquid mass transfer by increasing the relative velocity between the phases. Denser solids are also easier to sediment but the density should not be too high relative to that of the liquid, or fluidisation will be difficult.

Liquid fluidised beds tend to be fairly quiescent but introduction of a gas substantially enhances turbulence and agitation. Even with relatively light particles, the superficial liquid velocity needed to suspend the solids may be so high that the liquid leaves the reactor much too quickly i.e. the solid-liquid contact time is insufficient for the reaction.

In this case, the liquid may have to be recycled to ensure a sufficiently long cumulative contact time with the biocatalyst. The minimum fluidisation velocity i.e. the superficial liquid velocity needed to just suspend the solids from a settled state-depends on several factors, including the density difference between the phases, the diameter of the particles, and the viscosity of the liquid.

14.10.4 Packed Bed Columns

A bed of solid particles usually with confining walls, constitutes a packed bed (Fig. 14.7). The biocatalyst is supported on, or within, the matrix of solids that may be porous or a homogeneous non-porous gel. The solids may be particles of compressible polymeric or more rigid material. A fluid containing nutrients flows continuously through the bed to provide the needs of the immobilised biocatalyst. Metabolites and products are released into the fluid and removed in the outflow. The flow may be upward or downward, but down flow under gravity is the norm. If the fluid flows up the bed, the maximum flow velocity is limited because the velocity cannot exceed the minimum fluidisation velocity or the bed will fluidise.

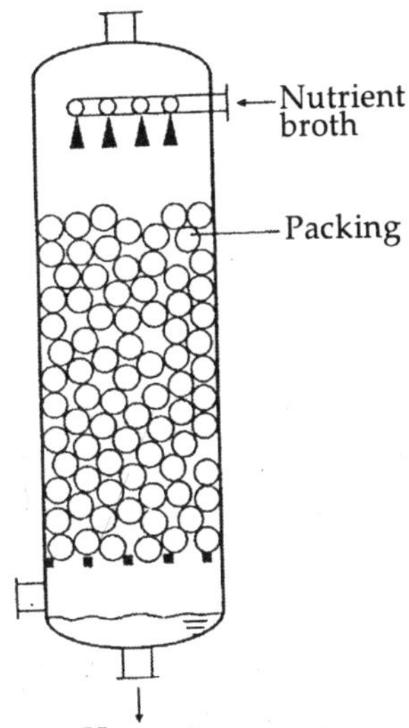

Fig. 14.7 : A packed bed bioreactor.

The depth of the bed is limited by several factors, including the density and the compressibility of the solids, the need to maintain a certain minimal level of a critical nutrient, such as O_2, through the entire depth, and the flow rate that is needed for a given pressure drop. For a given void volume (i.e. solids-free volume fraction of the bed) the gravity-driven flow rate through the bed declines as the depth of the bed increases. Nutrients and substrates are depleted as the fluid moves down the bed. Conversely, concentrations of metabolites and products increase. Thus, the environment of a packed bed is non-homogeneous but concentration variations along the depth can be decreased by increasing the flow rate. Gradients of pH may occur if the reaction consumes or produces H^+ or OH^-. Because of poor mixing, pH control by addition of acid and alkali is nearly impossible. Beds with greater void volume permit greater flow velocities through them but the concentration of the biocatalyst in a given bed volume declines as the voidage (void volume) is increased. If the packing, i.e. the biocatalyst-supporting solids, is compressible, its weight may compress the bed unless the packing height is kept low. Flow is difficult through a compressed bed because of a reduced voidage. Packed beds are used extensively as immobilised enzyme reactors. Such reactors are particularly attractive for product inhibited reactions: the product concentration varies from a low value at the inlet of the bed to a high value at the exit; thus, only a part of the biocatalyst is exposed to high inhibitory levels of the product.

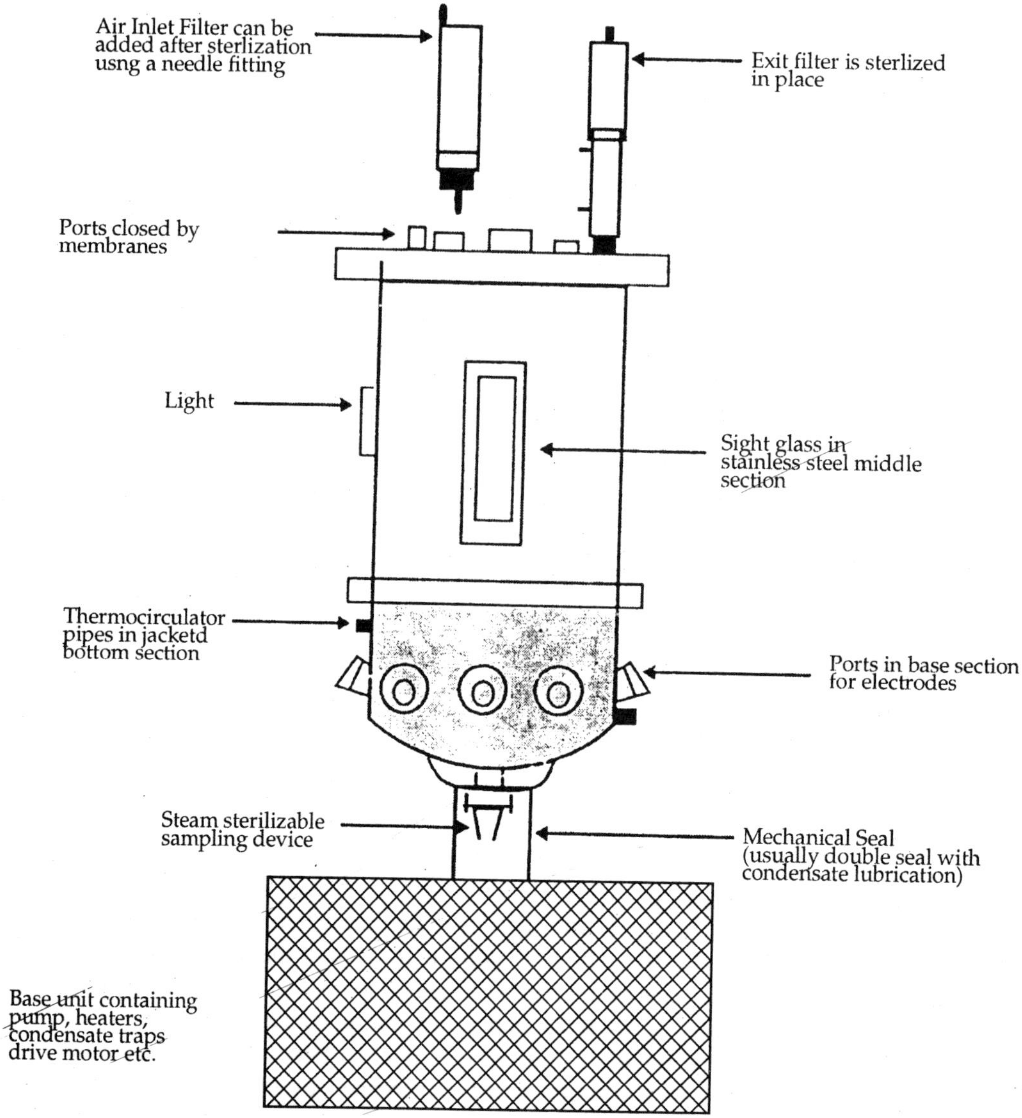

Fig. 14.8 : Illustration of major parts of an *in situ* sterilizable fermenter vessel.

14.10.5 In situ Sterilizable Fermenters

For safety and insurance purposes, vessels above the recommended size are usually made of stainless steel and are sterilized *in situ* using house steam or an electrical steam generator, which may be built into the base unit. The heating for the vessel is normally done through a double jacket, which can either be the full length of the vessel or cover just the bottom. The bottom section contains large port fittings for electrodes and usually has some kind of steam-sterilizable sampling device/harvest valve. The mechanical seal and drive shaft usually enter from the bottom on this size of vessel. In some fermenters, the air inlet filter and connecting pipe to the air sparger are also provided. Major parts of an *in situ* sterilizable fermenter are shown in figure 14.8.

14.10.6 Hollow Fibre

In this cells are embedded in fibres contained in a cartridge, which are bathed in circulating culture medium. This is often used for mammalian cell culture, where anchorage-dependent cell lines can be perfused with oxygenated medium. An extension of this method is to use cartridges containing two different bundles of fibres as a separation 'membrane' between a pair of fermenter vessels and allow dissolved gases and metabolites to be exchanged without cells crossing the barrier to allow, for example, it is used for the study of toxicity or interactive studies.

14.10.7 Novel Seesaw Bioreactor

The design of this type reactor has taken into consideration certain problem inherent i.e., foam formation due to aeration and agitation. This not only disturbs the physical system blocking the flow passages, but the cell mass attaches to the foam, thereby reducing the cell density in the broth. Cells in contact with bubbles may be destroyed. In such cases aeration may be done outside the reactor but, at the increased cost of the system. Novel 'seesaw' bioreactor aeration of broth is done by surface diffusion. This avoids bubbling and reduces the cause of foam formation. Shear effect is also negligible.

This reactor comprises a minimum of two bioreactor chambers and two gas chambers. Each bioreactor chamber is connected to gas chamber. By controlling the gas pressure inside the chamber by means of valve and blower, the gas is alternately forced into each bioreactor chamber, thereby forcing the liquid levels inside the bioreactor to move up and down alternately. The system is a closed circuit one and mainly make-up gas needs to be supplied from outside and, if necessary, excess gas removed. Additional solid surfaces are provided inside the bioreactor chamber for improving the aeration efficiency. Provision for control of pH, dissolved O_2, CO_2 level and temperature can be incorporated.

14.11 PHOTOBIOREACTORS

Certain micro-algae and cyanobacteria provide important chemicals, such as astaxanthin and β-carotene in addition to being used as aqua-culture feed for bivalves and fish hatchlings. Cyanobacteria, such as *Spirulina,* are also grown as human health-foods. Photosynthetic cultures require sunlight or artificial illumination. Although some algae may be grown heterotrophically, i.e. without sunlight, this type of growth requires an alternative organic energy source, usually glucose. Heterotrophically grown cultures often lack photosynthetic pigments and may not yield the same products as a photosynthesising population. However, much progress is being made in developing heterotrophic strains for production of commercially valuable products.

Artificial illumination is impracticably expensive; only outdoor photobioreactors appear to be promising for large-scale production. Open ponds and 'raceways' are often used to culture micro-algae, but mono-septic culture requires fully closed photobioreactors. Because it needs light, photosynthesis can occur only at relatively shallow depths. Algal ponds are typically no deeper than 0.15 m. However, too much light causes photobioreactos; a situation in which slightly reducing the light intensity will actually improve the rate of photosynthesis. With increasing cell population, the self shading effect of cells further limits light penetration. In addition to light, photosynthesising algal cells need a source of carbon, usually carbon dioxide. Part of the carbon may be derived from dissolved bicarbonate species. Cells convert CO_2 to carbohydrate and other cellular components. Too high a concentration of CO_2 can reduce photosynthetic productivity.

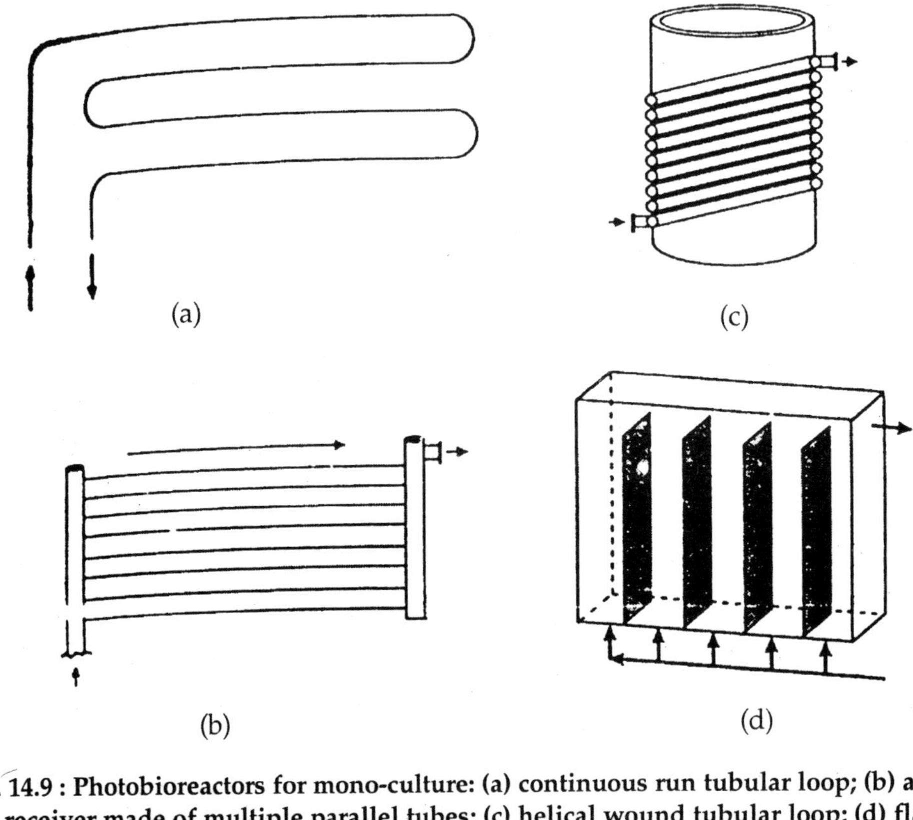

Fig. 14.9 : Photobioreactors for mono-culture: (a) continuous run tubular loop; (b) a solar receiver made of multiple parallel tubes; (c) helical wound tubular loop; (d) flat panel configuration. Configuration (a) and (b) may be mounted vertically, or parallel to the ground.

Closed photobioreactors for monoculture consist of arrays of transparent tubes that may be made of glass, or more commonly, a clear plastic. The tubes may be laid horizontally, or arranged as long rungs on an upright ladder; see Fig. 14.9. A continuous single run tubular loop configuration is also used, or the tube may be wound helically around a vertical cylindrical support. In addition to the tubes, flat or inclined thin panels may be employed in relatively small-scale operations. An array of tubes or a flat panel constitutes a 'solar receiver'. The culture is circulated through the solar receiver by a variety of methods, including centrifugal pumps, positive displacement mono pumps, Archimedean screws and airlift devices. Airlift pumps perform well, have no mechanical parts are easy to operate aseptically and are suited to shear sensitive applications.

The flow in a solar receiver tube or panel should be turbulent enough to aid periodic movement of cells from the deeper poorly lit interior to the regions nearer the walls. Generally, a minimum Reynolds number value of 10^4 is recommended. While turbulence is needed to improve radial mixing, too much turbulence can be harmful. The velocity everywhere should be sufficient to prevent sedimentation of cells. Typical linear velocities through receiver tubes tend to be 0.3-0.5 m s^{-1}.

Because of the need to maintain adequate sunlight penetration, a tubular solar receiver cannot be scaled up by simply increasing the tube diameter. The diameter should not exceed 6 cm, although this constraint may be relaxed somewhat by deploying specially designed static mixers that improve radial mixing inside the tube. Without the mixers, reducing the tube diameter from 6 cm to 5 cm and further to 2 cm noticeably improves culture performance. Light penetration depends on biomass density, cellular morphology and pigmentation, and absorption characteristics of the cell-free culture medium.

Photobioreactors are normally operated in continuous mode. Micro-algae grow slowly; the doubling time of *Chlorella pyrenoidosa* is approximately 9 hours. Consequently, the dilution rate is kept low; about one culture volume per day, confined to daylight hours. Although the biomass grows only during daylight, certain products are produced predominantly during the dark hours. Biomass productivity of outdoor cultures is generally less than 2 gl^{-1} day, often no more than 1.5 g l^{-1} day. Culture is generally carried out at 22-37°C. During daylight hours, the solar receiver tubes need to be cooled to prevent temperature rise to damaging levels.

A solar receiver is more amenable to scale-up by increasing the length of a continuous run tube; however, the maximum permissible length should not exceed 50-100 m because the photosynthetically generated O_2 builds up along the tube and high oxygen concentrations inhibit photosynthetic productivity. An alternative to a long continuous run tube is an array of multiple parallel tubes. For minimal costs, each tube of the multi-tube arrangement should be as long as possible but within the threshold where O_2 levels become inhibitory. The rate of photosynthesis and, consequently, the rate of O_2 generation of the amount of sunlight available. With *Spirulina platensis* under intense artificial illumination O_2 production rates have been estimated at 0.35-0.5 g l^{-1}h for radiation intensity levels of 1500-2600 μE m^{-2}. In a long tube, the rate of photosynthesis varies along the tube because of several factors: (i) the light intensity declines because of increasing cell concentration; (ii) the CO_2 is depleted; and (iii) O_2 concentration may increase to inhibitory levels.

14.12 FERMENTER PREPARATION AND USE

14.12.1 Disassembly of the Vessel

For carrying out the fermentation process, it is necessary to disassemble the vessel. The fermentation unit is shut down from the control unit and transfer lines and cable connections are removed. The vessel should be re-autoclaved and the culture should be disposed of as laid down by departmental safety procedures. The complete vessel including its each and every part is disassembled.

14.12.2 Cleaning

The vessel should be rinsed several times in distilled water to remove any loose culture residues. During cleaning, an examination of any chips or cracks in the vessel glass can be carried out and a replacement can also be made, if necessary. Vessels must be stored clean and dry. In use, any spillages of reagents or medium should be wiped up immediately with a damp cloth and not be allowed to dry out. The pump heads and covers must be thoroughly cleaned if a tube breaks and reagent leaks out.

14.12.3 Preparations for Autoclaving

Before proceeding for the autoclaving, the vessel seal should be removed and lightly greased with suitable silicone grease. On replacing the seal, it must be correctly located so that there is no chance of any part lifting or kinking. All fittings such as pipes etc. are fitted. Ports not in use have 'stoppers' fitted, and their O-ring seals should have a light coating of silicone grease applied. The pH and dissolved oxygen electrodes should be fitted, taking care not to damage them by careless insertion into the port. Both must be tightly capped to prevent moisture getting into the electrical contacts. Reagent bottles are prepared in a similar way to the fermenter vessel.

The exit gas cooler should be fitted to one of the wider ports. The air outlet line must be kept open during autoclaving. A short length of silicone tubing must be fitted to the air sparger inlet pipe with a 0.2 μm disposable filter mounted on top. The tubing between the sparger pipe and the filter must be clamped shut during autoclaving.

14.12.4 Autoclaving

On a steel tray or in an autoclave basket the vessel and reagent/sampling bottles already connected by silicone tubing are assembled together. It should check that all the outlets and inlets are properly closed. A quantity of liquid is certain to be lost during autoclaving (approximately 10%) so either the medium is over-diluted to compensate for this or sterile, distilled water is added afterwards to restore the volume. Some form of indicator such as autoclave tape should be included to provide a warning if the correct sterilization procedure has not been carried out. Autoclaving at 121°C for 30 min is normally considered adequate for vessel sterilization to minimize damage to the constituent chemicals of the medium. The autoclave used must have good pressure equalization during the cooling down phase of operation to prevent medium being boiled off. The vessel and any accessory bottles must be allowed to cool completely before handling.

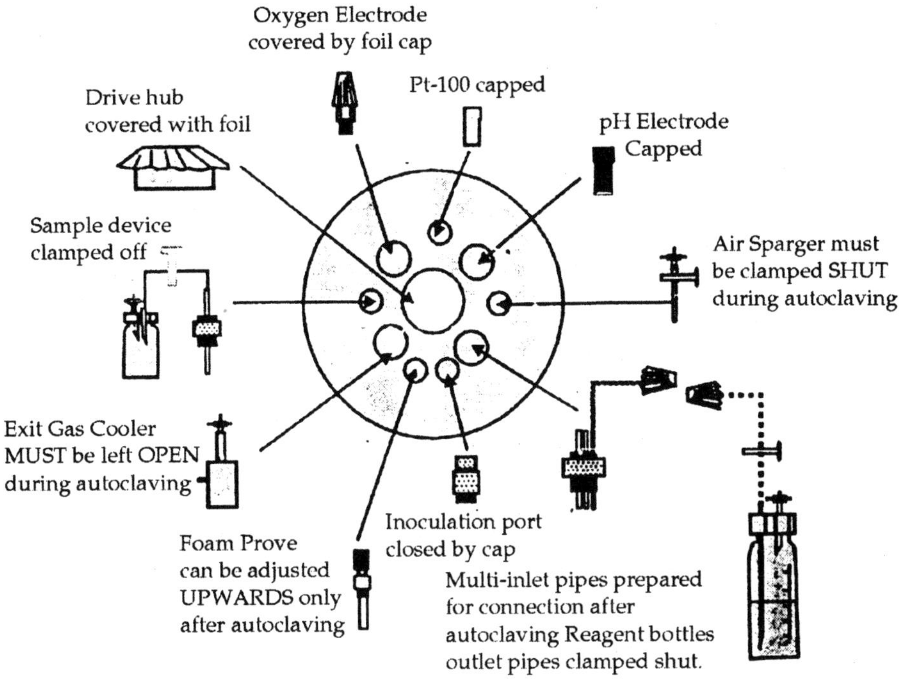

Fig. 14.10 : Illustration of vessels and accessories which are prepared for autoclaving process.

14.12.5 Set-up following Autoclaving

The air sparger is connected to the rotameter by a piece of silicone tubing from the top of the filter to the air outlet of the rotameter. The air sparger line is unclipped between the metal pipe and the air filter. The exit gas cooler is connected to the water supply, either directly or via the fermenter base unit. The tubing for water in, out and drain is connected to the vessel jacket for a water system, and the water turned on so that the vessel jacket is filled. Alternatively, any pads or heater cartridges are connected to the base unit or temperature control modules.

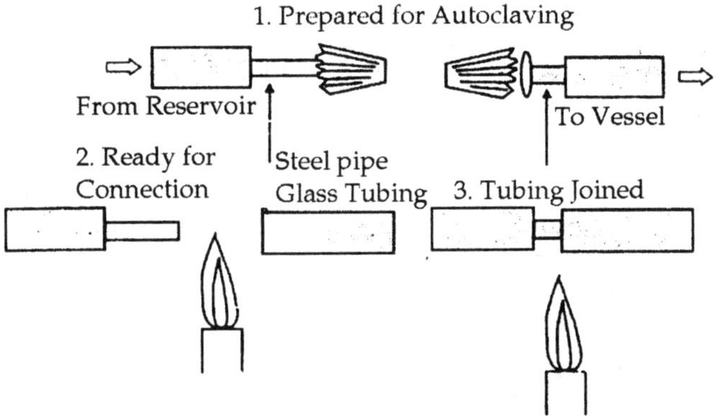

Fig. 14.11 : Illustration of how tubes are connected aseptically.

Aseptic connections (Fig. 14.11) must be made first if the reagent bottles were autoclaved separately from the vessel. The clamps are removed so liquid can flow freely. The Pt-100 temperature sensor is connected to the control module and the pH electrode by removing the shorting cap and screwing in the cable. The dissolved oxygen electrode is connected to the appropriate cable. Connections to the foam probe are made, usually one wire into the electrode and one on the vessel top plate to make a circuit. Setting the temperature control at this stage will ensure that fermenter is ready to inoculate. The maximum stirrer speed to be used and the maximum airflow required on the rotameter are set. The fermenter is now ready to inoculate.

14.12.6 Inoculation of a Fermenter Vessel

The inoculum, which should normally be not more than 5-10% of the total culture volume, is aseptically transferred (Fig. 14.12) to a sterile, disposable syringe of a suitable size. The port fitting is removed and held vertically to prevent contamination of the bottom end. The syringe needle is quickly pushed through the membrane and the inoculum transferred into the vessel. The vessel may be actively aerated during this procedure to minimize the risk of a contaminant getting into the vessel. The syringe needle is quickly withdrawn and the silicone membrane reseals. The port fitting is now replaced. Aseptic connection can be made to an inlet pipe.

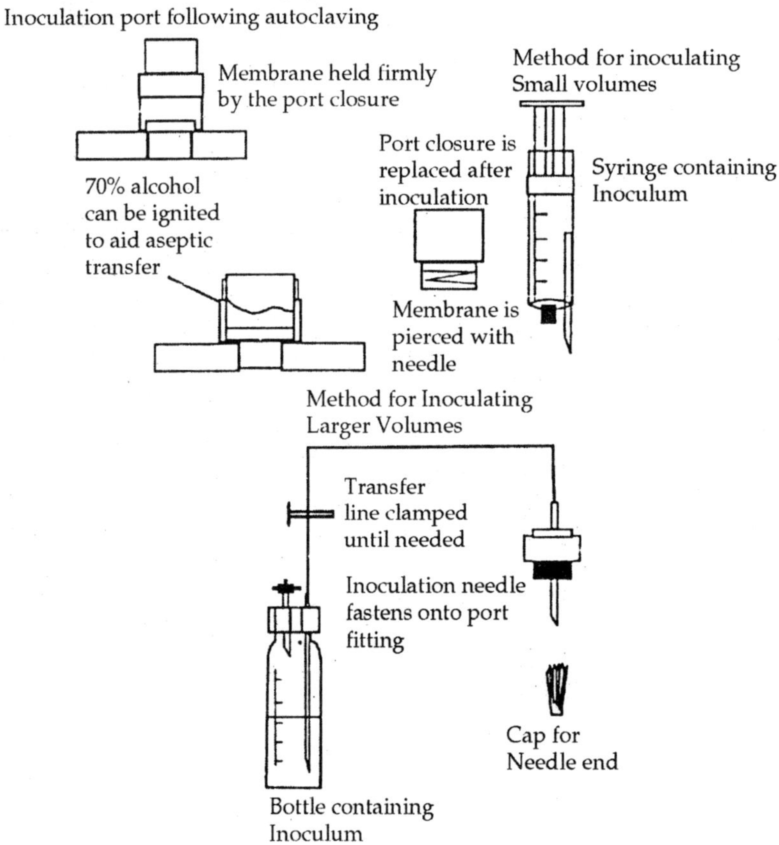

Fig. 14.12 : Illustration of inoculation process of a fermenter vessel.

14.12.7 Sampling from a Fermenter Vessel

Aeration is stopped during this process to prevent 'surging' of culture into the syringe. This method can be prone to contamination and is not suitable for organisms with any risk to operators. This sampling device is a bottle, which has two metal needles/pipes permanently, fixed through both the metal and rubber seals of its cap. A syringe is fitted to the air filter after autoclaving. The sample device is usually attached to the vessel top plate so that the glass bottle hangs down vertically beneath the cap and a supply of bottles of the same size are autoclaved ready for use. When a sample is to be taken, the clamp on the sample pipe tubing is released, and the syringe pulled back to create a partial vacuum in the sample bottle. Culture will flow into the sample battle. When the required volume of sample has been taken, the syringe is pushed into clear the line and the sample tubing is clamped shut again. The sample bottle with the culture is quickly removed and capped with a sterile cap from a new glass bottle, which replaces it under the sampling device. Aseptic sampling is explained in figure 14.13.

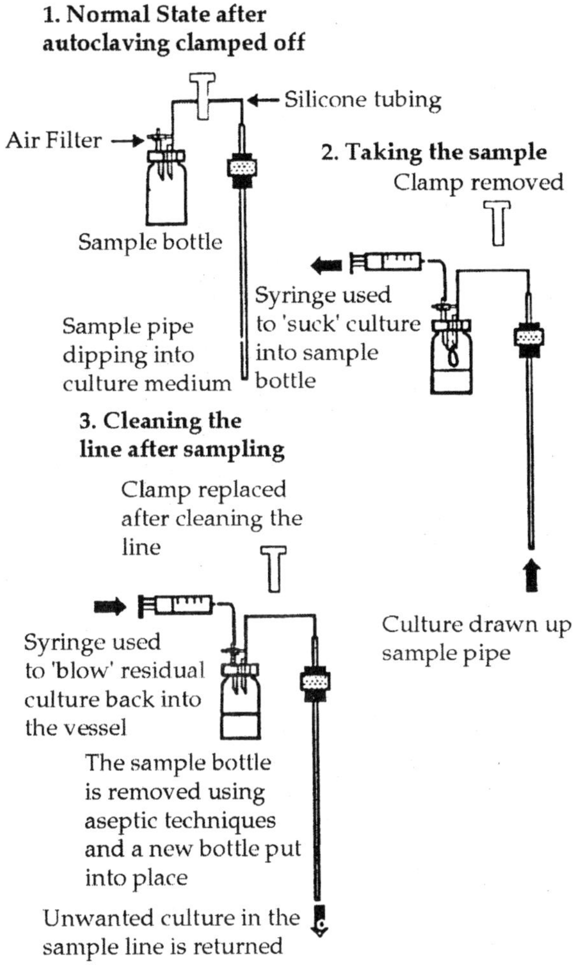

Fig. 14.13 : Illustration of sampling.

14.13 MAINTENANCE OF ASEPTIC CONDITIONS

Bioreactor should remain contamination free. The contamination is caused by microorganisms and results in a variety of consequences e.g., depletes nutrients, nature of the final product, reduction in yields, degradation of the desired metabolite, interference with downstream recovery etc. Therefore, the bioreactor should be designed in such a way so that there should be minimum or no contamination occurs. Integrated efforts should be made while designing bioreactor, use of materials and layout, effective operating procedure, scheduled checks and trained personnel. Contamination may be minimized by using sterile inoculum; adequately sterilized nutrient medium, the design, material and fabrication of the bioreactor (aeration and agitation systems, sampling and monitoring ports, and foaming) air liquid transfers. The operation of sterilization and aseptic additions of industrial plants involve complex procedures and perhaps, manipulations of hundreds of valves. The following are some of the important factors to be considered for an aseptic fermentation:

1. For proper aseptic condition in the bioreactor, regular preventive maintenance of vessels is essential.
2. The fermenter must be designed properly in such a way that the fermenter and its contents are steam sterilizable.
3. Entry and exit of steam should be proper.
4. The agitated drive shaft should have double mechanical aseptic seal. Seals should be of best quality to withstand with host water lubrication.
5. Filtered and sterilized air should be supplied to bioreactor.
6. The vessel should be operated at positive pressure.
7. Addition or transfer of liquid systems in a vessel should be designed in aseptic manner.

14.14 DESIGN FOR STERILE OPERATION

Most commercial fermentation processes are mono-cultures. To establish and maintain aseptic conditions are vital for the success of these processes. Hence, a bioreactor must be sterilised prior to inoculation and contamination during operation must be prevented. Contamination during culture is a common cause of process failure.

14.14.1 Sterilisation-In-Place [SIP]

A bioreactor intended for *in situ* sterilisation requires a complex arrangement of pipework, valves, and filters to enable initial sterilisation and maintenance of sterility. A typical arrangement for *in situ* sterilisation is shown in Fig. 14.14. Because almost all biopharmaceutical production processes involve aeration, the figure includes aeration and exhaust groups that must also be sterilised. The air inlet and exhaust lines have *in situ* steam-sterilisable gas filters. Typically, hydrophobic membrane cartridge filters are used. These filters are rated for removing particles down to 0.45 µm or even 0.1 µm. Often the gas streams have two filter cartridges in series; with the first serving to protect the final filter.

A good system is designed so that the different sections can be sterilised independently of any of the others, thus sterilisation of any section during fermentation can be carried out if, and when, required. Saturated clean steam (1.1-1.4 bar gauge) is used for sterilisation. The air inlet and exhaust groups are sterilised first, and then, in a second step, the bioreactor. The system is designed so that the filters, valves and the associated pipe-work reach sterilisation temperature (~121°C) very quickly (~1 minute), and are held at the temperature for the required time (25-30 minutes).

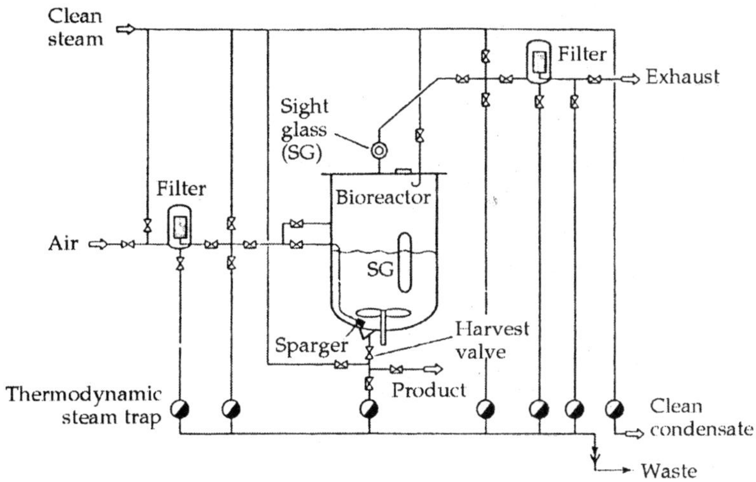

Fig. 14.14 : A bioreactor (fermenter) with air inlet and exhaust groups arranged for in-place sterilization with steam.

Apart from the harvest valve, all other valves shown in Fig. 14.14 should either be diaphragm or pinch valves. The harvest valve is usually a piston valve with a metal bellows sealed stem. The valve closes flush with the internal surface of the bioreactor and there is an unobstructed flow path through the valve body. The valves may be operated manually, but pneumatic operation under automatic control is more efficient and reproducible. The bioreactor is sterilised either filled with the medium or without. Empty sterilisation is the norm in cell culture applications where the media are invariably heat sensitive. In this case, filter sterilisation is used to sterilise the medium.

Proper closing and opening sequence of the various valves is important for attaining sterility and preventing recontamination from the adjacent non-sterile areas. Once the sterilising steam supply is shut off, the bioreactor is immediately pressurised with sterile air through the air inlet filter so that any leakage from the outside to the sterile vessel is prevented. In bioreactors with stirrer or foam breaker shaft penetrations, the shaft seals require suitable piping and valves for steam sterilisation and maintenance of a sterile barrier fluid between the contents of the fermenter and the outside.

14.14.2 Clean-In-Place considerations [CIP]

Industrial bioreactors and much of the other processing equipment are cleaned in-place using automated methods. Automation ensures consistency of cleaning and reduces down-time (i.e. unproductive time of a machine). Attaining an acceptable state of cleanliness is essential to prevent contamination and cross-contamination of biopharmaceuticals and food products. An effective and trouble-free cleaning capability requires attention to design of the bioreactors and the clean-in-place (CIP) systems. At any given time a plant may have several bioreactors at different stages of processing and some empty reactors which need to be cleaned along with any associated transfer piping. The CIP devices and procedures must be matched to the specific configuration of the bioreactor and to the fermentation process to ensure satisfactory cleaning. Generally, a bioreactor which has processed hybridoma or other animal cell culture broth is far easier to clean than one which has processed broths of *Streptomyces* or mycelial fungi such as *Penicillium.*

Design aspects

To ensure removal of solid particles and avoid sedimentation, the minimum flow velocity through piping should be 1.5 ms^{-1}, but a higher value of 2.0 m s^{-1} is preferred. In addition, the piping should be free of dead spaces as much as possible; if unavoidable, the depth of the dead zone must be less than two pipe diameters to ensure adequate cleaning using CIP techniques. Only valves with a metal-bellows-sealed stem, or diaphragm and pinch valves are recommended as all other valves carry a significant risk of contaminating reactors with accumulated debris during the final rinse cycle. For adequate cleaning, the CIP solutions are sprayed into the reactor through one or more removable, static or dynamic spray balls, or dynamic spray nozzles (see Fig. 14.15). In addition, the piping for air exhaust, which is upstream of the exhaust gas filter, and the air inlet piping, should also receive the cleaning solutions. For cleaning with jet spray, pressures of 308 to 377 kPa (absolute) are optimal. Permanently installed spray heads are not recommended for bioreactors because of potential difficulties with sterilisation. These devices must be inserted into the reactor through one of the ports on the head plate. Often, the spray heads are designed to spray the upper one-third of the tank and the falling liquid film irrigates the remaining surface.

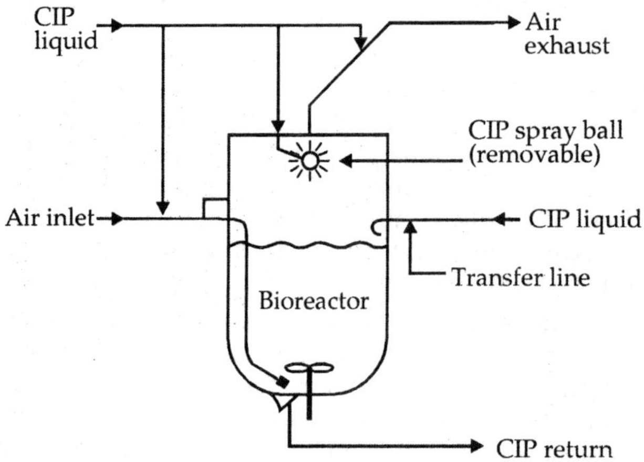

Fig. 14.15 : Delivery of the clean-in-place (CIP) liquid to the bioreactor (fermenter). The flow of CIP solutions is sequenced through the transfer line, the air inlet and exhaust groups and the spray ball of the bioreactor.

For bioreactors for parenteral (injectable) products and other biopharmaceuticals, potable quality deionised water is recommended for all pre-rinsing and detergent formulations. Pre-rinse should be on a once-through basis without recirculation. A five minute pre-rinse, is usually sufficient for bacterial, yeast and animal cell culture reactors. Following pre-rinse, 1% (w/v) NaOH at 75-80°C should be circulated through the equipment so that all product contact surfaces are exposed to this solution for 15-20 minutes. The alkali should be discarded afterwards. Dilution, contamination with soil and microbial spores that can survive for long periods and loss of quality definition of the starting material for the next cleaning, are some of the arguments against re-use of cleaning chemicals. A deionised or reverse osmosis water rinse at 25-35°C is used to remove all alkali from the system. Process equipment for products that are injected into the body must undergo a final wash with hot water-for-injection grade water. This ensures that all residual water complies with the requisite quality standards.

In mechanically agitated bioreactors, the spray of cleaning solutions may be unable to achieve proper cleaning of the agitators, magnetic couplings, mechanical seals and the lower portions of baffles. Therefore, filling of the vessel to at least above the level of the lowermost impeller and agitation at impeller. Reynolds numbers of 10^8-$10^{8.5}$ is recommended during pre-rinse, alkali recirculation and the final rinse. Agitation for 2-3 minutes is sufficient to dislodge adhering dirt or soil. These recommendations assume that reactors are being cleaned in-place soon after use and caking of dirt has not occurred.

14.15 SUMMARY REPRESENTATION OF FERMENTERS (BIOREACTORS)

The wealth of fermenter (bioreactor) designs can be referred to basic types. Figure 14.16 gives a summary representation of fermenter (bioreactor) variants that have become known.

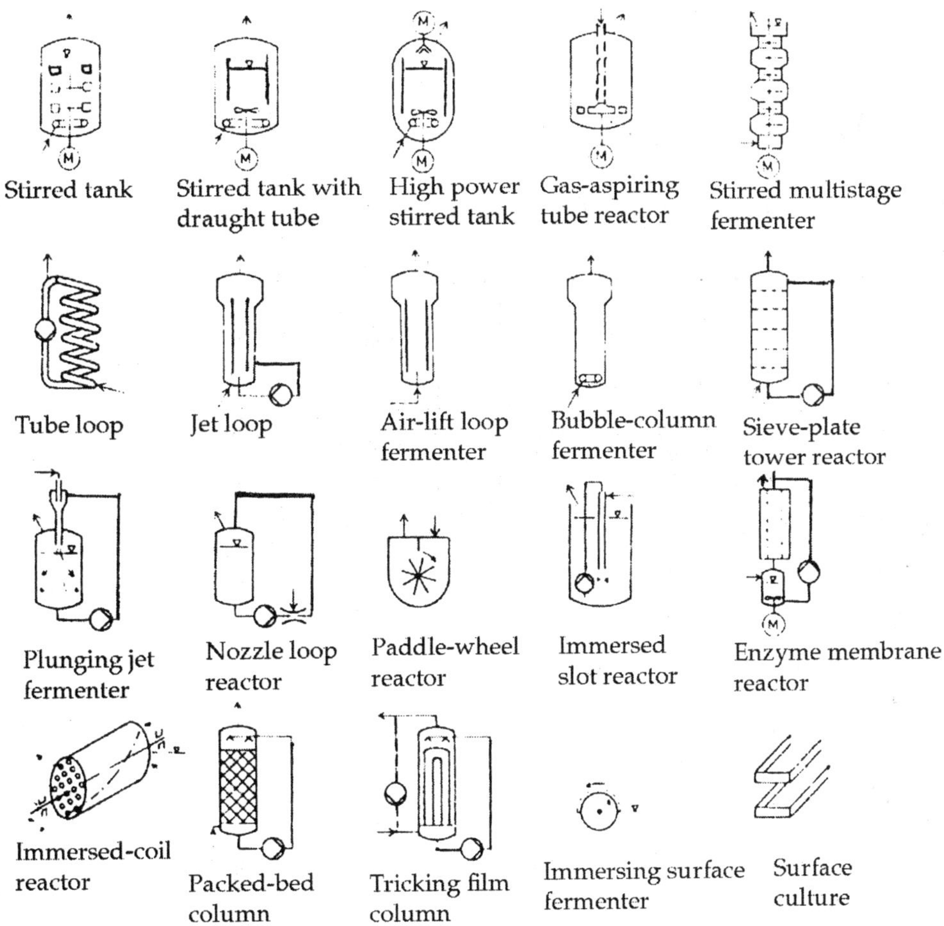

Fig. 14.16 : Types of reactors.

15

MEASUREMENT AND CONTROL OF BIOPROCESS PARAMETERS

15.1 INTRODUCTION

For any fermentation in a fermenter, measurement and control of various types of bioprocess parameters are essential to take care of smooth product formation. Offline and online measurements and control methods for many fermentation variables have been developed through years. In aerobic fermentation the bioprocess parameters which need measurement and control include: aeration-agitation, pH, temperature, foam, CO_2, redox level etc. The obvious need for regulation and control of oxygenation in oxic bioprocessing is well known.

In all processes there arises the need of keeping flows, pressures, temperatures, compositions etc. within certain limits for reasons of safety or specification. This is accomplished simply by measuring the variable it is required to control (the control variable), comparing this measurement with the value at which it is desired to maintain the controlled variable (the desired value or the set point) and adjusting some further variable (the manipulated variable) which has a direct effect on the control variable until the desired value is obtained. Thus automatic and efficient process operation is obvious.

The objectives of applying instrumentation to fermenters are:

(1) to analyze the process status, and

(2) to establish an optimum set of environmental conditions in order to maximize specific cellular or enzymatic activities.

15.2 TYPES OF INSTRUMENTATION USED IN BIOPROCESS ANALYSIS

Analysis of conditions in a biological process is performed by two general methods; namely, off-line operating instruments and on-line operating instruments.

15.2.1 Off-Line Operating Instruments

Off-line operating instruments are not connected directly to the fermenter or biochemical processor; consequently, sampling, sample preparation or treatment, and sample analysis are performed and the analysis results provide information on the process status. However, the elapsed time between the sampling and the assay results can be significant. Information obtained by this method is for the most part useful for postexperimental evaluation of the process but cannot normally be used for process-control purposes. Current examples of off-line operating instruments include spectrophotometers, mass spectrometers, gas chromatographs, enzyme activity analyzers, and electron microscopes, as well as specialized wet chemical and microbiological procedures for product analyses (e.g., reducing sugar or dry weight measurements). Recently, there has been a trend toward connecting certain of these instruments to fermenter units by means of automatic samplers and in-line sample treatment devices, using electronic conversion of the analyzer signals for data logging. A typical example of such an arrangement is the application of a Technicon Auto Analyzer for measuring in-process sugar levels where the elapsed time between sampling and results (approximately 20 minutes) may allow the data to be used for process-control action in certain biological processes.

15.2.2 On-Line Operating Instruments

On-line operating instruments are directly connected to the fermenter or biochemical processor unit and have the capability to sense a specific variable which directly or indirectly reflects a process condition and/or the performance characteristics of a controller. It is noted here that there are relatively few on-line direct sensing devices which give indication of the process status at the cellular level (e.g., NAD—$NADH_2$ measurement belongs to this category). The predominant types of on-line operating instruments indirectly reflect a process condition in the whole broth culture; e.g., a dissolved oxygen sensor connected to an analyzer indicates the dissolved oxygen level in the broth, which is a consequence of the metabolic activity of the culture and the oxygen transfer capacity of the fermentor system. Most of these instruments therefore are reflecting the performance characteristics of a controller which interacts with a process variable. A dissolved oxygen sensor connected to a dissolved oxygen controller reflects the performance characteristics of the method by which the transfer of oxygen from the gas phase into the liquid phase is performed while the culture consumes the available oxygen.

15.3 SIGNIFICANCE

The yield of microbial products depends significantly on the strict control of technological parameters and a stringent adherence to optimum technology. Laboratory, pilot-plant and industrial establishments have therefore to be equipped with reliable measuring, recording and control devices. Measuring and control equipment has considerable economical importance since it makes it possible to optimise processes, saves labour, and increases the demands for technical qualification of the personnel.

Significant changes have recently taken place in the research and development of microbial processes. Large series of poorly equipped laboratory and pilot-plant fermenters are being replaced by small series of highly automated fermenters. This trend requires better knowledge of the mechanism of microbial processes and the development of suitable sensors and measuring devices. The current technological level practically excludes manual regulation and evaluation of a large number of diverse quantities. Contemporary fermenters should meet the following demands:

a) ensure full evaluation of conclusions from control and experimental cultures;
b) facilitate interfacing the fermenter to a computer which may control the process and automatically evaluate all data;
c) reduce operation costs and increase the applicability of the acquired information.

The efficient control and further development of any microbial process requires knowledge of the maximum possible number of parameters, each with a statistically valid set of data and with minimum delay. This can be achieved only by automating the process, or at least by mechanising most of the analyses. Automation may still be limited by the lack of suitable sensors and detectors (Humphrey, 1971, 1977). Sensors should have the following features: a) accuracy, b) reliability, c) reproducibility; the last two features are the most important. The experimental assembly should be flexible enough to ensure both relatively simple measurements accomplished by a small number of mechanised sensors, and fully automated processes including automatic detection of multiple parameters by sensors connected to a computer.

Sensors can be classified according to a number of criteria (Röhr and Hampel, 1980) with regard to the

a) location of sensor—in the bioreactor, at the bioreactor, outside the bioreactor;
b) mode of data acquisition—discontinuous, continuous;
c) kind of environmental factors—physical, chemical, biological, molecular-biological;
d) kind of data acquired—primary (direct) data, secondary (derived) data.

A survey of physical, chemical, biological and molecular-biological factors and modes of their measurement is given in Tables : 15.1, 15.2 and 15.3.

Table 15.1 : Physical environmental factors (adapted from Röhr and Hampel, 1980)

	Measured factor	Mode of measurement
1.	Culture liquid volume (V_L)	a) Calibrated sight glass (V_L via H_L)
	Culture liquid mass (M_L).	b) Capacitance probe
	Culture liquid density (Q_L)	c) Differential pressure cell (V_L via Δ_p)
		d) Pressure gauge (M_L)

contd. ...

contd. ...

	Measured factor	Mode of measurement
2.	Foam level (H_F)	Foam electrode
3.	Agitator speed (n)	a) Tachometer b) Stroboscope
4.	Agitator shaft power (P) Agitator shaft torque (M)	a) Watt meter b) Torque meter c) Strain gauge
5.	Pressure (p)	a) Bourdon-type manometer in exit line b) Diaphragm manometer
6.	Temperature of culture fluid (T) Temperature of cooling fluid (T)	a) Mercury thermometer b) Thermocouple c) Thermistor d) Metal resistance thermometer
7.	Culture fluid dynamic viscosity (μ_L)	No on-line measurement used yet
8.	Gas (air) flow rate (FA)	a) Rotameter b) Mass flow meter c) Calibrated orifice plate
9.	Liquid flow rate (s)	a) Various flow meters b) Weight measurements at addition vessels c) Mechanical signals of metering devices (pumps, etc.)

Table 15.2 : Chemical environmental factors (adapted from Röhr and Hampel, 1980)

	Measured factor	Mode of measurement
1.	pH	pH meter with sterilisable
2.	rH	electrodes
3.	Dissolved oxygen tension DOT (PO_2) Dissolved oxygen concentration (CO_2)	a) Steam sterilisable dissolved oxygen electrode-polarography b) Steam sterilisable dissolved oxygen electrode —voltammetry c) Tubing-method
4.	Dissolved carbon dioxide tension (pCO_2) Dissolved CO_2 concentration (CCO_2)	Dissolved carbon dioxide electrode

contd. ...

contd. ...

	Measured factor	Mode of measurement
5.	Concentration of carbon source (s)	a) Automatic analyser systems b) Enzyme electrode
6.	Total organic carbon in cell-free broth (TOC) Chemical oxygen demand (COD) in cell-free broth	Automatic analyser system
7.	Concentration of nitrogen source	a) Automatic analyser systems b) Ion-selective electrodes
8.	Concentration of products in broth	a) Automatic analyser systems b) Enzyme electrodes
9.	Oxygen concentration in exhaust gas	a) Paramagnetic oxygen analyser b) Oxygen electrode
10.	Carbon dioxide concentration in exhaust gas	a) Infrared carbon dioxide analyser b) Carbon dioxide electrode
11.	Concentration of organic substances in exhaust gas	a) Mass spectrometer b) Gas chromatographic detectors c) Catalytic sensors

Table 15.3 : Biological and molecular-biological environmental factors (adapted from Röhr and Hampel, 1980)

	Measured factor	Mode of measurement
1.	Cell concentration (X) Cell density (N)	a) Turbidimeter b) Spectrophotometer c) Fluorimeter d) Nitrogen analyser e) Carbon analyser f) Phosphorus analyser
2.	Concentration of nucleic acids in cells	No sensor available to date
3.	Concentration of protein in cells	Automatic analyser
4.	Concentration of NAD/NADH in cells	Reflexion fluorimeter
5.	Concentration of nucleoside phosphates in cells	No sensor available to dale

15.4 TEMPERATURE

(A) Temperature Measurement—Temperature may be measured by (a) thermometers, (b) thermocouples, (c) thermistors, (d) resistance thermometers.

Among these instruments, the most often used are mercury thermometers. In most cases they have to be placed in wells or pockets with contact liquid such as mineral oil which causes a considerable delay between the thermometer reading and the actual attainment of a given temperature in the culture. In addition, the thin glass of the thermometers breaks easily and drops of mercury may then contaminate the medium.

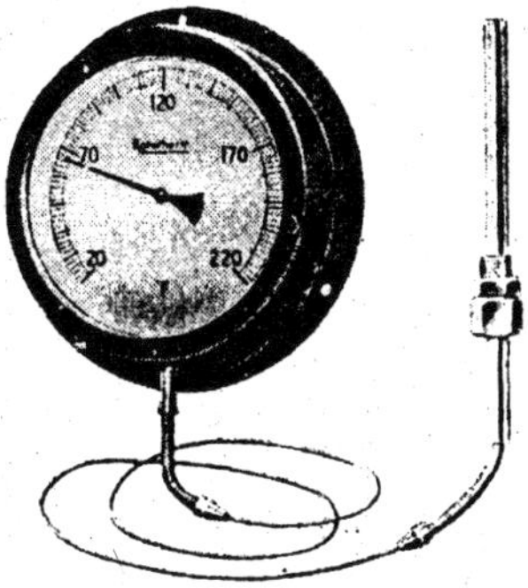

Fig. 15.1 : Direct reading thermometer

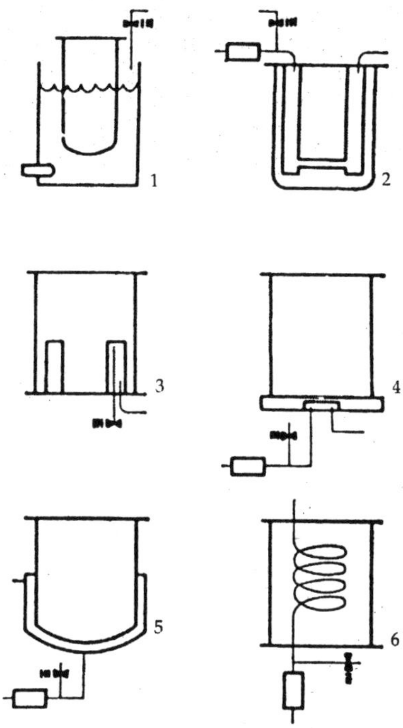

Fig. 15.2 : Methods of temperature regulation in fermenters (Solomons, 1969). 1. Water bath, 2. Water flow through baffles, 3. heating and cooling chamber, 4. heat exchanger in fermenter bottom, 5. jacket, 6. coil.

(a) A direct-reading thermometer (Fig.15.1) has a more robust construction and may be placed directly in the fermenter or in a pocket according to the sterilisation procedure used (direct or autoclave sterilisation). Bi-metal thermometers have a similar construction; they are small, are encased in stainless steel sheaths and increasingly replace glass thermometers since they may be easily detached from the culture device. The measurement precision is 0.5-1 %.

(b) Thermocouples are not used very often for temperature measurements in microbial industries since they usually require a room temperature. Moreover, their output voltage is only 25-40 μV while resistance sensors have a voltage of 2 mV.

(c) Thermistors used for temperature measurements are very cheap and small. With increasing temperature their resistance rises considerably but the rise is not linear. They have a low long-term stability and must therefore be frequently calibrated.

(d) Resistance thermometers (nickel or platinum) display a lower resistance increase with temperature rise than thermistors. However, they have a high stability and the change in resistance is linear. They are routinely used in pilot-plant and industrial fermenters.

Regulation of temperature is usually accomplished by means of various heating elements combined with cooling water pipes controlled by a solenoid valve. Some of the types of temperature regulation used in fermenters are shown in Fig. 15.2.

(B) Temperature control—A thermocirculation system around a vessel jacket has been used for the temperature control because it is the most complex of all the methods. For simple direct systems such as a heater pad, it is simply a matter of fixing the heater, setting the desired temperature and switching on. Cooling is normally via a cold finger and flow of cooling water is controlled via the action of a solenoid valve.

The Pt- 100 sensor provides the feedback signal, which causes the controller to take one of the following actions:

1. Heat at full power as the actual temperature is some way below set point;
2. Pulse the heater power as the actual temperature is close to set point;
3. Turn on the cooling valve as the actual temperature is above set point.

Indicator lights usually show which action the controller is taking at any given moment. A low-voltage DC heater system (e.g. 50 W) is often chosen for safety reasons. A circulation pump and pipework are added to the system for water circulation and any heating is indirect, i.e. on the water circulating in the vessel jacket. In this case, the connections to cold tap water must be made (securely, using jubilee clips or cable ties) and a drain pipe provided from the overflow point to a sink with a clear fall to the drain, i.e. the sink must be the lowest point for the whole length of this pipe.

The water should be delivered from the mains at a minimum pressure of 2 bar and a flow rate of greater than 5 litres per minute. The water hardness should be no more than 50 ppm. The vessel must be connected to the circulation loop (normally by rapid coupling connectors and flexible pressure tubing). Water is first supplied by opening a manual valve until the jacket is filled. The manual valve is then used to maintain just a trickle of water going through the jacket to the overflow to compensate for evaporation losses, e.g. 30-50 drops per minute. (Fig. 15.3).

The heating and cooling is controlled in exactly the same way as a directly heated system. The jacket provides a large surface area in contact with the vessel wall for heat exchange. Good temperature control can be achieved from approximately 5-8°C above the ambient temperature or 5°C above the temperature of cooling water. Counter-cooling with water ensures stable temperature control when operating near ambient temperatures. Measured range is typically from 0 to 60°C (exceptionally upto 90°C).

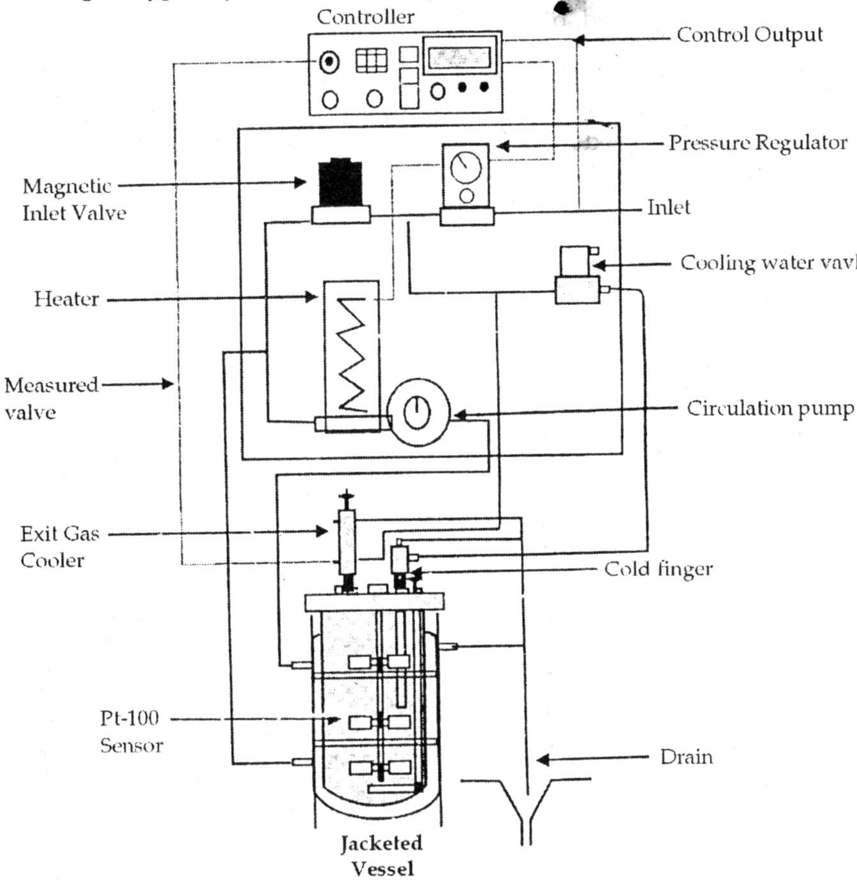

Fig. 15.3 : Illustration of temperature control system using water circulation.

15.5 PRESSURE MEASUREMENT

Static head pressure inside the bioreactor influences the partial pressure conditions and, hence, solubility of gases (O_2, CO_2, in particular) and volatile compounds. A positive head pressure (about 1.2 aim absolute) can also contribute to the maintenance of sterility inside bioreactors operating under aseptic conditions. In addition it is important to measure internal vessel pressure conditions during sterilization as well as the pressure condition of the main supply services (steam, water, and air). Depending on the mode of application, there are two pressure measuring systems generally accepted for biochemical reactors: (1) pressure devices without electrical signal outputs, and (2) pressure transducers. In their applications there are several points to note.

1. At nonsterile points, generally Bourdon manometer gauges are used. For steam pressure measurement a gauge protector is usually employed.

2. For determination of vessel pressure under aseptic conditions a diaphragm is placed between the sterile environment and the gauge.

3. For fermenter head pressure measurement, which also provides an electronic signal useful for on-line analysis as well as for automatic control of a given head pressure when gas flow rates are changed, pressure transducers are required.

(a) Bourdon Manometer

The pressure of steam, water and compressed air may be measured in simple way by a Bourdon manometer. If used for measuring steam pressure, the siphon has to be placed between the manometer and the tubing to protect the Bourdon tube.

The pressure of compressed air is measured only in fermenters for sterile fermentation processes carried out under air overpressure. In small fermenters the measuring set up is simple since it includes Bourdon manometers placed beyond the air outlet filters. The filter must not be too wet since then the manometer reading would deviate from the actual pressure inside the fermenter.

Pressure measurement under sterile conditions is conveniently done by means of **membrane manometers** placed in the blind end of a Bourdon tube. They have a membrane made of stainless steel, placed between two flanges one of which may be situated directly on the fermenter lid. The membrane is easily sterilised. The other face of the membrane is attached to a mechanical system with a pointer moving on a dial, or the pressure from the membrane is transmitted to a reading device *via* a liquid-filled tube.

(b) Pressure Transducers

An example of a typical pressure transducer with an electronic output is shown in Fig. 15.4. Pressure entering the port moves the diaphragm, causing a frictionless floating cone to be repositioned within the magnetic field of a linear variable differential transformer (LVDT). Movement of the transformer cone alters the inductive coupling between the primary and the secondary coils, resulting in a voltage change proportional to the displacement caused by variations in the input pressure. Typical specification for such a device are (a) a range of 3-15 to 0-3000 psi, (b) a linearity of ±0.25%, (c) a stability of 0.01%/°F (from +10°F to 140°F), and (d) a response time of 100 msec. One source of pressure transducer devices is Schaevitz Engineering.

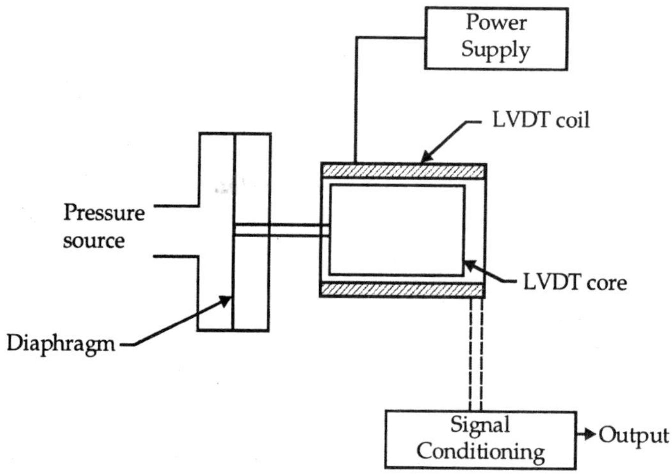

Fig. 15.4 : Schematic of a diaphragm pressure transducer LVDT = Linear variable differential transformer.

15.6 SPEED CONTROL

Speed control basically relies on the feedback from a tachometer, which is located within the drive motor and determines the power delivered by the speed controller to maintain the speed set-point value, which is set by the user. A digital display shows the actual speed in rpm, as determined by the tachometer signal.

A power meter is sometimes included which indicates how hard the motor has to work to maintain the set speed and thereby, indirectly, the viscosity or 'density' of the culture fluid. A DC, low voltage (24-50 V) motor is often used for safety reasons. Speed range is typically from 50 to 1500 rpm (exceptionally up to 2000 rpm).

Where speed is used to control the level of dissolved oxygen, an external signal from the oxygen controller can have an effect on stirrer speed. In this case, an absolute maximum and minimum value for speed can be set on the speed control module to limit the effects of the oxygen controller (which could set either too low a speed and impair mixing or too high a speed and cause excessive foaming) (See figure 15.5).

Agitation Speed Measurement

The measurement of agitation speed can be by either direct or indirect monitoring devices. Indirect monitoring devices include both magnetic and optical methods. Magnetic sensors are noncontact transducers whose output voltage reflects the rate of change that is taking place in their magnetic field. The sensors are easily mountable adjacent to the external drive shaft (in a nonferrous housing) and opposite special gear teeth devices mounted directly to the drive shaft. These devices come in both analog and digital outputs. A typical supplier of such magnetic sensor systems is Electric Corporation, Inc.

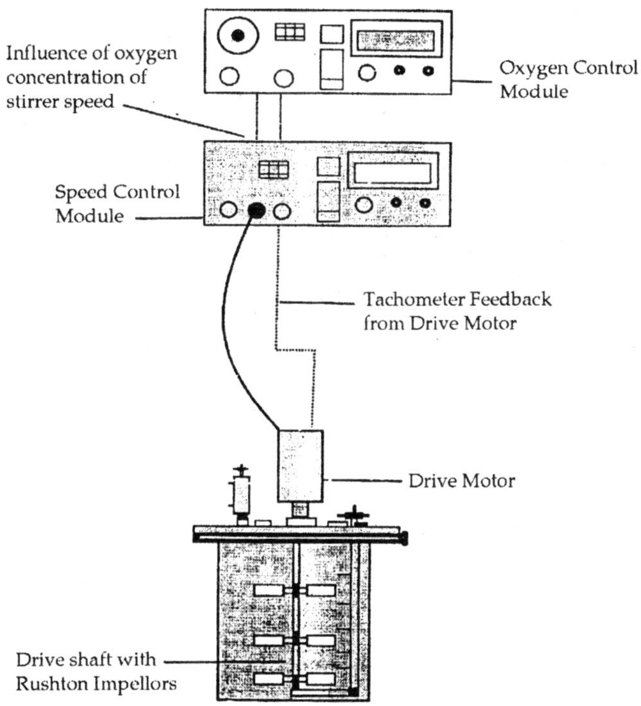

Fig. 15.5 : Illustration of speed control system.

Direct gear or belt-driven tachometers are also available for monitoring of agitator speed but are only advisable for use at low rpm (<250). For higher speed monitoring, indirect magnetic devices are preferred. Note that output meters generally read in revolution per minute (rpm). Caution must be exercised in comparing agitation speed literature data from different experimenters in view of the fact that the diameter of the impeller will determine the absolute rotation speed of the tip of the impeller, which in turn defines the liquid shear rate and the liquid turnover rate in the culture vessel.

15.7 POWER INPUT

The measurement of power input is important in the scale-up of fermentation processes from bench to industrial scale. The pertinent literature mentions only a few fermentation systems equipped with devices for input measurement. In industrial fermenters wattmeters are successfully used for the purpose but in laboratory or pilot-plant fermenters electrical measurements are not accurate. The inaccuracy is mainly due to energy losses in the stuffing box chamber and unfavourable characteristics of the loading factor in small motors. The power input in small fermenters is usually measured by a torsion dynamometer and extensometer.

A **torsion dynamometer (torque meter)** measures the torque moment between the motor shaft and the stirrer shaft. The chief component of the device is usually a shaft placed in a floating bearing connected with a spring or spiral. The torque moment causes a deviation of the spring which is then read on a dial. A commercially manufactured torque meter is shown in Fig. 15.6.

Fig. 15.6 : Torsion dynamometer (Torque meter).

The conductivity of a resistance wire changes with its longitudinal or cross-sectional mechanical stress. This principle is utilised in the construction of extensometers for power input measurement. In contrast to torque meters extensometers are mounted on the shaft inside the fermenter. The measurement of torque moment is accomplished by four strain gauges attached to the shaft and tilted at an angle of 45°. This angle is used since with this arrangement the shear deformation attains a maximum value. The inlet wires to the bridge are brought out of the fermenter through the hollow shaft *via* a system of collector rings.

The theoretical basis of power input measurements has been developed by Aiba *et al.*, (1973).

15.8 FOAM LEVEL

Foam formation complicates considerably the course of fermentation processes. Foam is removed either mechanically or by the addition of antifoam agents (The details are discussed in Chapter 18). Foam detection by sensors is based on **conductance** or **capacitance** principle.

15.8.1 System Based on Conductance Principle

The control of foam is based on its detection by a conductance probe (sensor) in the vessel head space, which leads to the controller delivering a dose of liquid antifoam agent *via* a peristaltic pump.

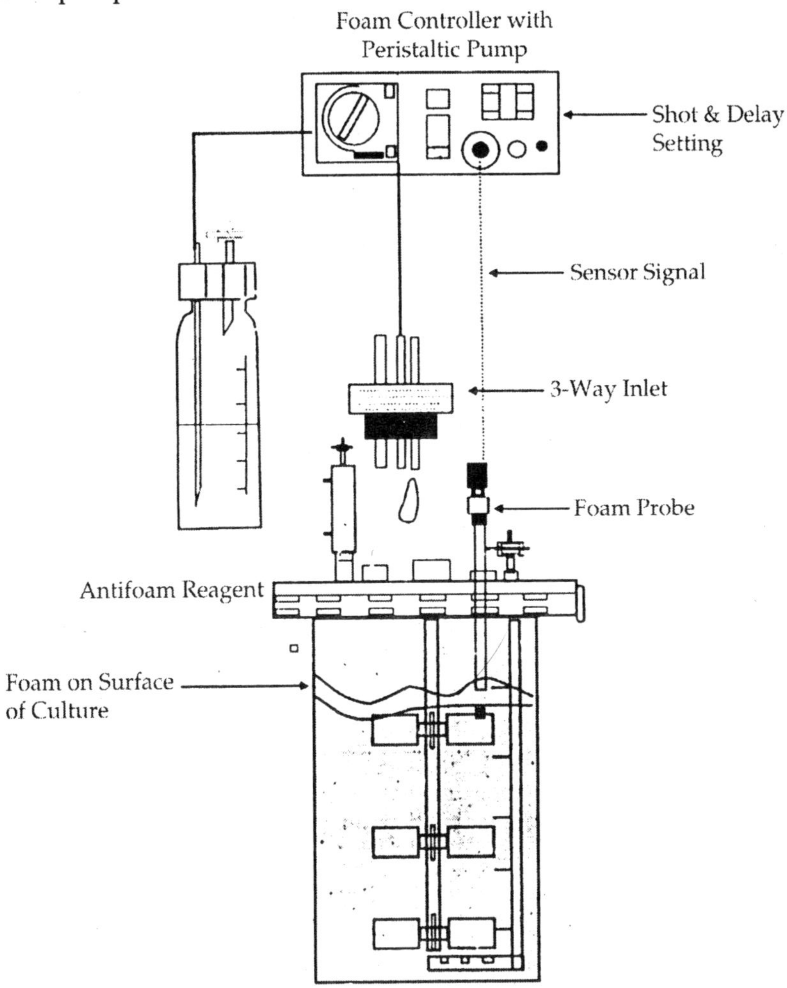

Fig. 15.7 : Illustration of foam control system based on conductance principle.

A delay timer ensures the antifoam agent has adequate time to reduce the foam level before another 'shot' of antifoam is added. The sensitivity of the probe to foam can be adjusted by the user to suit the conditions prevailing in the fermenter. A sheath of inert material around the probe prevents splashes of foam thrown up from the culture from giving 'false positives'. Normally, the metal top plate is used to provide the electrical circuit for the probe to operate so a flying lead is provided which fits into a socket somewhere on the top plate. This principle of foam control system is illustrated in figure 15.7.

If foam is allowed to build up unchecked it can eventually reach the exit gas filter so blocking it and providing a path for contamination, so antifoam reagents are used. Antifoam reagents like mineral oils, vegetable oils or certain alcohols are commonly used in the fermentation process. Commercial preparations are available for use with cultures, which ultimately provide materials for pharmaceutical manufacture. The key thing with using oils is that they can form a 'skin' on the surface of the culture and interfere with gas transfer at the liquid/for interface.

15.8.2 System Based on Capacitance Principle

The probe based on capacitance consists of a metal rod covered by a teflon layer. When the foam reaches up to the lower part of the probe, a solenoid valve or a pump is switched on and an antifoam is fed into the fermenter. These probes are more expensive but less prone to short-circuit than conductance probes.

The antifoam dosage must be sufficient to break foam but must not be too high since it may cause a drop in oxygen transfer rate and complications during the isolation of fermentation products. After the first dose of antifoam the foam may still reach the end of the probe for some time, stimulating further antifoam dosage. For this reason the next antifoam dose is usually added with some delay to prevent a spurious overdosage of the agent. A simple dosage system is shown in Fig. 15.8. Some fermenters are fitted with a reflector plate onto which the antifoam is injected; this arrangement gives rise to a better dispersion of the antifoam on the foam surface.

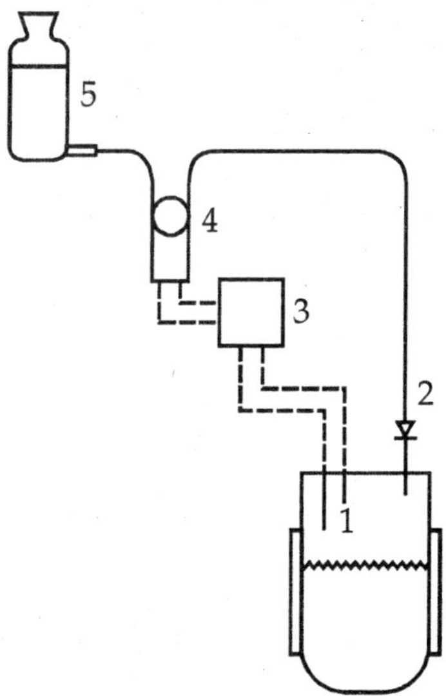

Fig. 15.8 : Scheme of foam control, 1. Probe, 2. antifoam feed, 3. time switch, 4. pump, 5. antifoam reservoir.

15.9 GAS FLOW/SUPPLY

(A) Measurement

The flow of gases such as air, oxygen, carbon dioxide and methane is usually measured by flow meters **(rotameters)**. They are glass tubes with an inner diameter tapering conically fitted with floats in the form of a small ball or hollow cylinder. The accuracy of measurement is usually ± 2% and the ratio of maximum to minimum flow rate is about 10 : 1. Although most flow meters can be sterilised by steam, water often condenses inside them and they should therefore be attached to the fermenter prior to air filtration. The accuracy of measurement depends on the stability of gas pressure in the main tube.

Another system, suitable for gauging flow rates in excess of 10 l min^{-1}, is the turbine meter. These compact devices work with an accuracy of ± 1% but their price is higher than that of rotameters.

The measurement of the flow of large volumes of air in industrial fermenters is often accomplished using so-called orifice gauges. In addition to an increased precision of measurement, these devices have a number of other advantages. They are not affected by pressure variations (automatic compensation) and are suitable for the attachment of instruments for remote flow reading, monitoring and regulation. They also ensure sterility since they are placed at the end of the outlet air tube. The capital costs and maintenance expenses are lower than with rotameters. As indicators we may use either liquid U-shaped manometers or differential mercury manometers fitted with an integrating meter and a remote control.

(b) Control of gas supply

The supply of gas (normally air) to the fermenter vessel is provided from a compressed air supply (oil-free) which may provide air for more than one vessel depending upon its size and output. A pressure regulation valve ensures that air reaches the vessel at a maximum of 1.5 bar.

The rotameter controls the actual flow rate of air through the fermenter. This should not exceed 1.5 vessel volumes per minute or droplets of water may become entrained in the stream of gas leaving the fermenter and so wet the exit gas filter, causing it to block. A valve at the bottom of the rotameter is turned and the indicator 'ball' in the rotameter tube rises or falls in proportion to the valve position. A scale on the tube gives flow rates in ml per minute or litre per hour.

Illustration of gas supply system is shown in figure 15.9. The air passes through the inlet air filter which prevents any microbes from entering the vessel *via* this path. The end of the sparger is typically a ring with small holes through which the air is forced. The bubbles are immediately broken up and dispersed by the impellers on the drive shaft and the baffles which can be fitted near the wall of the vessel. The use of several impellers ensures all regions of the vessel receive good aeration.

A 'head space' of around 20% is normally left between the culture level and the vessel top plate. Sometimes, gas can also be introduced into this region *via* a short pipe in the fermenter top plate. For certain type of fermentations (e.g. mammalian cell cultures), a gas mixing station can be used to pre-mix several gases before they are introduced into the fermenter.

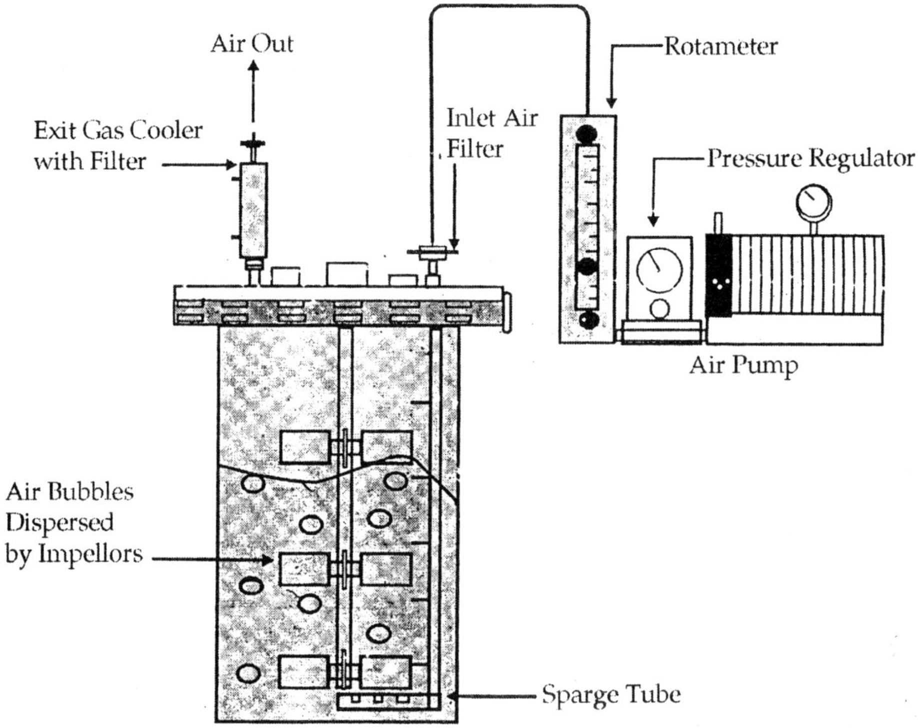

Fig. 15.9 : Illustration of Gas supply system.

15.10 LIQUID FLOW

Dosage of fluids and measurement of their flow belong to the most important operations in fermentation industry; dosage devices are used for the addition of inoculum, nutrients, alkalies and acids for pH adjustment, and antifoam agents. Continuous culture processes require particularly precise fluid dosage and flow measurement.

Dosage devices used in the fermentation industry have to meet the following demands:

a) constant rate of feed into fermenter;
b) elimination of contamination of dispensed fluid by undesirable microorganisms
c) possibility of dosage against a pressure;
d) low effect of pressure and temperature changes on dosage;
e) easy sterilisation;
f) construction from nontoxic materials that do not release undesirable substances on sterilisation or fluid flow;
g) reliability of dosage over long periods of time.

The constant feed rate may be ensured by the dosage of either volume or mass of the delivered liquid. Devices based on the measurement of the volume of the dispensed fluid are more frequent since they are simple and cheaper. The dosage based on fluid mass measurement is more precise since it is independent of temperature, pressure, viscosity and other physical factors. The volume dosage of fluids offers two possibilities: either the fluid feed is ensured separately (by overpressure, pump, etc.) and the volume of the delivered fluids is merely measured, or a device is used that fulfils both these roles, transporting and measuring the fluid at the same time.

Two types of pumps, viz. membrane and peristaltic pumps, are frequently used in the fermentation industry. Piston pumps have been successfully used in the dosage of small volumes of fluids but membrane pumps are more convenient than piston pumps since they are closed and do not leak. The pumped fluid is separated from the device by an elastic membrane made of an inert material. The motion is transmitted to the membrane either directly without a piston, or the piston movement is transmitted to the membrane *via* another auxiliary liquid which also serves as a lubricant in the piston chamber (Ricica, 1964).

Peristaltic pumps are based on the principle of a gradual compression of the walls of an elastic tubing through which the fluid passes. When the compression ceases, the elasticity of the tubing restores its original shape and this causes the suction of a new portion of the fluid. Dosage reproducibility depends on the quality of the tubing; frequent compression adversely affect its material and shorten its service life. The dosage rate is determined by the frequency of compression and the tubing inner bore. Depending on the way the tubing is compressed, peristaltic pumps can be divided into rotary (Fig. 15.10) and piston peristaltic pumps (Fig. 15.11). Liquid flow can be measured by a variety of devices: (1) Graduated burette, (2) air bubble, (3) slit flow meters, and (4) rotameters.

Fig. 15.10 : Peristaltic pump (rotary) (New Brunswick-Scientific).

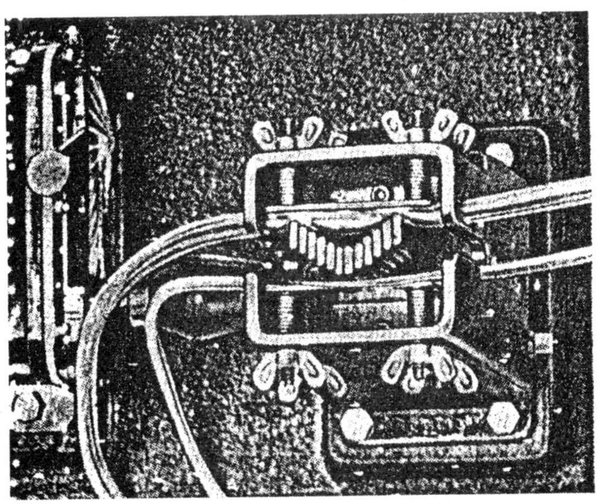

Fig. 15.11 : Peristaltic pump (piston) (Sigma).

15.11 VISCOSITY

Viscosity measurement in the microbial industry has so far been given undeservedly little attention; yet it may be used as an indication of cell growth and morphology and could also shed some light on mechanical changes in processes using filamentous microorganisms. As yet no system has been developed for viscosity measurement inside fermenters since the results depend strongly on the character of the fermenter. Fermenters of the same volume but different shapes and inner equipment give rise to different viscosity values, as also do different flow rates and filament shapes (Fig. 15.12).

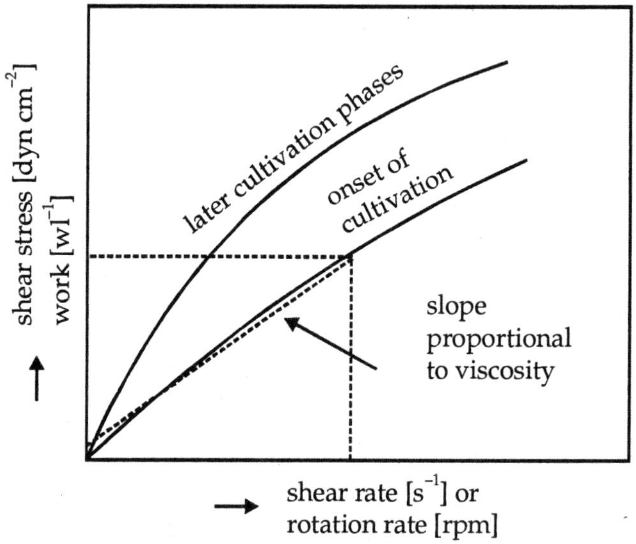

Fig. 15.12 : Viscosity features of culture fluids (Humphrey, 1971).

Cultures of unicellular microorganisms resemble Newtonian fluids whereas cultures of many-celled microorganisms are non-Newtonian. Non-Newtonian fluids change their "viscosity" at different shear rates. A number of viscometers with known shear rates are available at present; however, many of them are unsuitable for use in the microbial industry as they have tubes with small inner diameter. The most suitable is the Couett viscometer based on the principle of a rotary coaxial cylinder. The shear rate of the inner cylinder may be defined but the varying shear in the inter-cylinder gap must be taken into consideration. The viscometer should make it possible to measure certain shear rates; simpler viscometers with only three or four rates are ineffective in distinguishing between plastic and pseudoplastic fluids. One of the most widely used types of viscometers is the Rheomat 15 manufactured by Contrave.

15.12 TURBIDITY

The measurement of culture turbidity inside the fermenter is subject to the same effects as the viscosity measurement. It is highly important since it ensures rapid monitoring of the growth of microorganisms. In its present form it is fraught with all the shortcomings of sampling techniques, especially concerning the acquisition of an average sample. The main obstacle in turbidity measurement inside the fermenter is the sedimentation of microorganisms in the measuring chamber. To avoid this factor, double measurement of the batch of fermenter contents must be carried out.

A reliable device for automatic turbidity measurement was developed by Biolafitte (Fig. 15.13). The measurement is done by a photodiode in a measuring chamber, in a closed circuit outside the fermenter. The circulation of the culture is ensured by a peristaltic pump with variable flow rate. To eliminate the sedimentation of microorganisms, the wall of the measuring chamber is thoroughly cleaned by a piston prior to measurement. The measured values are automatically recorded.

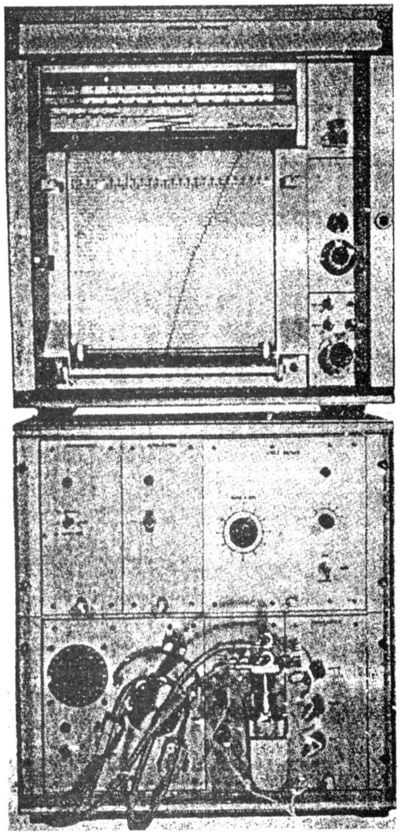

Fig. 15.15 : Device for automatic turbidity measurement (Biolafitte).

15.13 VOLUME MEASUREMENT

The first important physical condition of a liquid culture is its volume or working volume. Information on liquid volume (or level) is essential in liquid flow control (e.g., filling up vessels, continuous culture feeds, or introduction of liquid nutrients) which ultimately affect metabolic activity. There are two systems which are generally used for on-line volume determinations: (1) liquid level sensors, and (2) weight or mass measurement devices.

For sensing of liquid level experience indicates fairly reliable performance can be obtained in small-scale equipment with capacitance probes. As an example, the principle of operation of the Robertshaw capacitance level probe is described.

The sensor uses an electronic circuit which compares the rapid cyclical rate of charge of two capacitor systems, one reference capacitance internal to the electronic system, and the other, the external sensor capacitance probe itself, located in the fermenter. When the liquid level in the fermenter changes, the sensor probes capacitance and hence rate of charge changes. This rate difference is measured and provides an output signal proportional to the change in liquid level. Experience has shown that capacitance probe systems will provide level measurement with an accuracy of $\pm$2-4% (full scale) depending on the liquid mixing surface turbulence and foaming conditions of the fermenter.

For larger scale systems (5000 litres and above) the ΔP measurement method is usually used, where a flush mounted diaphragm pressure cell is located in the base of the vessel for liquid height, and hence volume measurement. Among suppliers of volume measurement systems are Ranco Controls, Drexelbrook, and Robertshaw Controls.

15.14 WEIGHT/MASS MEASUREMENT

Weight or mass of the liquid can be determined by scales of various types. In this method the vessel is suspended on a scale and the combined weight of the vessel and liquid is electronically measured, usually by strain-type gauges. Several methods for electronically or mechanically eliminating the tare weight of the system can be used. Advantages of this method are that (a) the system is external to the vessel and need not be steam sterilizable, (b) its calibration is relatively simple, and (c) its accuracy is within ± 1—2% (based on field experience). Among the disadvantages are the following: (a) Special engineering is needed for suspension of the vessel on the scale; consequently, this approach cannot be readily applied to existing installations. (b) There is a lower volume limit (approximately 250 litres) below which measurement accuracy falls off rapidly as the dynamic load becomes less than the tare weight. (c) Certain systems are sensitive to vibration and ambient temperature changes and compensation must be allowed for this in design. (d) The system is more costly than the simpler level probe method by a factor of four to five times. Suppliers of strain gauge systems are BLH Electronics, Transducers, Inc., and Micro-Strain, Inc.

This parameter is useful for continuous culture or fed-batch culture where the rate of addition of feed must be known and controlled very accurately. The 'deformation' of the load cell provides an output which can be transduced into an electric signal. Control is normally by setting a maximum and/or minimum weight such that a feed pump operates whenever the value is less than the maximum. At the maximum value, a second pump operates to remove culture until the minimum weight is reached.

15.15 pH MEASUREMENT AND CONTROL

(A) Monitoring

Automatic monitoring of pH involves either simply the measurement and recording of change of pH, or pH measurement and regulation (usually its maintenance at a constant value). The system comprises:

(a) measuring (glass) and reference (calomel) electrodes,

(b) devices for measurement, recording and regulation,

(c) reservoir for acid or alkali solutions with pumps.

To make the regulation precise, these solutions have to be delivered slowly and into the vicinity of the electrodes.

(B) Control system

pH control is achieved by the addition of either acid or alkali to correct changes in the pH of the culture during growth. The controller uses a pH electrode (either glass or gel) to sense these pH changes and provide a feedback signal. The pH measurement system is identical to a bench pH meter and is calibrated in the same way before the electrode is autoclaved. Temperature compensation during operation may have to be set manually or may be automatic. The pumps are used to supply the acid and alkali are normally built into the control module or the base part of the fermenter. Measured range is typically pH 2-12.

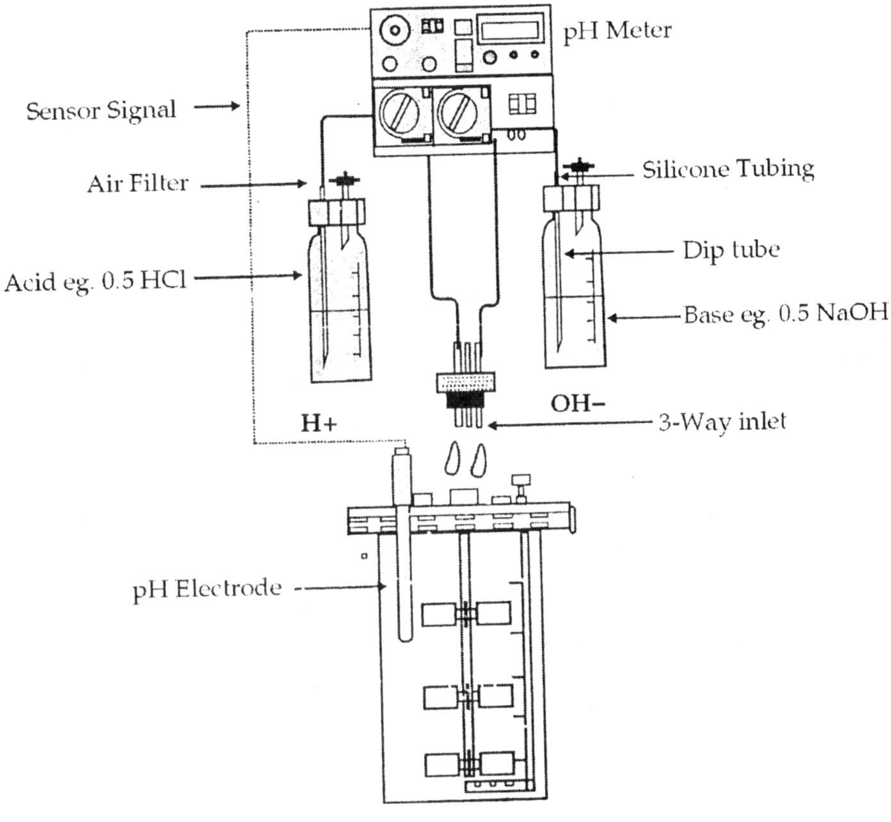

Fig. 15.14 : Illustration of pH measurement and control.

The reagent bottles are connected to the fermenter via transfer lines of silicone tubing. The bore of tubing selected will determine the volume of acid or base added by each activation of the relevant pump by the controller. Normally, solutions between 0.2 and 0.5 M of acid and base are used as a starting point. Figure 15.14 shows illustration of pH measurement and control.

A set-point value is set on the controller and/or an upper and lower limit to provide a 'dead band' range in which the controller is not active. This band should normally be around 0.5 pH units each side of the desired value to prevent overdosing of acid and alkali sequentially. A proportional band adjustment may be present to widen or tighten the range of pH value over which the controller acts and even a switch to give a non-linear control if good control cannot be achieved by routine adjustment.

15.16 OXYGEN MEASUREMENT AND CONTROL

(A) Offline Measurement of Dissolved Oxygen

Until the development of steam sterilizable gateway sensors/probes for dissolved oxygen measurement it was measured offline either by nonsterilizable probes or by chemical method (Winkler's method). So, perhaps the most significant advancement in the area of fermentation instrumentation has been the development of the *in situ* (steam sterilizable) dissolved oxygen sensor.

The Winkler's chemical or iodometric method and its modification are the standard procedure for determining dissolved oxygen. In principle it involves the fact that oxygen oxidizes Mn^{2+} to a higher state of valence under alkaline conditions and that manganese in higher state of valence is capable of oxidizing I^- to free I_2° under acid conditions. Thus the amount of free iodine released is equivant to the dissolved oxygen originally present. The iodine is measured with standard sodium thiosulfate solution and interpreted in terms of dissolved oxygen. The flow diagram of the method is shown below (Fig. 15.15). The demerit of the method relates to colour the solution in which d.o. is to be measured. As in most cases fermentation liquid is coloured. This method is difficult to use.

Liquid containing dissolved oxygen

↓ $MnSO_4$

$Mn(OH)_2$

↓ [O]

$MnO(OH)_2$

↓ Alkaline KI

Liberated I_2

↓ titration with standard $Na_2S_2O_3$ solution.

Free I_2 estimated

$$\text{D.O.} = \frac{\text{Normality of thiosulfate} \times \text{Volume of thio used}}{\text{Volume of sample used}} \times 200 \text{ ppm}$$

Fig. 15.15 : Flow diagram of Winkler's method.

Both offline and online dissolved oxygen measuring probes work on the same principle. In the principle of the design of an oxygen probe a distinction is made between two different types of oxygen probe/electrode e.g. electrodes without membrane and electrodes with gas permeable membrane.

(B) Online Measurement

Dissolved oxygen is one of the most important and yet one of the most difficult parameters to control properly during a fermentation cycle (on line).

Types—The electrodes used to measure on-line dissolved oxygen are of two types:

(1) Galvanic electrodes—They are generally simple, cheap, slow and have a limited life.

(2) Polarographic electrodes—They are generally more complex, accurate, rapid and robust but expensive. The polarographic electrode is used more commonly. The key point is that it requires a voltage to polarize the anode and cathode of the detecting cell. During polarization, the electrode must be connected to relevant module which in turn must be switched on.

Preparative process—The membrane of the polarographic electrode needs periodic replacement. The electrode can be 'zeroed' prior to autoclaving using a special zeroing gel provided by the manufacturer. The electrode is simply inserted into a sachet of gel, left for a short time and then the displayed value adjusted until it reads zero.

The 100% value is a relative setting made after autoclaving and polarization of the electrode. The air flow and stirrer speed are turned to the maximum needed for the fermentation and the vessel left for some minutes. The display is now adjusted until it reads 100%.

Working process—One of the widely used steam sterilizable membrane covered dissolved oxygen electrode example is W. Ingold PO_2 electrode. The tip of this electrode is kept immersed in the fermentation broth in the fermenter. Next the electrode cable is connected through the amplifier to the recorder (as shown in figure 15.18), which in turn is connected to the supply switch recorder. Chart speed is selected by the selector switch on the recorder as per need. After aseptic transfer of inoculum in the fermenter and with aeration and agitation, the amplifier connected recorder pointer is adjusted at 100% saturation point. Now it can give on line measurement of dissolved oxygen.

(C) Control

Two main mechanisms have been employed to control dissolved oxygen tension (d.o.t.) in fermentation, explained as follows—

(i) altering the feed gas composition, and

(ii) altering agitator speed.

The aim of both (i) and (ii) methods is to control dissolved oxygen tension (d.o.t.) by manipulating oxygen mass transfer driving force and by manipulating volumetric oxygen mass transfer coefficient.

(i) D.O. Control based on altering the feed gas composition

The first method of d.o. control as depicted in Fig. 15.16 basically consists of a d.o. probe/electrode to measure online the level of d.o., the level being controlled by altering the ratio of oxygen to nitrogen in the feed gas by means of a proportionating valve while maintaining a constant total gas flow rate. This method involved primarily manipulation of the oxygen mass transfer driving force as mechanism for controlling dissolved oxygen tension.

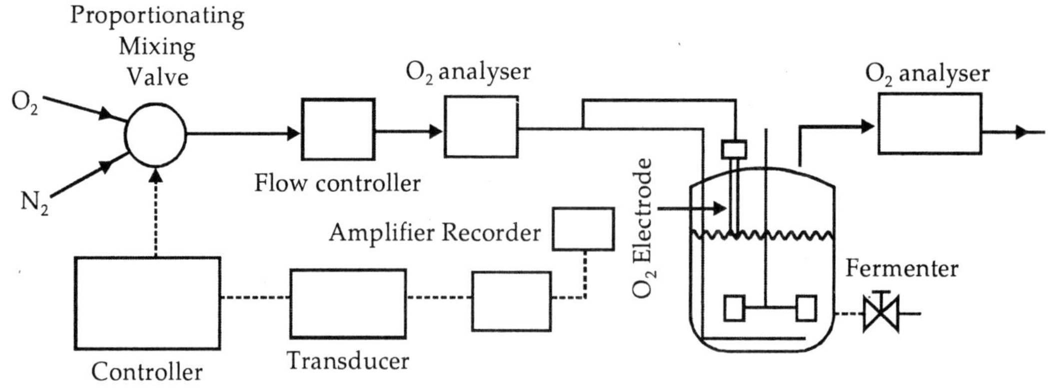

Fig. 15.16 : Schematic diagram of online measurement and control of D.O. concentration based on altering the feed gas composition.

(ii) D.O. Control by altering agitator speed

In the second method output of the probe governs the speed of the agitator motor, whilst the air flow rate is held constant. It manipulates the volumetric oxygen mass transfer coefficient (K_La) as mechanism to control dissolved oxygen tension. A schematic of this control arrangement is shown in Fig. 15.17.

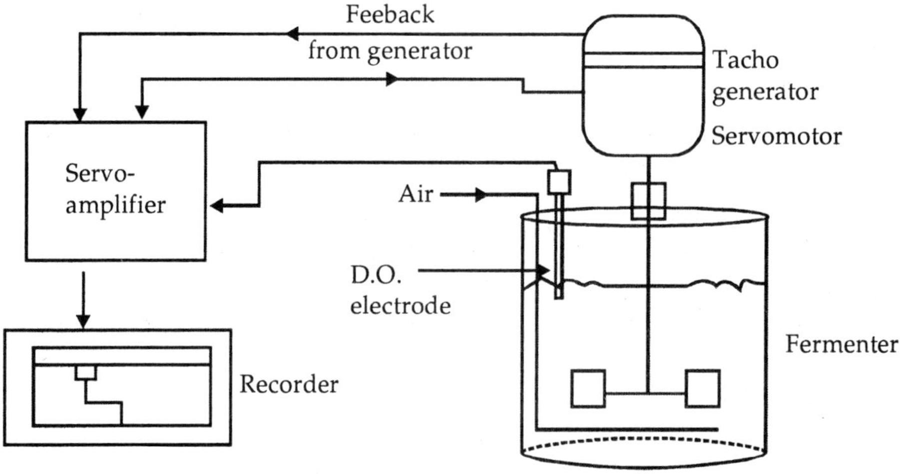

Fig. 15.17 : Schematic of D.O. control by regulating the speed of the agitator in fermenter.

(iii) D.O. Control using combination of both—controlling agitator speed and proportional air

In such system, control of dissolved oxygen is achieved by adjusting agitator speed as well as by controlling proportional air valve. Only speed adjustment may interfere with good mixing, while only adjustment of air flow rate may increase foaming. The most accurate form of flow control is to use a thermal mass flow control valve which measures and controls air flow based on the cooling effect the gas exerts when passed over a heated element. A simple option is the use of a solenoid valve with a manually set "bleed" value to simply vary air supply between a maximum and a minimum flow rate (see figure 15.18). Range is typically 0-120%.

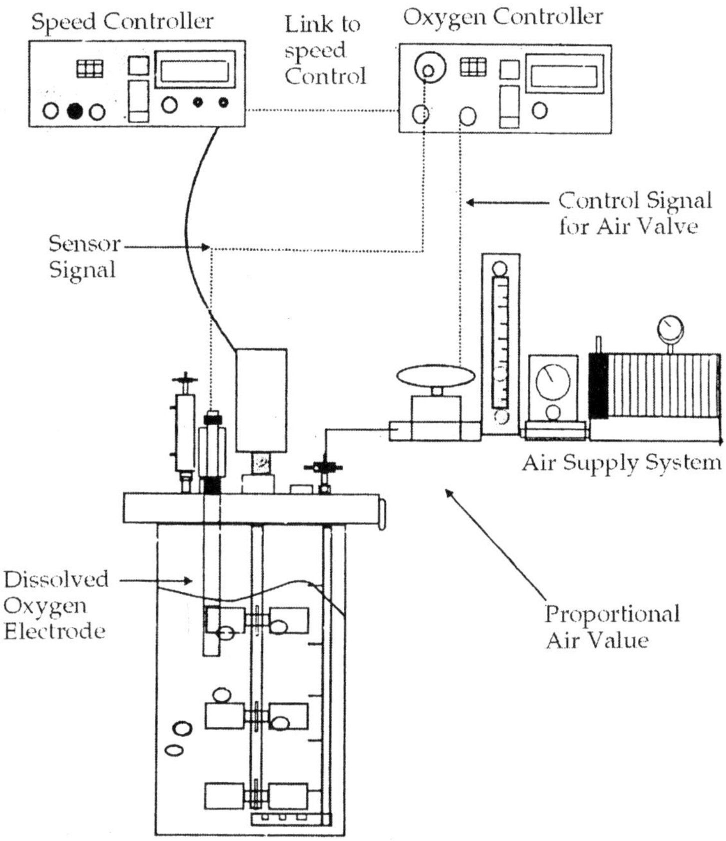

Fig. 15.18 : Illustration of Dissolved oxygen measurement as well as control, using speed controller and oxygen controller.

15.17 REDOX POTENTIAL MEASUREMENT AND CONTROL

Many aerobic microbial processes take place at concentrations of dissolved oxygen that are impossible to measure using commercial probes for dissolved oxygen. However, the supply of oxygen to the cultures is essential to the microorganisms. Since the range of the oxygen-limited state may vary from the almost completely anaerobic state through the so-called partially aerobic state to a state where oxygen can be measured by means of a commercial probe (i.e., above 10 μM O_2), it is important to have a tool that can provide information about the degree of oxygen limitation. Such a tool is the measurement of the redox potential in the growth liquid (Jacob, 1970).

In principle there are two ways of measuring a redox potential: (1) by redox dyes and (2) by electrodes (Kjaergaard, 1977). Measurement of the redox potential by dyes is not exact and requires a number of different dyes to obtain even a semiquantitative measurement; furthermore, many of these dyes may be toxic to the cells or may inhibit enzyme activities in biological liquids. Therefore, this method is not used in biochemical engineering except in classification assays.

The redox potential can also be measured using a device similar to that for pH monitoring. The main difference is in the electrodes used: the potential is measured using an inert metal electrode, usually platinum, and is recorded in millivolts, usually in the range of 0-800 mV or -500 to + 500 mV. Regulation is effected by feeding gaseous nitrogen or oxygen, cystine, ascorbic acid or sodium thioglycolate into the fermenter. Measurement of the redox potential during aerobic microbial processes is considerably complicated by dissolved oxygen: the measurement is therefore performed only rarely.

15.18 MEASUREMENT AND REGULATION OF OTHER DISSOLVED GASES

Other gases dissolved in the culture liquid, which require measurement or regulation during culture, are carbon dioxide, methane and hydrogen. Severinghaus and Bradley (1958) described a modified glass electrode for measuring the concentration of dissolved carbon dioxide. It differs from the conventional pH-measuring electrode in that it is filled with bicarbonate buffer and its end is formed by a wetted cellophane film encased in a teflon membrane. The membrane prevents the penetration of ions but carbon dioxide may diffuse into the cellophane. The electrode measures the changes in pH of the bicarbonate buffer due to changes in the concentration of dissolved carbon dioxide. The electrode cannot be sterilised by steam. Sensors for measuring the concentration of dissolved methane and hydrogen have not yet been developed.

Gases dissolved in culture liquid can be measured analogously to dissolved oxygen using appropriate gas analysers (Fig. 15.18). The level of dissolved gases may be regulated as shown in Figs. 15.19.

15.19 EXIT GAS ANALYSIS

Measurement of amounts of different gases leaving a fermenter vessel can provide valuable information about the metabolic processes taking place under the conditions in which the culture is growing. For example, the ratio of oxygen and carbon dioxide entering and leaving the vessel, or respiratory quotient (RQ), can be used to determine whether a yeast is producing biomass or alcohol.

The RQ can be calculated by most fermentation software packages and can then be used as part of a control algorithm so that the flow rate of a feed pump can be altered accordingly. The entry gas does not need to be anlysed providing it is air, because the amounts of O_2 and CO_2 will be those of the atmosphere. The exit gas may need to be conditioned (e.g. moisture removed) before going into the analyser, depending on the type of instrument used.

Determination Methods of outlet gases

(A) Oxygen. An important requirement in the study of aerobic microbial processes is the automatic and continuous determination of oxygen in the gas mixture leaving the fermenter, as it supplements the data on the overall oxygen consumption.

(i) One of the first methods used for measuring the content of oxygen in gases was based on changes of electrical conductivity. A resistance body (a platinum wire placed in the air stream) changes its electrical properties in a stream of oxygen; if the level of oxygen in the gas stream drops, the resistance of the wire decreases due to a higher cooling effect.

(ii) Paramagnetic oxygen analyser—The best results concerning the measurement of oxygen in a gas mixture have been obtained by methods based on the paramagnetic properties of oxygen. A **paramagnatic oxygen analyzer** is a light dumbbell suspended on a silicon filament between magnet poles. Each ball of the dumbbell is in a nonhomogeneous magnetic field. The dumbbell system has a defined constant magnetic susceptibility. When the oxygen-containing gas surrounds the suspension system, it gives rise to a force depending on the difference between the magnetic properties of the gas and those of the system. The force causes a rotation of the dumbbell around the silicon fibre. The magnitude of the force and, hence, of the deviation, is proportional to the content of oxygen in the gas. The suspension of the dumbbell is covered with rhodium and is thus conductive. In the vicinity of one of the dumbbell balls is situated a pair of electrodes or flaps each of which is kept at a certain potential. This holds the ball in an electrical field; an electrostatic force may also be brought to bear on the dumbbell by altering the potential of the ball. The variable potential brought to the suspension system is actually a stabilising or balancing potential. Any force acting on the suspension system may be counterbalanced by the electrostatic force so that the dumbbell does not move. The magnitude of the balancing potential is therefore proportional to forces acting on the suspension system and these in turn are directly proportional to the oxygen content in the gas. The balancing potential is measured as follows: A light ray reflected by a mirror is left to impinge separately on two photodiodes which have the same resistance if the suspension is in zero position. If the suspension system starts rotating due to a change in the oxygen content, the light incident on the two photodiodes becomes unequal and the resistance of one of them increases while that of the other decreases. This results in a rise or drop of voltage on the central terminal from the two valves depending on the direction of rotation. This voltage is amplified and is counter-balanced by the voltage imposed on the suspension. The suspension system is thus brought to zero position. The balancing potential is measured and may be recorded by a suitable device. The recordings obtained by the method are linear and proportional to the oxygen content in the gas.

(iii) A cheaper method of determining oxygen concentration includes the bubbling of the outlet gas through a water column in which the concentration of dissolved oxygen is continuously monitored. With suitable temperature regulation, this system is as reliable, and simpler to construct, than paramagnetic analysers.

(B) Carbon dioxide. The study of microbial processes and balance calculations necessitate the determination of carbon dioxide in outlet gases. The following methods can be used:

a) Measurement of conductivity changes caused by temperature changes in a resistance body due to stream of CO_2. The principle is analogous to that used in O_2 determination but the increased cooling of the body brings about an increased resistance.

b) Sound analyser. The measurement is carried out in a resonance cavity kept in a state of oscillation by a transistor amplifier. The change in frequency due to the presence of CO_2 is linearly related to the CO_2 content.

c) Photometric determination (Maxon and Johnson, 1952). The equilibrium between CO_2 and water, giving rise to HCO_3^- and H^+, depends on the concentration of an aqueous solution of bicarbonate and on the partial pressure of CO_2. This equilibrium may be determined by measuring the concentration of hydrogen ions using, e.g., phenol red.

d) Absorption spectrometry (Telling *et al.*, 1958). The measurement is performed using commercially available automatic continuous absorption spectrometers such as those manufactured by M.S.A. Ltd., Beckman and Hartmann and Braun.

e) Measurement of the difference in oxygen and CO_2 content after absorption of the latter (Phillips, 1963). The system comprises a pump, timer, CO_2 absorber and flow meter (Fig. 15.19). The outlet gas mixture is intermittently passed through a CO_2 absorber and the resulting change in the reading on a paramagnetic oxygen analyser is used to assess the content of CO_2.

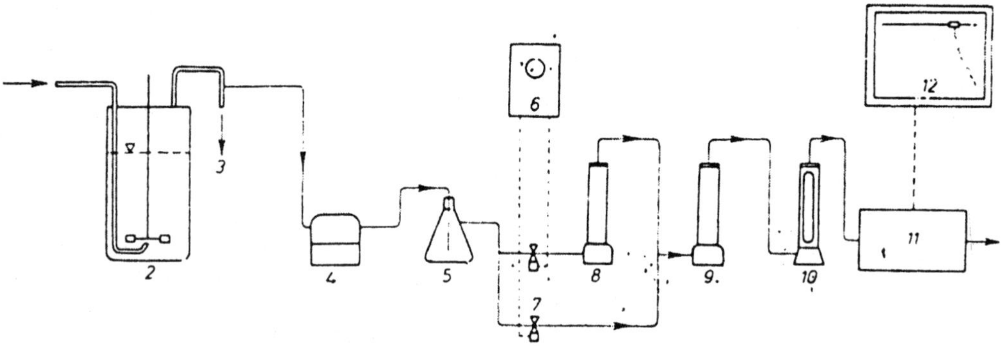

Fig. 15.19 : Scheme of oxygen and CO_2 determination in a gas mixture by paramagnetic oxygen analyser (Phillips, 1963). 1. air inlet, 2. fermenter, 3. gas outlet, 4. peristaltic pump, 5. separator, 6. time switch, 7. cocks with screw joints, 8. CO_2 absorption, 9. drier, 10. flow meter, 11. O_2 analyser, 12. recorder.

f) Infrared CO_2 analyser—A hot wire is used to generate a source of infra-red radiation which passes through the gas which comes from the fermenter into the sample chamber. An infra-red detector measures the amount of radiation reaching it after some is absorbed by the gas. An optical filter is used to make sure the detector only responds to the gas of interest. A "chopper" or rotating shutter is used to allow the detector or to see a reference source at regular intervals.

This type of detector allows continuous measurement with good sensitivity, accuracy and selectivity. Output signals are normally provided either as analogue signals (e.g. 0-10 V) or a string of ASCII characters for printing or transfer to computer software.

15.20 ION-SELECTIVE ELECTRODES (ISE)

The key property of every ion-selective electrode (ISE) is that it possesses a membrane which exhibits selective permeability to a restricted number of ionic species, ideally only to the one required to be analysed. Only a few examples of each type of ISE are included although many more ionic species may be determined directly using the appropriate electrode.

(A) Classification

It has become conventional to classify ion-selective electrodes into following types:

(1) **Solid State**

(a) Glass (H^+, K^+, Na^+, NH_4^+, etc.)

(b) Other e.g. Ag/AgX for Cl^-, Br^-,
I^-, Ag^+, $S^=$
LaF for F^-
MS_2 for Pb^{++}, Cu^{++}

(2) **Liquid Membrane**

(a) With dissolved Ion-Exchanger e.g. Ca^{++}, NO_3^-

(b) With neutral carriers, e.g. Valinomycin for K^+

(3) **Gas Sensing**

CO_2, SO_2, NH_2, NO_X

(4) **Enzyme Electrodes**

e.g. glucose oxidase for glucose, penicillinase for penicillin, urease for urea.

(B) Design

In each case it is necessary to have a reference electrode against which the potential of the ISE may be determined. For most applications the calomel or Ag/AgCl electrode exhibits a potential independent of ionic concentration and is therefore suitable. A typical electrochemical "cell" containing an ISE is shown in Fig. 15.20.

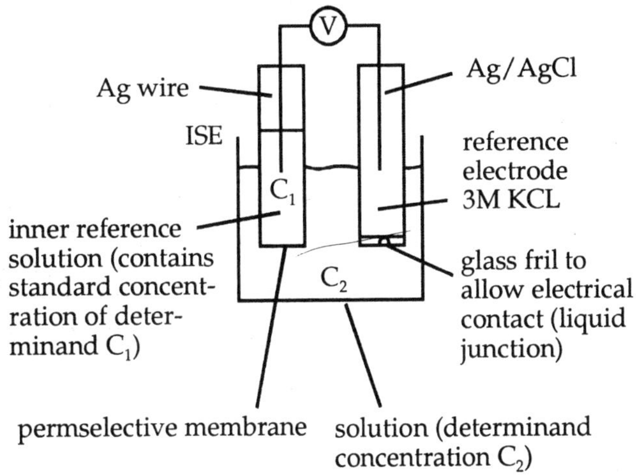

Fig. 15.20 : Design and principle of operation of an ion-selective electrode.

(C) Working principle

When two phases containing electrically charged particles come into contact an electrical potential difference develops at their interface. The potential across the membrane is given by the familiar Nernst equation:

$$E = \text{constant} + \frac{RT}{zF} In \frac{c_{i(1)}}{c_{i(2)}}, \tag{15.1}$$

where *R*, *T* and *F* have their usual thermodynamic meaning, z is the charge of the selected ion and the concentrations (strictly, activities) of ions in phase 1 and 2 are respectively $C_{i(1)}$ and $C_{i(2)}$. It is clear that the potential exhibited by the ISE relative to its reference electrode will vary with $C_{i(2)}$ if $C_{i(1)}$ is held constant and the potential of the reference electrode remains unchanged. This is the principle of the use of ISE's. The factor *RT/F* has a value of approximately 60 mV at 30°C. Thus the potential of an ISE changes linearly with the logarithm of the required ion concentration. This is indeed one of the many advantages of the use of ISE's, that a very large concentration range may readily be encompassed, typically 6 orders of magnitude.

(D) Advantages

Some of their other merits are listed in Table : 15.4, and these alone justify the belief that ISE's may be useful devices in the control of microbial processes.

Table 15.4 : Advantageous features of ion-selective electrodes for control of microbial processes (Kell, 1980)

1. Continuous, real-time assay.
2. Sensitive (usually to < 10^{-6} mole l^{-1}).
3. Electrodes biologically inert.
4. Non-destructive assay: no added reagent.
5. Many electrodes, using same equipment.
6. Selectivity: good to poor, depending on need.
7. Relatively rapid response.
8. Good lifetime.
9. Low cost.
10. Response to thermodynamically important activities rather than concentrations.
11. Possibility of use in optically opaque and turbid suspensions.

15.21 ENZYME ELECTRODES [ENZYME SENSORS/BIOSENSORS]

(A) Advantages

Electrochemical enzyme sensors, so-called enzyme electrodes, are used for simple and rapid determination of different metabolites. They combine the advantages of electrochemical sensors with the specificity of enzyme action and are finding an ever wider use in biology and biochemistry. The analysis usually requires no reagents other than a buffer and a standard. In comparison with conventional enzyme assays, biochemical sensors can be used repeatedly or continuously, which considerably lowers the costs of the enzyme preparation used (Barker and Somers, 1978).

(B) Working Principle

An enzyme electrode consists of an electrochemical sensor which carries on its surface a thin layer of immobilised enzyme. The substrate present in the medium into which the electrode is immersed diffuses into the enzyme layer and the resulting reaction product is detected by the sensor. The ensuing recorded electrical signal is proportional to the concentration of the substrate which is then determined from a calibration curve. (See Fig. 15.21)

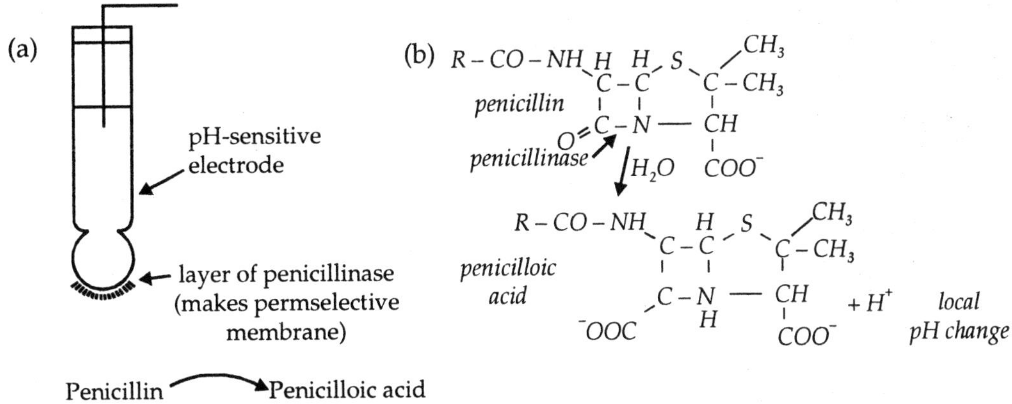

Fig. 15.21 : Principle of an enzyme electrode (penicillin electrode) (a) diagrammatic representation, (b) reactions leading to electrode response (Kell, 1980).

(C) Construction

The construction of an enzyme electrode involves the following steps:

(1) Selection of a suitable enzyme reacting, if possible, specifically with the substrate(s) to be determined; if the enzyme is not commercially available, it has to be isolated, purified and concentrated to the required degree;

(2) Suitable immobilization of the enzyme in a thin layer; the procedure should not cause any considerable losses in enzyme activity;

(3) Fixation of the enzyme layer on the measuring part of the indication electrode sensitive to the reaction product or cosubstrate, i.e. to certain ions (NH_4^+, H^+, I^-, CN^-) or dissolved gases (O_2, CO_2).

The enzyme layer usually contains a single enzyme catalysing a simple transformation of the substrate, e.g. oxidation, hydrolysis or decarboxylation. Electrodes with two enzymes acting in a sequence as a metabolic chain are rather rare. In this case the product of the first enzyme reaction is the substrate of the second (indication) reaction which produces readily measurable substrate or consumes a cosubstrate (e.g. O_2). The enzymes most suitable for the purpose are those which do not require a coenzyme or which have the coenzyme firmly bound. Recently, however, procedures have been developed even for the immobilisation of nicotinamide coenzymes.

(D) Application example

Table : 15.5 gives a list of enzyme electrodes in current use with some of their characteristics. It demonstrates that a substrate can sometimes be determined by two or more different electrodes.

The enzyme electrodes are of two main categories:

(a) Electrodes for product assay.

(b) Electrodes for cosubstrate assay.

Table 15.5 : Characteristics of currently used enzyme electrodes (Biosensors)

	Assayed substrate	Immobilised enzyme	Sensor	Concentration range (mol/l)	Electrode stability
1.	β-D-Glucose	Glucose oxidase	O_2	10^{-4}—10^{-2}	3 weeks
			$Pt(O_2)$	10^{-5}—10^{-1}	> 4 months
			$Pt(H_2O_2)$	10^{-4}—2.10^{-2}	> 14 months
			pH	10^{-3}—10^{-1}	1 week
		Ditto + peroxidase	I^-	10^{-4}—10^{-3}	7-30 days
2.	Urea	Urease	NH_4^+	10^{-4}—10^{-2}	> 4 months
			NH_3	5.10^{-5}—5.10^{-2}	> 4 months
			pH	5.10^{-5}—5.10^{-3}	3 weeks
3.	D-Amino acids	D-Amino acid oxidase	NH_4^+	5.10^{-5}—10^{-2}	1 month
4.	L-Amino acids	L-Amino acid oxidase	NH_4^+	10^{-4}—10^{-2}	> 1 month
5.	L-Phenylalanine	Ditto + peroxidase	I^-	5.10^{-5}—10^{-3}	> 3 weeks
6.	L-Asparagine	Asparaginase	NH_4^+	5.10^{-5}—10^{-2}	1 month
7.	L-Tyrosine	Decarboxylase	CO_2	10^{-4}—10^{-1}	3 weeks
8.	L-Lysine	Decarboxylase	CO_2	2.10^{-4}—$1.5.10^{-3}$	1-2 weeks
9.	L-Lysine (L-Arginine)	Ditto + Diamine oxidase	O_2	2.10^{-5}—2.10^{-4}	> 6 months
10.	Uric acid	Uricase	$Pt(O_2)$	10^{-4}—10^{-2}	> 4 months
11.	Hypoxanthine, xanthine, purine (aldehydes)	Xanthine oxidase	O_2	2.10^{-5}—4.10^{-4}	> 3 months
12.	Monophenols and polyphenols	Polyphenol oxidase (*o*-Diphenolase)	O_2	7.10^{-6}—7.10^{-5}	> 6 months
13.	L-Ascorbic acid	Ascorbate oxidase	O_2	2.10^{-5}—10^{-3}	> 7 months

... contd.

contd. ...

Assayed substrate	Immobilised enzyme	Sensor	Concentration range (mol/l)	Electrode stability
14. Diamines, histamine, polyamines	Diamine oxidase	O_2	7.10^{-6}—3.10^{-4}	1 year
15. Lactic acid	Lactate dehydrogenase	$Pt[Fe(CN)_6^{4-}]$	10^{-4}—2.10^{-3}	1 week
16. Alcohols	Alcohol oxidase	$Pt(O_2)$	0.5—100 ppm	> 4 months
17. Penicillin	Penicillinase	pH	10^{-4}—3.10^{-2}	2 weeks
18. Amygdalin	Emulsin	CN^-	10^{-5}—10^{-1}	1 week
19. L-Glutamic acid (pyruvic acid)	Glutamate + lactate dehydrogenases	NH_4^+	10^{-4}—10^{-3} (2.10^{-5}—8.10^{-4})	15 days
20. Inorganic phosphate (Glucose-6-Phosphate)	Alkaline phosphatase + glucose oxidase	O_2	10^{-4}—10^{-3}	3 months

(a) Electrodes for product assay—An example of such an electrode is that used for the determination of urea. A potentiometric sensor is covered by immobilised urease and it responds to the presence of the product of urea hydrolysis, i.e. NH_4^+ (NH_3) or CO_2. The calibration curve demonstrating the relationship between the monitored potential change and the logarithm of urea concentration is sigmoid, with a linear part in the range of 10^{-2}-10^{-4}. When the enzyme layer is sufficiently active, a tenfold change in urea concentration corresponds to a potential change of 60 mV at 30°C. The time constant, i.e. the time necessary to attain a steady electrode response, is 1-5 min depending on the type of sensor and the thickness of the enzyme layer.

(b) Electrodes for cosubstrate assay—An example of this type of assay is the amperometric determination of glucose by the Clark oxygen electrode coated with immobilised glucose oxidase. The electrode indicates the consumption of O_2 for glucose oxidation by a drop in the electrode current which is proportional to the concentration of glucose in the surrounding solution; this proportionality is maintained over a wide range, almost to the exhaustion of oxygen diffusing into the enzyme layer. Stationary electrode response is reached within 1-2 min.

Electrodes for glucose and urea determination are commercially available, but only as integral parts of appropriate instruments (Yellow Springs Instrument Co., Radiometer, Owens-Illinois, etc.). In principle enzyme electrodes make it possible to measure not only specific substrates but also enzyme activities. For instance, the urease electrode can be used for determining the activity of arginase or the phenoloxidase electrode may serve to assay the activity of alkaline phosphatase, β-glucosidase, β-galactosidase, β-glucuronidase, aryl esterase, etc., when an appropriate phenol conjugate is used as substrate.

Enzyme electrodes are judged by their sensitivity, stability and, in some cases, vulnerability to interference. Most of the electrodes described so far can be used over a concentration range of 10^{-5}-10^{-2} M. Linearity of calibration curves is achieved at $[S] < K_m$, when the enzyme reactions obey first-order kinetics. The lower limit is given by the sensitivity of the electrochemical sensor used. Since the electrode response is determined chiefly by the diffusion of the substrate into the enzyme layer and the diffusion of the product to the surface of the indication electrode, thorough stirring and a constant temperature are essential for the reproducibility of the assay (Enfors and Molin, 1978).

15.22 ON-LINE MEASUREMENT OF BIOMASS

A direct measurement of the number of organisms in a culture is clearly desirable for the control of feed rate, oxygenation and general process optimization. Normally, measurements of total cell numbers, wet weight, dry weight and viable cell counts would be made by taking samples of the culture at set times and performing the relevant laboratory analysis. The results would be entered some minutes, hours or even days later, when they would be of little value for real-time control. Fluorescence microscopy has also been used as a measure of viable cell density either with a probe or externally through the vessel glass.

Two probe-based systems which are reasonably accurate and can be used with bench-scale vessels are described by way of examples below.

15.22.1 Mettler-Toledo FCS turbidity system

This is actually a nephelometer as it measure the scattering of light by particles in the culture (i.e. microbes). The advantage over a standard nephelometer is that the sensor is directly mounted in the vessel and connected to the electronics via a fibre-optic cable. A two-point calibration procedure is used and result can be expressed in a variety of units. One of the restrictions of such a system is that if the culture is very thin, then shiny, reflective components close to the sensor could affect the readings. Conversely, too thick a culture would not allow the light scattering to be detected properly. However, a range upto 250 gram per litre dry weight is measurable. The probe is resistant to fouling over time.

15.22.2 Capacitance/Conductance-based Biomass Monitor (Aber Instruments)

This uses a totally novel measurement system to give viable cell numbers for all types of organisms (bacteria, yeast, filamentous fungi and animal cells).

A probe in the vessel uses a radio-frequency electrical field to measure the natural capacitance of living cells with an intact plasma membrane. This build-up of charge is proportional to the cell concentration and specific to different types of cells. The signal is processed and can be expressed as dry weight or concentration in cells per ml. An antifouling system is used to maintain the ability of the probe to give accurate readings over long time periods.

This type of instrument makes possible the control of feed pumps by biomass concentration, and could form the basis for an automated transfer system for delivering seed culture to a larger vessel after an inoculum of sufficient concentration has been produced in a small fermenter.

15.23 OFF-LINE AUTOMATIC ANALYSERS

Difficulties in the development of steam sterilizable on-line reading sensors for detection of various metabolites have resulted in the use of off-line analytical devices utilizing repetitive and/or continuous sampling of exhaust gases to monitor specific parameters. Sampling devices usually consist of a pipeline (sterilizable) through which gases are released by the culture are connected to the particular analyser.

15.23.1 Gas Chromatography

Off-line determination of various volatile compounds (e.g., methane, acetaldehyde, ethanol) using gas chromatographs was successfully tested in biochemical processes by several researchers.

One of the most elegant applications was reported by O'Brien and Cecchini (1970), who analyzed the metabolic patterns of bacteria by determining their end products (alcohols, di- and tricarboxylic acids) and using this "finger print" for taxonomic identification. It is interesting to note that this approach might also have practical significance for the detection of certain types of contamination in pure culture, though such application has not yet been reported in the literature.

Recent developments in gas chromatograph (GC) technology have made it possible to connect them to the exit gas lines of fermentors. As an example, there is Hewlett-Packard's Series 5830A, Reporting Gas Chromatograph. This instrument can be connected to a maximum of 16 sample lines with an automatic sampler. It can be programmed to control oven temperature, has a dual thermal conductivity detector, and automatically integrates the data and reports them in alphanumeric and/or graphical formats. Signals can also be obtained in binary coded decimal (BCD) form for computer interfacing. An HP5830A GC has been used in the verification of the effects of C/N feed ratios on protein and ethanol biosynthesis for yeast grown on molasses.

The advantage of gas chromatography is the ability for automatic determination of a relatively wide range of compounds.

The disadvantages are (1) the discontinuous signal produced and the time length between signals, since the reporting frequency is controlled by the retention time and the number of compounds being analyzed; (2) the relatively high price which, however, can be partially offset if several fermenters are to be monitored; and (3) the practical problem which requires investigation for each scale of use—to what extent do the volatile compound concentrations in the exit gas reflect the actual concentration in the culture broth? For the example shown (acetaldehyde and ethanol) the liquid-gas distribution ratio could be determined for a given gas flow rate.

15.23.2 Mass Spectrometer

In recent years reports have appeared about the use of mass spectrometry for monitoring low-molecular-weight compounds in physiological fluids, e.g., ethyl alcohol and dimethylsulfoxide in blood. This practice was currently extended to the field of fermentation technology.

Weaver and co-workers (1976) reported the mass spectrometric measurement of volatile organic chemicals in aqueous solutions whereas Reuss *et al.*, (1975, 1976) applied mass spectrometer for determination of dissolved oxygen, carbon dioxide, and methanol in fermentation broth.

The, advantages of application of mass spectrometer (MS) to monitor dissolved volatile compounds in biological fluids can be listed as follows:

1. Response time is rapid (<1 minute).
2. Sensitivity (detection limit) is in the range of 10^{-5} *M*.
3. There is a linear response to compound concentration over wide range (e.g., between 10^{-5} *M*—10^{-1} M).
4. It is possible to determine the compound concentration both in liquid and gaseous phases.

In the case reported by Reuss *et al.* (1975, 1976) a quadruple type mass spectrometer (QMS) was connected to a bacterial culture broth for the analysis of dissolved oxygen, dissolved carbon dioxide, and methanol concentrations. The physical connection between the culture liquid and the QMS sensing element was established by a sensor covered by a silicon rubber membrane. The thickness of the membrane made it possible to uncouple the pressure conditions existing inside the fermenter and in the high vacuum of the ion source in the mass spectrometer.

Application of mechanically supported semipermeable membranes as separators makes it possible to transfer the dissolved, low-molecular-weight (LMW) volatile compounds into the MS, which at the same time can also be connected to the exit gas phase of the fermenter. This technical approach makes it possible to determine the concentration difference of compounds existing between the dissolved and gaseous phase (e.g., HCO_3^-/ CO_2 proportion).

The application of mass spectrometry to monitor concentration changes of important compounds in fermentation is in its early stage. Certain technical problems are anticipated (e.g., proper resolution of compounds to be measured, limitation of measurement speed by the diffusion rates of compounds through the membranes, preparation of sample to volatilize the compounds); however, these problems can be solved by technical improvements (e.g., selection of suitable membranes, application of selected ion-peak monitoring systems for resolution, and microprocessors for calibration—peak holding as well as tuning).

In view of the fact that more than one fermenter can be connected to one MS, the relatively high unit cost of the mass spectrometer is eliminated. In addition, it is anticipated that with the development of the methodology less expensive and more versatile mass spectrometers will appear on the market, bringing this analytical technique into the realm of widespread applicability.

15.23.3 Dialysis Systems

Development of plastic membranes with various molecular weight cutoff values has opened up the possibility for selective separation of molecules from the culture broth (on-line) and analysis off-line in readily available equipment. This same principle has been applied in mass spectrometers. In a recent study Zabriske and Humphrey (1978) described a dialysis system which is attached to a modified baffle in the fermenter. Compounds with low molecular weight are continuously dialyzed from the culture fluid into the water-buffer solution circulating in the dialysis tube. Glucose was measured by a glucose analyzer with a response time of 4-7 minutes to step changes in the culture broth's glucose content. The system is a modified version of a device developed by Glaxo Ltd. (England) for continuous determination of penicillin.

Advantages of the dialysis approach are (1) selectivity for the compound of interest, (2) continuous signal production (analog signal), and (3) relative simplicity of the sampling system. The potential disadvantage of the approach is the relatively long response time which is a function of the distance between the dialysis system and the analyzer and the assay procedure selected.

15.23.4 Automatic Wet Chemical Analyzers

A general purpose and sophisticated device which can automatically and semicontinuously determine a variety of organic and inorganic constituents of a culture broth is the Technicon AutoAnalyzer. This instrument consists of a train of modules which can pretreat, dilute, sample, react, and analyze the sample broth by a specific quantitative detection device (colorimeter, flame photometer, flowmeter, potentiometer, etc.) for the compound in question. Results of the analysis can be either recorded and/or sent to an interface in the form of a digital signal for further data processing. Typical of the assay procedures for fermentation processes which have been reported for the AutoAnalyzer are mono- and disaccharides, inorganic and organic nitrogens, phosphorus, amino acids, vitamins, and certain antibiotics. Although the AutoAnalyzer is considered an off-line device because of the relatively long elapsed time (typically 10-30 minutes) between the sampling and availability of results, it will find increased "on-line" application because of its accuracy, reliability, and versatility of available measurements. Process information obtained by this method has the potential for process monitoring and possible control response as well as post-experimental comparison of data obtained with on-line operating sensors and other wet-chemical analysers.

❑❑❑

16

AERATION AND AGITATION

16.1 INTRODUCTION

The primary purpose of aeration should be to provide microorganisms in submerged culture (or fermentation) with sufficient oxygen for metabolic requirements. So, supply of sterile air (oxygen) at constant rate throughout the vessel is known as **aeration.**

The primary purpose of aeration should ensure that an uniform suspension of microbial cells is achieved in a homogenous nutrient medium so that each cell can get equal opportunity to satisfy its substrate and oxygen requirements. So, uniform mixing of all the contents present in a fermenter is known as **agitation** (mixing).

The type of aeration–agitation system used in a particular fermenter depends on the nature of fermentation process and type of organism involved in fermentation process. Aeration and agitation are crucial processes to carry out successful fermentation process.

Mechanically agitated fermenters are used for fermentations carried out by filamentous fungal (e.g. **Penicillin**) and actinomycetal (e.g. **Streptomycin**) organisms. Production of Single–cell Protein [SCP] is an example of "**Non–agitated fermentation**" carried out by yeasts. In SCP fermentation, fermenter vessel (e.g. Tower fermenter) is aerated with fine bubble aerators without mechanical agitation. Here, aeration also provides sufficient agitation or mixing.

16.2 STRUCTURAL COMPONENTS INVOLVED IN AERATION–AGITATION SYSTEM

The major structural components of the fermenter involved in aeration and agitation processes are.

(a) **Impeller** (also know as Agitator or stirrer),

(b) **Stirrer glands, bearings** and **seals,**

(c) **Baffles,**

(d) **Sparger** (Aeration system).

16.2.1 Impeller/Agitator/Stirrer

An impeller is mounted to a shaft extending through a bearing in the lid of the fermenter and driven by an external power source. The external power source may be a motor or by direct drive. In recent designs, the impeller is driven by a magnetic coupling to a motor mounted beneath the fermenter.

(A) Impeller categories—Depending on the positioning of impeller blades in the fermenter, impellers are of two categories—

(I) **Horizontal Impeller**—Here, the positioning of the impeller blades is in a horizontal position in reference to the bottom of fermenter. These are of convenient nature.

(II) **Vertical Impeller**—Here, the positioning of the impeller blades is in a vertical direction in reference to the bottom of fermenter. In such category, the blades have to rotate in opposite direction of the gravitational force.

(B) Functions of the Impeller—

(I) To reduce or diminish the size of air bubbles to give a bigger interfacial area for oxygen transfer and to decrease the diffusion path. So that each microbial cell can get maximum possibility of acquiring oxygen.

(II) To facilitate complete mixing of contents like nutrients, substrates, oxyben (air) and microbial cells so that an uniform environment throughout the fermenter vessel can be achieved.

(C) Types of Impellers—Some commonly used various types of impellers are shown in figure 16.1

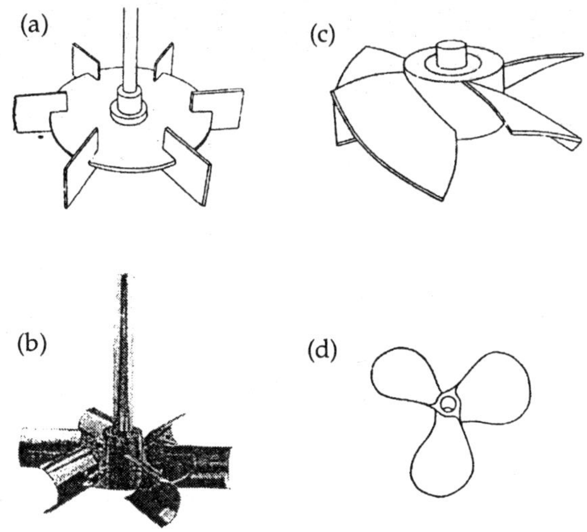

Fig. 16.1 : Some commonly used impellers—(a) Rushton disc turbine; (b) a concave bladed turbine; (c) a hydrofoil impeller; and (d) a marine propeller.

(I) Rushton disc turbine type—(Fig. 16.1a)—This is the most suitable type of impeller in a fermenter because it breaks up a fast air stream without itself becoming flooded in air bubbles.

(II) Concave bladed turbine—(Fig. 16.1b)—The impeller diameter is about 1/3 of the vessel diameter for gas dispersion.

(III) Hydrofoil impeller—(Fig. 16.1c)—Larger hydrofoil impellers with diameters of 0.5 to 0.6 times the vessel diameter are especially effective bulk mixtures and are used in fermenters for highly viscous mycelial broths.

(IV) Marine propeller type—(Fig. 16.1d)—Animal cell culture vessels typically employ a single, large diameter, low–shear impeller such as marine propeller. The marine propeller produces axial flow rather than radial flow. So, it is less efficient in breaking up a stream of air bubbles. It is also being flooded at a lower Vs (velocity of air based on cross–sectional area of vessel) than the disc turbine type.

(D) Size of Impeller—It is also important to consider size of the impeller, and where to position in the fermenter vessel.

In tall vessels, more than one impeller is needed for adequate aeration and agitation.

The ideal diameter of impeller (L) is one–third to one–half of vessel diameter above the base vessel (D). If L/D ratio is increased above 1.00, additional impeller must be incorporated into the vessel. Though at 1.00 diameter (D) distance along the impeller shaft, multiple impeller blades are spaced. In a tall narrow vessel, there may be a series of impellers on horizontal shaft. This helps in achieving high speed agitation.

Bottom entry and use of stirrer shaft also helps to achieve high-speed agitation.

Multi–rod-mixing impeller is used in novobiocin fermentation, fitted in turbine–stirred fermenter capacity of 15,000 dm^3. Same yield of novobiocin is procured but at about half of power consumption.

16.2.2 Stirrer glands, Bearings and Aseptic Seals

The entry of stirrer shaft in fermenter vessel may be from the top or bottom or side of the vessel. Top entery is most common found in majority of industrial fermenters. Generally, bottom entry is undesirable. Bottom entry is advantageous if more space is needed on the top plate for entry ports and the shorter shaft permits higher stirrer.

The satisfactory sealing of the stirrer shaft is very necessary to prevent contamination. The impeller and other devices are introduced in the fermenter by making holes on the lidtop of the fermenter. These holes should be sealed properly to prevent the entry of contaminants present in the external environment.

(A) Bearing—A porous bronze is fitted in the centre of the fermenter's top and another in a yoke directly above it for making the shaft to be rotated.

The bearings were passed into steel bushings, which screwed into position in the yoke and fermenter top.

(B) Aseptic Seals—Following four basic types of assembly are being used—

(I) Packed–gland seal [Stuffing box]—In this type, the shaft is sealed by several packing rings of asbestos or cotton yarn, pressed against the shaft by a gland follower. This packing may be difficult to sterilize properly because of unsatisfactory heat penetration and is necessary to check and replace the packing rings regularly.

(II) Bush seals—These seals have been developed especially for laboratory–scale fermenters. These simple bush seals are usually made from oilite or teflon. These seals may be used with or without oil seals.

(III) Mechanical Seal—The mechanical seal assembly is commonly used in large industrial fermenters. The seal is composed of two parts. One part is stationary in the bearing housing while the other part is rotating on the shaft. Both the components are pressed together by springs. Steam condensate is used to lubricate and cool the seals during operation.

(IV) Magnetic drives—The problems of providing a satisfactory seal when the impeller shaft passes through the top or bottom plate of the fermenter may be solved by use of a magnetic drive.

A magnetic drive consists of driving as well as a driven magnet. The driving magnet is held in bearings in a housing on the outside of the head plate and connected to a drive shaft. The internal driven magnet is placed on one end of the impeller shaft and held in bearings in a suitable housing.

Multiple ceramic magnets are also available. It has become possible to transmit power across a gap of 16 mm. Water can be agitated at speeds of 300 to 2000 rpm in baffled vessels of capacity up to 300 dm^3. Adequate power transmission between magnets is needed, when it is necessary to stir viscous mycelial broth or bacterial cultures with high air.

16.2.3 Baffles

Due to continuous aeration and agitation for a longer time in fermenter and the impeller action in the tank tends to spin the liquid in a circular motion which results in the formation of a vortex above the impeller.

Under the vortex formation, it is possible for the liquid to spin in such a manner that suspended solids or cells tend to remain in the same relative position in relation to the moving liquid without actually being mixed through it. The vortex further result in relatively small amount of medium surface is exposed to the atmosphere in the head–space of the tank. This condition causes decreased aeration.

To correct this situation or break vortex, vertical baffles (either 4 or 6 or 8) are often attached to the inside wall of the fermenter. These baffles disrupt the flow pattern of the liquid so that much greater mixing will occur. Baffles will break the vortex and also help in improving aeration efficiency. Baffles are metal strips or may be of wooden. Baffles are roughly one–tenth of the vessel diameter and attached radially to the fermenter wall.

Examples—(i) Wegrich and Shorter designed a fermenter (10,000 dm^3 capacity) containing four baffles–each of 10 cm wide and placed 3.5 cm away form the wall.

(ii) In a 300–dm^3 Chernap fermenter having 51 cm diameter, four baffles were fitted–each of 5 cm wide and 5 mm away form the wall.

16.2.4 Sparger (The Aeration System)

(A) Design—A sparger may be defined as a device for introducing air into the medium in a fermenter. It is important to know whether the sparger is to be used on its own or with mechanical agitation.

The sparger consists of a pipe with minute holes (1/64 to 1/32 inch) which allow the air under pressure to escape as tiny air bubbles into the liquid medium. Sparger holes smaller than this require too great an air pressure for economical bubble formation. Spargers in fermenters for growth of mycelia–forming organisms often utilize 1/4 –inch holes to prevent plugging of the holes by hyphal growth.

The air bubbles from the sparger are picked up and dispersed through the medium by the action of the impeller blades mounted above the sparger. Thus, the rotation rate of the impeller blade should be great enough to allow thorough mixing of the air in the medium.

The smaller the bubbles of air produced by the sparger, the greater is the bubble surface area and the more likely is the oxygen of that air to pass across the bubble boundary and dissolve in the liquid of the medium. However, small air bubbles require greater air pressure for their formation through the fine holes of the sparger.

In some fermentation tanks, an impeller is not utilized. The medium is stirred by the directed rush of air bubbles from a sparger at the bottom of the tank.

(B) Types of sparger—

(I) Porous Sparger—The porous sparger consists of sintered glass, ceramics or metal and has been used primarily on a laboratory scale in non–agitated vessels. The bubble size produced from such spargers is always 10 to 100 times larger than the pore size of the aerator block. Throughout air is low due to pressure drop across the sparger.

Problem of blocked time holes due to microbial growth is a limitation of porous sparger.

(II) Orifice Sparger—In small stirred fermenters, the perforated pipes are arranged below the impeller in the from of rings, approximately three–quarters of the impeller diameter. In most designs, the air holes are drilled at the under surfaces of the tubes, making up the ring or cross.

Brown and Peterson used a ring sparger of 17–cm diameter with six holes of 0.75 mm (1/32") in the ring below a 22–cm diameter impeller in a 30–dm^3 fermenter.

Such type of spargers without agitation have been used in yeast manufacture, effluent treatment and also for production of single–cell protein (fitted in pressure–cycle fermenter).

(III) Nozzle Sparger—Most modern mechanically stirred fermenters have a single open or partially closed pipe as a sparger to provide the stream of air bubbles. Ideally the pipe should be located centrally below the impeller and as far away as possible to ensure that the impeller is not flooded by the air stream.

The single–nozzle sparger causes a lower pressure loss than any other sparger and does not get blocked.

Nozzle sparger has limited application especially in mechanically stirred fermenters, e.g. vinegar process.

(IV) Combined Sparger–Agitator—On a small scale, Herbert *et al.* developed the combined sparger–agitator design, introducing the air *via* a hollow agitator shaft and emitting it through pores between the blades of a disc–turbine.

16.3 SIGNIFICANCE

Suitable aeration and mixing of microbial cultures is one of the main factors ensuring successful aerobic microbial processes (Brown, 1920, Tsao and Lee, 1977). Despite considerable success achieved in solving pertinent problems, many theoretical and practical questions still need to be answered. The rate of oxygen consumption by aerobic microorganisms may be so high that even at maximum saturation of the medium with air the oxygen present would be consumed within 15 s. Thus if oxygen supply is interrupted for a mere 15 s, the respiration of highly aerobic microorganisms such as *Acetobacter* may be seriously impaired. Such a dramatic effect resulting from the interruption of oxygen supply is not very frequent; however, insufficient oxygen always causes reduced yields of microbial products such as organic acids, enzymes or antibiotics and reduced growth of microorganisms.

Considerations of oxygen supply to microbial processes usually include simultaneous aeration and mixing since both processes affect the oxygen level and are necessary for ensuring a sufficient oxygen concentration in microbial cultures. Some microbial processes using yeasts and bacteria do not always require mixing of the culture by a separate agitator provided the fermenter contains correctly placed air spargers. On the other hand, the cultivation or filamentous microorganisms such as actinomycetes and moulds requires mechanical mixing.

16.4 THEORY OF OXYGEN TRANSFER

The transfer of oxygen from the gaseous phase proceeds by molecular diffusion. The double-film theory proposes the formation of a gaseous and a liquid film at the gas—liquid interface; all the resistance to the transfer of oxygen from the gaseous into the liquid phase is then concentrated in the liquid film (Fig. 16.2).

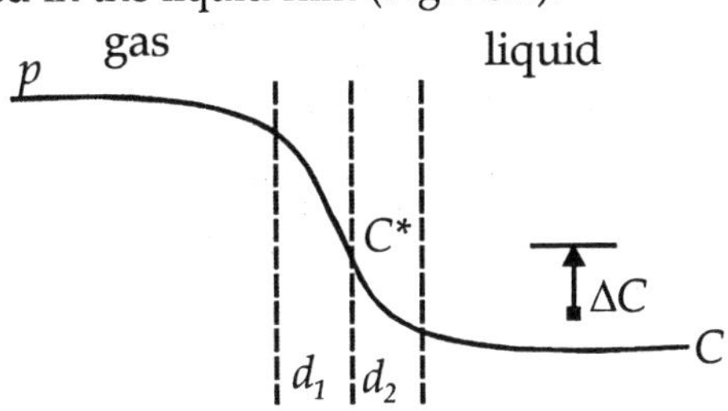

Fig.16.2 : Scheme of oxygen transfer from gas into liquid. d_1— film thickness, d_2 — liquid film thickness, C* — equilibrium oxygen concentration (function of T, PO_2 and physical properties of liquid), C — concentration of dissolved oxygen, ΔC — driving force.

If the oxygen concentration profile in the film is linear, no reaction is seen to take place in the film and the absorption of oxygen is physical in nature. In the case of chemisorption the profile is not linear. The magnitude of resistance of the liquid film depends on the physico-chemical properties of the liquid and the film thickness is affected by the mixing intensity. The double-film theory presumes an immediate equilibration between the oxygen concentration in the gas and in the liquid when the two phases come into contact.

$$P/He = C \qquad \text{Henry's law} \qquad (16.1)$$

where He — Henry's constant.

The solution of the diffusion process by the double-film theory is based on the first of Fick's laws

$$N = -D\left(\frac{dC}{dx}\right)_{x=d} \quad (16.2)$$

where N — diffusion flow

D — oxygen diffusivity

$\frac{dC}{dx}$ — gradient,

x — distance

The solution of the formula yields an equation

$$N = \frac{D}{d}(C^* - C) \quad (16.3)$$

$$N = k_L(C^* - C) \quad (16.4)$$

where k_L — coefficient of oxygen transfer.

When the rate of oxygen absorption is referred to the total surface area of suspension A in a volume V_L,

$$V_L \frac{dC}{dt} = k_L A(C^* - C) \quad (16.5)$$

$$\frac{dC}{dt} = k_L a(C^* - C) \quad (16.6)$$

where k_La — volume coefficient of oxygen transfer characterizing the aeration efficiency of the system. Here a = specific interface area in mass transfer, C^* = equilibrium oxygen concentration. C = actual concentration of dissolved oxygen at a given moment.

In a biological system consuming oxygen the equation has the form

$$\frac{dC}{dt} = k_L a(C^* - C) - rX \quad (16.7)$$

where r — respiration rate of the microorganisms

X — cell concentration.

If the rate of oxygen consumption exceeds the rate of absorption, the whole system will be oxygen-limited. Equations (16.6) and (16.7) hold for an ideally mixed system, i.e. homogeneous gaseous and liquid phases; this is not often fulfilled since in fermenters the oxygen concentration differs in different places.

If the oxygen concentration in the liquid is zero, equation (16.6) is simplified to the form

$$\frac{dC}{dt} = k_L a \,.\, C^* \quad (16.8)$$

which gives the maximum rate of oxygen transfer under given conditions.

The value of K_La is influenced by various factors such as the design of the vessel, the degree of agitation in a stirred vessel, the air flow rate, the viscosity of the culture broth and the presence of antifoam agent in the medium.

The design of the fermenter and operating conditions employed must be such that the organisms oxygen demand is met. The K_La of the system must be such that the desired amount of oxygen can be transferred per unit time (to meet the oxygen demand of the organism) in the presence of certain oxygen concentration in the medium.

16.5 FACTORS AFFECTING K_La AND OXYGEN TRANSFER RATE

16.5.1 Effect of Dissolved Oxygen Concentration

As seen from equation (16.6) the rate of oxygen transfer is maximal if the concentration of dissolved oxygen is nil. However, a drop of oxygen concentration below a certain limit, the so-called critical concentration C'c, brings about a decrease in the respiration rate which becomes, below this value, a function of oxygen concentration (Fig. 16.3). The critical oxygen concentrations for some microorganisms are shown in Table : 16.1. Oxygen limitation is demonstrated, like other substrate limitations, by a decline in the growth rate. In addition, a prolonged oxygen limitation in a medium may cause irreversible changes in the respiratory system of microorganisms. For practical purposes oxygen limitation is therefore undesirable and the oxygen concentration has to be kept at a level higher than the critical concentration. The critical concentrations are relatively low (of the order of one per cent of the saturation value); moreover, in a non-homogeneous environment, e.g. in cultures of filamentous microorganisms, local oxygen shortages arise which are reflected in an increased value of the critical concentration (the so-called apparent critical concentration, C'_C). The values of apparent critical oxygen concentration in some microbial processes may be quite high. Phillips and Johnson (1961) give the following values for *Aspergillus niger* in a synthetic medium (compared with values for *Escherichia coli*):

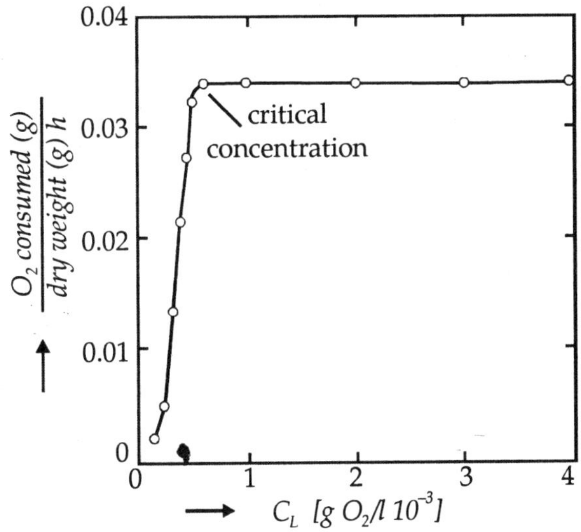

Fig. 16.3 : Dependence of yeast respiration rate on oxygen concentration (Hixson and Gaden, 1950)

	C (μmol 1^{-1})	C'_C (μmol 1^{-1})
Aspergillus niger	81	455
Escherichia coli	65	65

Table 16.1 : Critical oxygen concentrations typical for some microorganisms (Finn, 1967)

Microorganism	Temperature (°C)	C'_C μmol l^{-1}
Azotobacter vinelandii	30.0	18-49
Escherichia coli	37.8	8.2
	15.0	3.1
Serratia marcescens	31.0	about 15
Pseudomonas denitrificans	30.0	about 9
Yeast	34.8	4.6
	20.0	3.7
Penicillium chrysogenum	24.0	about 22
	30.0	about 9
Aspergillus oryzae	30.0	about 20

The apparent critical concentration C'_C responding to the 455 μmol l^{-1} represents, at the solubility of oxygen in the medium $C^* = 5.8$ mg l^{-1}, a value of $C = 1.93$ mg l^{-1}, or 33.3% saturation.

The driving force may be increased in two ways: by using oxygen-enriched air or by using increased air pressure during culture, for example by increasing the height of the liquid column. Both approaches lead to an increased value of C* and have identical effects.

16.5.2 Other Factors

Mixing has a positive effect on the oxygen transfer rate since it increases the value of k_La. The increase in k_La is caused by the following factors:

(a) A finer dispersion of air serves to increase the interface area, i.e. the value of a;

(b) Increased turbulence causes a reduction in the thickness of the liquid film at the interface and thus an increase in k_L;

c) The bubble hold-up in the fermenter increases.

The value of k_La in stirred vessels increases with rising superficial air velocity, i.e. the amount of air supplied referred to the fermenter cross-section, to reach a certain critical value. This value corresponds to the point of impeller choking, i.e. to the state when the impeller rotates virtually only in an air cushion.

The value of k_La within a certain range does not depend significantly on the bubble size since large bubbles with a relatively small surface create-turbulence in their vicinity, which reduces the thickness of the interface liquid film and enhances k_L; on the other hand, small bubbles create a lower turbulence but have a relatively larger surface (Finn, 1954). Only with very small bubbles (smaller than 1 mm in diameter) is an increase in k_La observed. Since the bubble diameter is about 10-fold larger than the diameter of the opening through which the air is forced into the fermenter, bubbles of this size can be produced only by using sintered glass or ceramic candles.

Surface-active substances may increase or reduce k_La. An increase in k_La may be due to a relative increase in the bubble surface, a drop in k_La may be explained by an accumulation of surfactants at the phase interface which increases the resistance to oxygen diffusion. Experimental data show that the addition of 0.02% sodium lauryl sulphate increases the rate of oxygen transfer about threefold whilst the addition of 0.01 % silicon oil or 0.02 % Tween 80 reduces the rate by half.

The presence of microorganisms, especially filamentous, brings about an increase in medium viscosity, thus causing a drop in turbulence and a decrease in k_La. The effect of viscosity on oxygen transfer was studied by Deindoerfer and Gaden (1955) and by Solomons and Perkins (1958), as well as by many other authors. They found that the transfer rate was considerably (40 to 85%) reduced with increasing viscosity of the medium. At higher viscosities the transfer rate does not rise even on increasing the impeller frequency. On the other hand, Phillips and Johnson (1961) found higher values of oxygen transfer rate in some bacterial and fungal cultures in comparison to the sulphite method. This fact may be explained, e.g., by the theory of direct absorption of oxygen by microorganisms from the gaseous phase (Tsao, 1968).

The medium viscosity may also be affected by various products of microbial metabolism, e.g. polysaccharides excreted into the medium; these products may also cause a decrease in oxygen solubility.

16.6 METHODS FOR DETERMINATION OF AERATION CAPACITY

The practical methods for determination of k_La may be divided according to their applicability to different systems, i.e. the system gas-liquid (model system) and the system gas-liquid-microorganism (culture system).

16.6.1 Methods Applicable to Model Systems

(A) Sulphite method (Cooper *et al*, 1944). The oldest method used for measuring k_La is based on determining the decrease in oxygen concentration chemically *via* oxidation of sulphite to sulphate in the presence or absence of a catalyst (Cu^{2+}, Co^{3+} — 10^{-3} to 10^{-5} kmol/m^3) according to the formula

$$SO_3^{2-} + \tfrac{1}{2}O_2 \xrightarrow[Cu^{2+}]{CO^{3+}} SO_4^{2-}$$

Since in this system the concentration of oxygen in the liquid C = 0, the aeration capacity is described by the formula

$$k_La = \frac{dC_{Na_2SO_4}/dt}{2C^*} \tag{16.9}$$

As the absorption rate is increased by the chemical reaction, it is necessary to know the accelerating factor of the method which may be determined as the ratio of values of chemical and physical absorption according to the relation.

$$f = \frac{k_L a}{k_L^{\circ} a} \qquad (16.10)$$

Since this factor is usually > 1, the measured $k_L a > k^o{}_L a$ and the calculated values of aeration capacity are overestimates.

The method is suitable only for comparing the efficiency of individual fermentation devices; owing to the considerable consumption of sulphite it cannot be used for testing large-volume fermenters.

(B) Static and volumetric method. Integration of equation (16.6) yields the formula

$$In \frac{C^*}{C^* - C} = k_L a.t \qquad (16.11)$$

and $k_L a$ is determined from an aeration experiment using a suitable measurement of concentration C throughout the aeration interval t. The two methods differ in the manner of determination of dissolved oxygen concentration C. In the static method the concentration is determined by an oxygen electrode; the air is fed in for a time t_1 and the steady-state value of oxygen concentration is measured after interrupting the aeration. The whole system is deaerated and again supplied with air for an interval t_2 ($t_2 > t_1$). In this way the whole range of saturation of liquid in the fermenter with oxygen is measured over the whole time t. In the volumetric method the amount of dissolved oxygen is assayed by desorption using a CO_2 stream and absorption of excess CO_2 into a KOH solution in a gas burette.

(c) Gassing-out method. This method is based on recording the exponential change in oxygen concentration in the liquid by an oxygen electrode after a transient gassing-out of oxygen by an inert gas, usually nitrogen. After the deaeration phase the nitrogen has to be removed from the space above the liquid by flushing with air; this prevents nitrogen inflow during the subsequent exponential change (Fig. 16.4)

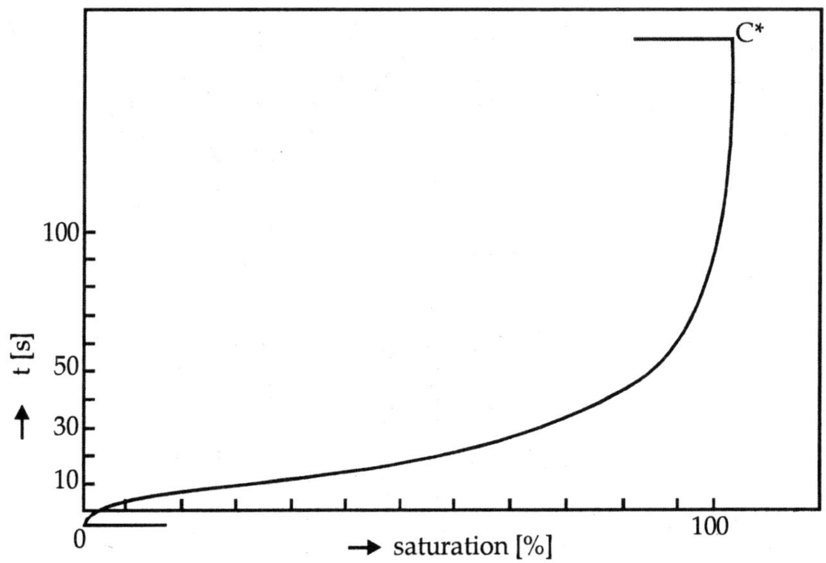

Fig. 16.4 : Time course of dissolved oxygen concentration in the gassing-out method.

Integration of equation(16.6) yields the formula

$$k_L a = \frac{1}{t - t_0} In \frac{C^* - C_0}{C^* - C} \tag{16.12}$$

16.6.2 Methods Applicable to Culture Systems

(A) Balance method. This method is based on the determination of changes in oxygen concentration in the gaseous phase. When the change in oxygen partial pressure in the liquid is also measured, the value of $k_L a$ can be calculated. Taking into account CO_2 respiration and referring the air flow to an inert gas (nitrogen), the rate of absorption, *Na*, may be defined by the relation

$$Na = \frac{V_{in}}{RV_L}\left[\frac{P_1 Y_1}{T_1(1 - Y_1 - Y_{1C})} - \frac{P_2 Y_2}{T_2(1 - Y_2 - Y_{2C})}\right] \tag{16.13}$$

where V_{in} — inert gas volume
V_L — liquid volume
R — gas constant
Y_1 — molar fraction of O_2 in the inlet
Y_{1C} — molar fraction of CO_2 in the inlet
Y_2 — molar fraction of O_2 in the out let
Y_{2C} — molar fraction of CO_2 in the outlet.

The rate of absorption is then a basic parameter in the calculation of other parameters of oxygen transfer, e.g.

$$k_L a = \frac{Na}{C^* - C} \tag{16.14}$$

C* has to be corrected for the oxygen concentration in outlet gases and a possible change in outlet gas pressure. It is also necessary to know the changes in oxygen solubility in the liquid due to the consumption of substrate and formation of products which may affect oxygen solubility C*.

The specific rate of oxygen consumption is defined by the relation

$$Q_{O_2} = \frac{Na}{X} \tag{16.15}$$

where *X* — concentration of cell matter.

Requirement for oxygen in the linear growth phase is defined by the formula

$$Y_{O/X} = \frac{Na}{\text{Pr}} \tag{16.16}$$

where *Pr* — productivity, and in the exponential phase of growth

$$Y_{O/x} = \frac{Q_{O_2}}{\mu} \tag{16.17}$$

Where μ — specific growth rate. The respiration quotient

$$R_Q = \frac{CO_2}{Y_1 - Y_2} \tag{16.18}$$

The advantage of this method is its wide applicability from laboratory to pilot-plant scale, with the exception of closed fermenters. It allows the measurement of oxygen transfer during culture without any interruption of the mixing and aeration of the culture.

(B) Integral balance method. A balance is performed in both the gaseous and the liquid phase with respect to the utilisation of dissolved oxygen by the microorganisms. The oxygen balance in the liquid phase is defined as

$$\frac{dC}{dt} = k_L a\left(\frac{P}{He}Y_2 - C\right) - Q(C, t) \tag{16.19}$$

The term $Q(C, t)$ must be assessed from growth experiments. The first right-hand term represents the change in oxygen concentration during growth, the second stands for its change during the formation of intermediates.

The aeration capacity may then be established by evaluating the parameters of the model on a computer using experimental data on biomass formation, dissolved oxygen concentration, substrate consumption, the time course of oxygen concentration in the outlet gases, or other quantities characterizing the cumulative oxygen consumption. Apart from the oxygen transfer parameters, the method permits the assessment of other kinetic parameters of growth and may also be used under conditions of oxygen limitation. In addition, it requires only the measurement of the quasi-steady oxygen concentration during the process and thus eliminates all the effects associated with the dynamic measurement by the oxygen electrode.

(C) Dynamic method. This method is based on determining the exponential change in oxygen concentration during culture using an oxygen electrode; the changes are brought about by switching-off and -on of the mixing and aeration in a fermenter (Fig. 16.5).

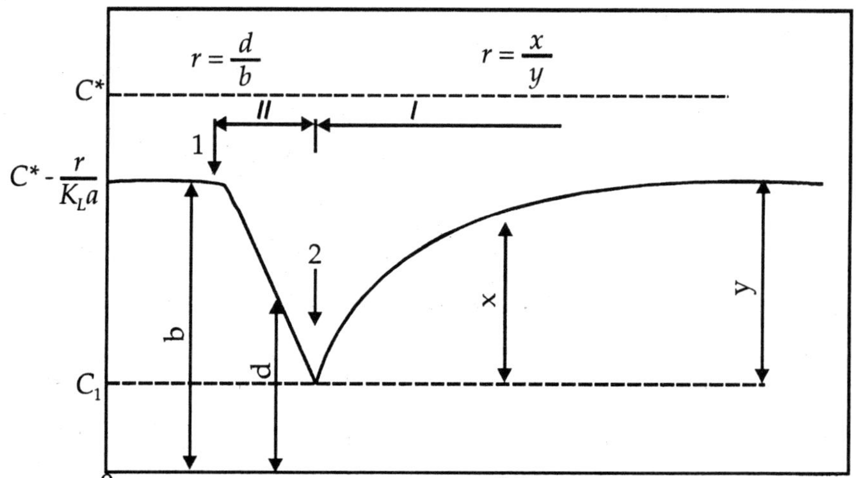

Fig. 16.5 : Time course of oxygen concentration in a culture after aeration switch-off (1) and switch-on (2)

The changes in oxygen concentration in the culture are described by equation (16.7) which is reduced in region II of Fig. 16.5 after aeration has been stopped, to the form

$$\frac{dC}{dt} = -rX \tag{16.20}$$

Equations (16.7) and (16.20) hold if oxygen concentration does not drop below the critical level for the microorganism, i.e. below the concentration at which the respiration rate begins to depend on dissolved oxygen concentration. The respiration rate may be determined from the slope of electrode recording in region II and aeration capacity may be calculated from the slope of oxygen concentration C plotted versus $-dC/dt + rX$. The limitations of this method are the same as those in the gassing-out method: it cannot be used in the range of oxygen limitation and is not suitable for testing large-volume fermenters because of the sudden energy loads on the stirrer during the switch-on.

16.7 MIXING (AGITATION)

The term mixing denotes the process of combining two or more different substances in order to obtain a homogeneous mixture. The term also denotes the maintenance of a given medium in an intensive motion which facilitates heat and mass transfer and the formation of suspensions. Mixing devices of conventional types transmit the supplied mechanical energy by means of stirrers into a flow of the mixed medium. It is advantageous to achieve high turbulence during mixing in order to ensure an effective transfer of heat and mass in the mixing device.

The measure of mixing efficiency is the degree of homogeneity, *I*, defined by the equation

$$I = 1 - \frac{1}{n}\sum_{i=1}^{n}\left|\frac{c_i - c_f}{c_O - c_f}\right| \tag{16.21}$$

The degree of homogeneity at a given moment is assessed by simultaneous withdrawal of samples from *n* representative points of the mixing vessel, determination of concentrations C_i of the component under study and insertion into equation (16.21) in which C_O is the initial and C_f is the final concentration of the component.

Mixing intensity is a term which is often used but still not precisely defined; the mixing intensities defined in different studies are valid only for a particular mixing device and process. The criteria for mixing intensity are usually the impeller frequency *n*, circumferential velocity of impeller blade tips v_c, the value of the Reynolds criterion for mixing Re_M or the specific power supply (power supply referred to unit culture volume) *P*/ *V*.

16.7.1 Agitators

The two principal types are low-speed and high-speed agitators. The peripheral velocity of low-speed agitators is in the interval $V_C \varepsilon < 0.5; 1.5 >$ ms^{-1} and revolution frequency $n\varepsilon < 20; 60 >$ rpm. Impeller diameter is close to the mixing vessel diameter *D*; this prevents the formation of incrustation on the walls and local overheating. Low-speed agitators are used for stirring highly viscous fluids.

Fig. 16.6 illustrates a range of low-speed agitators. Their dispersing efficiency is very low and their application in microbial industry is limited even though they have been tested in the culture of filamentous microorganisms (Steel and Maxon, 1966). Low-speed agitators include also worm stirrers which work at higher angular velocities ($n \varepsilon < 30; 2500>$ rpm). They are suitable for stirring highly viscous fluids; they need a low power input for medium circulation because they generate a low level of turbulence. Another type, paddle agitators, are formed usually by two paddles parallel with, or at an angle to, the stirrer shaft, of a diameter $d \varepsilon <0.5; 0.8 > D$. Their main disadvantage is a low mixing intensity and a high proportion of tangential component of fluid circulation.

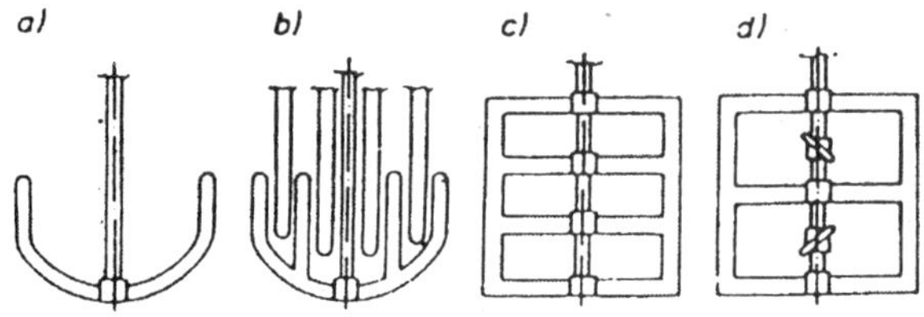

Fig. 16.6 : Low-speed agitators a) horseshoe, b) rake, c) gate, d) gate with tied paddles.

The circulation of fluid during mixing by high-speed agitators depends on the type of impeller used, (radial, axial, disc) and on the presence or absence of baffles at the mixing vessel walls.

A typical representative of radial impellers is a turbine with a separating disc and straight perpendicular blades, the so-called Rushton turbine impeller or, in short, turbine impeller. Its diameter is usually $d \varepsilon <0.25; 0.33> D$ and revolution frequency is chosen so that the circumferential speed v_c is in the interval $v_c \varepsilon <3; 9>$ ms^{-1}. Turbine impeller can be used in a wide range of viscosities of culture fluids ($\mu \varepsilon <10^{-3}; 20>$ kg m^{-1} s^{-1}); the maximum viscosity value decreases in large-volume vessels. The turbine impeller has an excellent dispersing efficiency and has therefore found wide application in industrial microbiology. The smaller the turbine impeller the higher the turbulence level generated by it; the dispersing efficiency also increases with decreasing size while circulation efficiency decreases.

A representative of axial impellers is the propeller stirrer which has the shape of a ship's propeller. Its dispersing efficiency is low whereas the circulation efficiency is very high. For this reason it is sometimes used in combination with a turbine impeller which acts as a disperser.

The propeller stirrer may also be used in a wide range of viscosities ($\mu \varepsilon < 10^{-3}; 10>$ kg m^{-1} s^{-1}) and its revolution frequency is adjusted so that the circumferential speed is in the interval $v_c \varepsilon < 6; 15>$ m s^{-1}.

Disc impellers formed by a rotating disc transmit energy into the culture fluid by viscous friction. They generate extremely high shear stress (high shear velocity gradients). In industrial microbiology they are therefore sometimes used for dispersion and foam breaking. They operate at high circumferential velocities ($v_c \varepsilon <5; 35>$ m s^{-1}) and cause only slight medium circulation.

16.7.2 Gas Supply Underneath the Stirrer

In small mixing vessels gas is usually introduced through an axially placed tube underneath the turbine impeller. Small diameter impellers have high revolution frequency and a very high dispersing efficiency; this efficiency decreases with increasing impeller diameter, i.e. with decreasing revolution frequency. In large-volume vessels the gas is therefore introduced through a distribution ring with a diameter equal to or slightly larger than the impeller diameter *d*. The number and diameter of distribution ring orifices can be determined from the equation (Leibson *et al.*, 1956)

$$\mathrm{Re}_t = \frac{V_g Q_S}{N_t d_B \mu_S}, \quad \mathrm{Re}_t \in <500; 2500> \tag{16.22}$$

where V_g is the gas volume flow through N_t nozzles with diameter d_t; index *S* denotes a fluid. The bubble diameter, d_B, in close vicinity to the nozzles is defined by the equation (Lehrer, 1971)

$$d_B = 0.8205\left(\frac{V_g}{N_t}\right)^{0.4} \tag{16.23}$$

The condition $d_B \leq 0.75\ X$ must be fulfilled; X_t is the distance between two neighbouring orifices in the distribution ring. If the condition is not met, the effective number of orifices, $N_{t,\,ef}$ has to be inserted into equation (16.23):

$$N_{t.ef} = \frac{0.75(N_t - 1)X_t}{d_B} + 1 \tag{16.24}$$

16.7.3 Dimensionless Criteria and Geometrical Similarity

Application of the theory of similarity to the dimensionless differential equations describing the liquid flow, and rearrangement of the equations, yields the following dimensionless criteria for characterizing conditions in the mixing vessel:

I. The Reynolds Number for mixing Re_M

$$\mathrm{Re}_M = \frac{nd^2 Q}{\mu} \tag{16.25}$$

where *n* is impeller frequency, d is impeller diameter, *Q* is specific weight of culture fluid and μ its dynamic viscosity. The calculation of the value of Reynolds' Number for mixing of non-Newtonian fluids requires the assessment of mean shear rate, D_M, for a given impeller frequency *n* in the mixing vessel according to the formula (Metzner and Taylor, 1960)

$$D_M = 11.5\, n \tag{16.26}$$

This correlation holds for d/D <.0.6 and $m < 1$. The apparent viscosity μ_D corresponding to the calculated D_M is obtained from the rheogram of the mixed fluid and is then inserted into equation (16.25).

II. The Euler Number for mixing Eu_M

$$Eu_M = \frac{P}{n^3 d^5 Q} \tag{16.27}$$

where P is impeller power input. Euler's Number reflects the dimensionless impeller power input.

III. The Froud Number for mixing Fr_M

$$Fr_M = \frac{n^2 d}{g} \tag{16.28}$$

which shows the ratio of inertial forces to gravitational forces; it is therefore used for describing processes in which gravitational forces play a significant part. Dimensional analysis yields

IV. Flow number Kp

$$Kp = \frac{V}{nd^3} \tag{16.29}$$

where V is the volume flow of the agitated culture fluid due to the stirrer. A number of K_p values have been published for turbine impellers; for the most part they are in the interval $K_p \varepsilon$ <0.5; 0.8>.

V. Flow number for aeration Kp_g is defined analogously to K_p:

$$Kp_g = \frac{V_g}{nd^3} \tag{16.30}$$

where V_g is flow of gas introduced underneath the stirrer.

16.7.4 BUBBLE HOLD-UP IN AN AERATED MIXING VESSEL

Many data and formulae have been published concerning the problems of determination of mean bubble diameter in the mixing vessel, d_{BM}, the gas hold-up ϕ and the interface area. They are usually restricted to a given geometrical arrangement and small-volume vessels, as well as to the physical properties of true Newtonian fluids, such as water or salt solutions. Of utmost importance is the absolute purity of the fluids since even the small addition of a surface-active substance alters substantially the surface tension and thus also the bubble size. The extrapolation of all such formulae to real devices has to be done with caution since their derivation is usually based on a pure liquid-gas system and even in this system deviations of several tens of per cent may be encountered.

The gas hold-up in a mixing vessel, ϕ defined as

$$\phi = \frac{V_g - 1}{V} \tag{16.31}$$

where V_g is the culture volume during aeration and V is the volume of liquid phase in the mixing vessel, can be estimated from various complicated formulae (Calderbahk, 1958) which hold for pure fluids of low viscosity.

Mean bubble diameter d_{BM} may be determined with the same restrictions; for fluids with properties close to those of water the determination may be done with the aid of the formula (Calderbank, 1958)

$$d_{BM} = 4.15\left(P_g / V\right)^{-0.4} Q_S^{-0.2} \sigma^{06} \phi^{0.5} + 0.0009 \qquad (16.32)$$

where Q_s is the density of the agitated fluid. However, in systems with different physical properties the d_{BM} value calculated in this way cannot be used because of a different bubble coalescence in the culture.

The range of bubble sizes during the aeration of non-Newtonian fluids, especially suspensions of filamentous microorganisms, is broader since very small bubbles are trapped among the filaments and, at the same time, large bubbles are formed due to increased coalescence.

The interface area a (the surface area of all bubbles in the mixing vessel referred to the volume of the aerated fluid) is often determined from the formula (Calderbank, 1958)

$$a = 1.44 \frac{\left(P_g / V\right)^{0.4} Q_S^{0.2}}{\sigma^{0.6}} \phi^{0.5} \qquad (16.33)$$

which is also, subject to the above limitations.

16.7.5 Turbine Impeller Hydrodynamics

Let us consider a turbine impeller rotating in a fluid. Fig. 16.7 shows the view of the rear face of the turbine blade relative to the direction of rotation. The fluid displaced by the front face of the blade fills the empty space behind the blade, forming vortices the centres of which can be connected by line *f*. These vortices are also subject to an excentric force which displaces them towards the outer blade edge. Hence two parallel lines of vortices arise behind the rotating impeller (Fig. 16.8). They are situated above each other, corresponding to the upper and lower end of the outer impeller blade edge. The direction of liquid flow in the vicinity of the impeller is shown by the streamlines *f* and *g*. Higher impeller frequencies may give rise to cavitation; the impeller blade displaces the fluid, causing a deformation of the streamlines while the return flow of the fluid forms vortices and the excentric force of their rotation gives rise to a negative pressure which releases the gases dissolved in the fluid in the form of bubbles. The impeller imparts its energy to the liquid-gas dispersion and its power input consequently drops; this is reflected in the intensity of stirring. The agitation of fluids must thus be performed under conditions preventing cavitation.

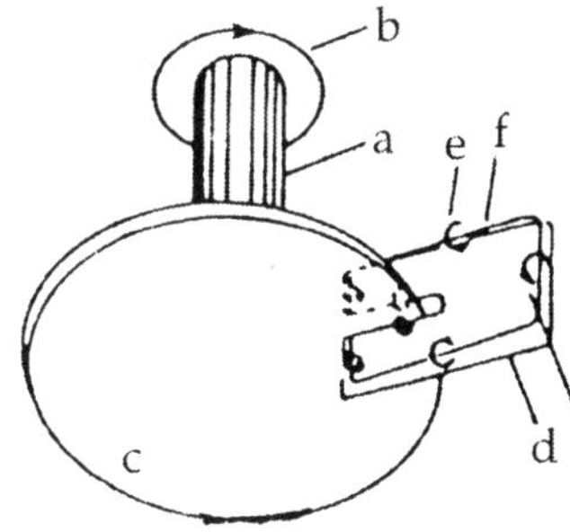

Fig. 16.7 : Scheme of turbine impeller a) shaft, b) direction of revolution, c) separating disc, d) blade, e) generation of vortices, f) line connecting vortex centres

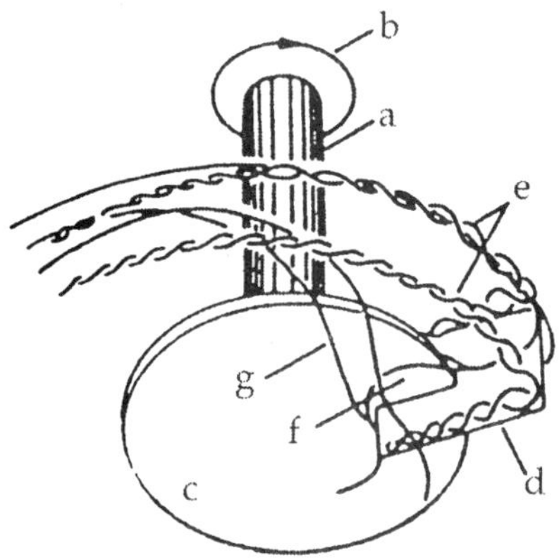

Fig. 16.8 : Scheme of turbine impeller action—generation of vortices a) shaft, b) direction of revolution, c) separating disc, d) blade, e) parallel lines of vortices, f) and g) streamlines.

When the gas is introduced below the turbine impeller (Fig. 16.9) it is displaced into regions with the lowest pressure, i.e. into the centre of rotation of vortices behind the impeller blade. The gas follows these vortices also during the outflow from the agitator. At very low flow rates the gas enters only the lower vortex while the upper vortex is formed by the fluid only. Only at higher flow rates does the gas also enter the upper vortex. With increasing gas flow the diameter of the two-phase region behind the stirrer also increases to give rise eventually to two parallel aerated layers *e* (Fig. 16.10). The bubbles are released at a greater distance from the impeller and the diameter of the bubbles also increases. At still higher gas flow rates the two layers merge and, on increasing the gas flow further, the impeller no longer disperses the gas sufficiently and, consequently, it rotates in a large bubble. This causes impeller choking and the gas passes vertically through the stirrer.

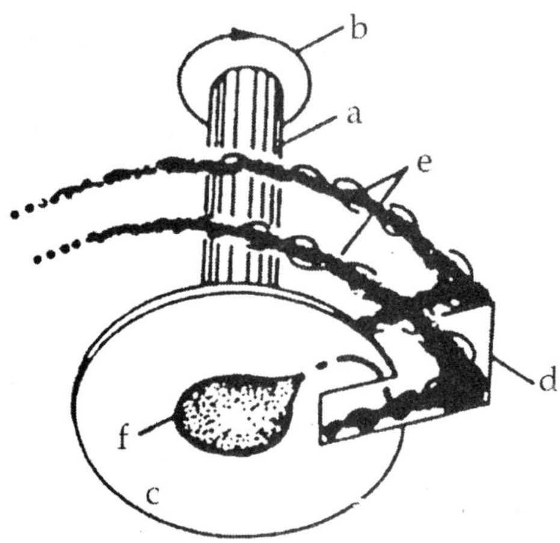

Fig. 16.9 : Scheme of turbine impeller action—gas supply underneath impeller a) shaft, b) direction of rotation, c) separating disc, d) blade, e) parallel lines of vortices, f) centre of vortex rotation.

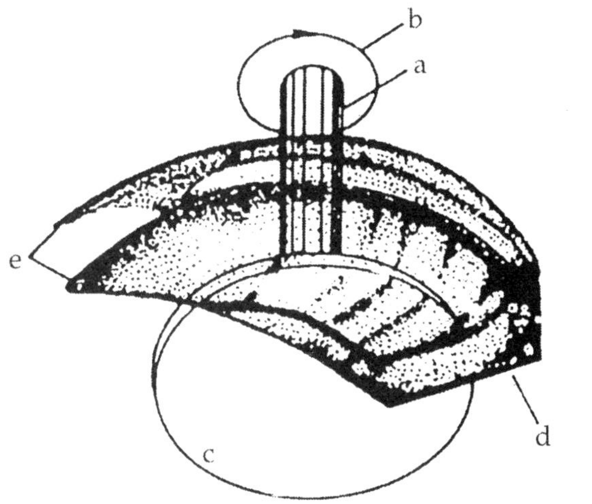

Fig. 16.10 : Scheme of turbine impeller action—generation of two aerated planes a) shaft, b) direction of rotation, c) separating disc, d) blade, e) parallel aerated planes.

16.7.6 Flow Pattern in a Mixing Vessel during Aeration

Let us consider the flow pattern in a mixing vessel with baffles (Fig. 16.11A). If no gas is introduced below the impeller *c* via the tube *e*, the agitated fluid forms streamlines *d* in the vessel. The stream of fluid from the stirrer towards the vessel walls carries with it large vortices generated by the impeller. In this region turbulence is highest and the conditions for mass transfer are optimal. The impact of the fluid on the vessel wall abolishes the pseudoturbulence and the fluid circulates in the vessel with a substantially lower level of turbulence. A mixing vessel with a turbine impeller is therefore modelled in a simplified manner as an ideal mixer (in which perfect homogenisation is achieved immediately) and a circulation zone in which there is practically no heat and mass transfer. The highest velocities of the fluid are attained in the stream from the rotating impeller and along the walls. In other places in the vessel flow velocities are lower, attaining a minimum in the centres of the circulation loops. From Fig. 16.11A we may easily see why the application of N turbine impellers on a single shaft does not yield a power input N times the input for a single impeller. The fluid flowing in a vessel with one impeller towards the centre of the vessel is retarded at the bottom owing to friction while at the surface the mechanical energy is spent on its uneven backstream. In a space between two impellers no such losses occur in the stream towards the centre of the vessel; moreover, the circulation loops of the two impellers cause a mutual acceleration of the streams. The energy expenditure during the deflection of the fluid stream from the wall towards the centre is also lower at the point of contact of the two flows than at other points. With two impellers, the total power input P_2 is only $\sqrt{2} = 1.414$ times the power input of a single impeller P_1; generally we may write

$$P_N = P_1(0.586 + 0.414\,N) \tag{16.34}$$

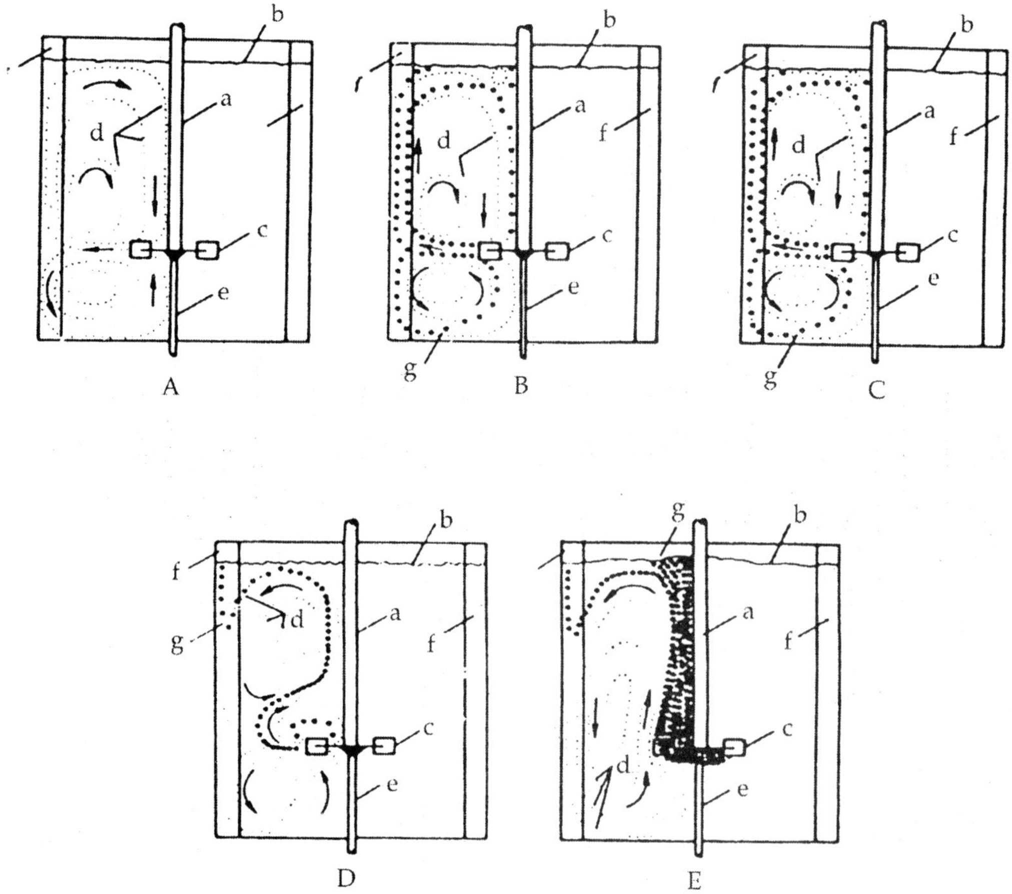

Fig. 16.11 : Flow pattern in a mixed vessel during aeration a) impeller shaft, b) liquid surface, c) impeller, d) streamline, e) gas supply, f) baffle, g) gas bubbles. Explanation see text.

If a small amount of gas is introduced below the impeller (Fig. 16.11B) it becomes completely dispersed, small bubbles continuously follow the particles of the fluid and are even carried with the upper circulation loop back to the impeller. The small bubbles are hardly affected by the excentric force due to the rotation of the fluid in the vessel; at higher gas flow rates this force (Fig. 16.11C) causes the displacement of the bubbles towards the centre of the mixing vessel due to the pressure of higher-density fluid. This process gives rise to another circulation loop at the fluid surface, *b*, with an opposite direction of rotation. With increasing gas flow rate the impeller creates bigger bubbles which are rapidly decelerated by the fluid. This even causes a back suction of the bubbles into the mixer (Fig. 16.11D) and the displacement of other bubbles towards the centre of the vessel is more pronounced; the circulation loop with the opposite rotation direction increases. The liquid flow velocity is so low that practically no gas is transferred into the lower loop. On further increase in the gas flow rate the impeller becomes choked (Fig. 16.11E) and the gas rises to the surface. The fluid gains momentum only from the rising bubbles which carry it along. There is therefore only one circulation loop in the vessel; the impeller has no effect on the mass transfer and the rate of mass transfer in a vessel with a choked impeller is thus very low.

Aeration aims at attaining the highest possible homogeneity of distribution of bubbles in the agitated fluid; the most suitable situation for the purpose is that shown in Figs. 16.11B and 16.11C. However, in most cases, the low gas flow (Fig. 16.11B) will not be sufficient and the flow pattern in an appropriately aerated vessel will closely resemble the schematic pattern shown in Fig. 16.11C. The situation shown in Fig. 16.11D is not convenient because of the low dispersing efficiency of the impeller and the shortage of dissolved gas in the lower circulation loop.

16.7.7 Homogenisation time

Homogenisation time T is defined as the time necessary for complete homogenisation of the culture. The simplest way to determine it is the addition of a tracer substance into the mixing device and studying the time course of its concentration changes. Depending on the type of the tracer substance, the momentary concentration at a representative point is assayed by conductometric, thermic or optical methods. The homogenisation time is defined by the formula

$$T = C\frac{V}{\dot{V}} = C\frac{V}{Kpnd^3} = CT_p \qquad (16.35)$$

where V is the volume of the mixing vessel, $\dot{V}$ is the volume flow of the agitated medium by the impeller, T_p is the mean circulation time, i.e. the mean time between two passages of a fluid particle through the stirrer region. For an ideal mixer a value of $C = 1$ may be expected (sufficient homogenisation is achieved by a single passage through the stirrer region), in practice a value of $C = 5$ is recommended. The mixing by turbine impellers in the high-turbulence region (Re_M 50000) may be described by the formula (Norwood and Metzner, 1960)

$$T = 4.1\sqrt{\frac{HD^3}{\left(n^4 d^{11}\right)^{1/3}}} \qquad (16.36)$$

17

Fluid Rheology

17.1 INTRODUCTION

In fermentation technology research, the broth rigidity or viscosity is important. Its importance relates to the power consumption in mixing of fermented mash.

The scale–up of microbial processes and the study of momentum, heat and mass transfer in industrial fermenters requires a knowledge of the flow properties of media as these properties fundamentally affect the process efficiency.

The flow of liquids is the subject of **rheology,** a discipline investigating the relations between different deformations of matter (elastic, plastic, flow) and their causes. Rheology is also concerned with the relations of macroscopic properties of fluids with their microstructure; on the basis of these relations it makes predictions concerning the kind and magnitude of deviations from Newtonian behaviour.

Fluids which obey Newton's law are called **Newtonian fluids** and their viscosity depends solely on the molecular structure of the fluid and on the properties of state (temperature and pressure). **Viscosity** is a measure of consistency of the liquid, i.e. of its resistance against flow (or the magnitude of the intrinsic friction between molecules) and, together with density represents a basic hydrodynamic parameter.

The current state of the rheology of non–Newtonian fluids does not allow us to describe their behaviour by a single flow equation. Instead, empirical formulae are used which show the best fit to experimental data. These formulae have no relation whatever to the basic understanding of deviations from Newtonian behaviour; this understanding is itself as yet only empirical.

17.2 RHEOLOGICAL BEHAVIOUR

Accurate viscosity/rheology measurement in such fluids is complicated by their non-ideal behaviour following Power law. Rheological behaviour of glucoamylase broth during the course of the fermentation is an example. The rheological characteristics of this fermentation broth exhibited apparent viscosity of about 500 c.p. Rheologists have assumed an empircial power law relationship of this fermentation broth fluid given by the following relation.

$$\tau = K\left(\frac{d\upsilon}{dy}\right)^n = K\upsilon^n \tag{17.1}$$

Here τ is the shear stress, $(d\upsilon/dy)$ is the velocity gradient, K is the fluid consistency index and n is flow behaviour index. The glucoamylase fermentation broth exhibited a value of K in the range 2-35 dyne/cm^{2}sec. and that of n varied from 0.3-0.9. This characteristic is shown in Fig. 17.1.

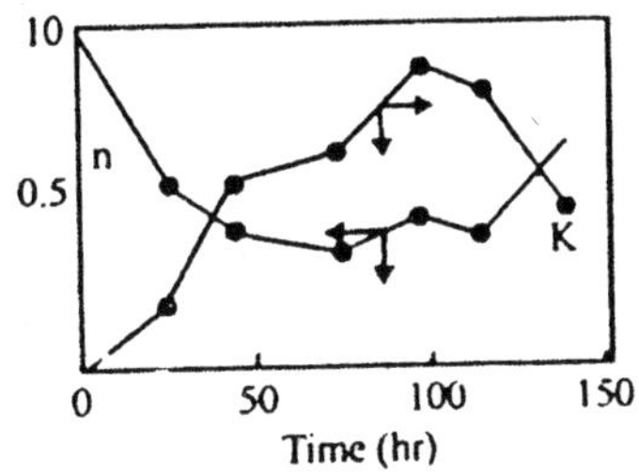

Fig. 17.1 : Rheological behaviour of a glucoamylase broth during fermentation course.

Gellan fermentation broth is an inelastic shear thickening fluid. Another typical fermentation manifesting power law fluid behaviour is novobiocin antibiotic production. Its behaviour is shown in Fig. 17.2. This behaviour has been reported to be of plastic fluid nature. However, the diversity of broth characteristics may be so great that no generalization is possible. From above examples and literature information it appears that fermentation fluid rheological variables are many and may be grouped as discussed in following sections.

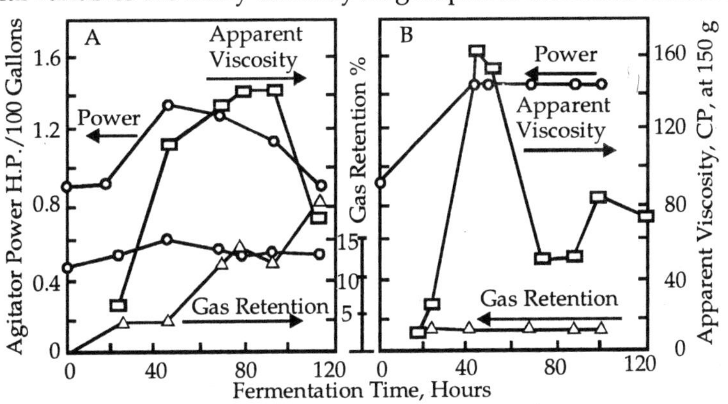

Fig. 17.2 : Changes in fluid physical properties during novobiocin fermentations in 66-gallon fermenter showed influence of power requirement. For high power, agitator speed = 266 r.p.m., air rate = 0.875 cubic feet per minute; for low power, agitator speed = 190 r.p.m., air rate = 0.612 cubic feet per minute. A. Normal fermentation. B. Two litres of antifoam added during first day.

17.3 RHEOLOGICAL VARIABLES

17.3.1 Cell Variables

Various microbial cell related variables may influence rheological characteristics of fermentation systems. Among them study on cell density, cell size, the electrostatic property of the cell surface, specific heat and heat conductivity of cells, properties of cell, suspensions like morphology, viscosity and surface tension etc has been felt to be of great significance depending on the nature of system. Flocculation property of cells is also an important variable.

17.3.2 Equipment Variables

The equipment design variables effecting rheology of fermentation fluid include: a) geometry of the vessel and internal fittings and b) maximum rating of agitator motor and air compressor.

17.3.3 Equipment Capacity Variables

These relate to mass and heat transfer and mixing characteristic related variables of the equipment hardware.

17.3.4 Fermentation Process Operating Variables

Rheology of fermentation fluid may vary depending on the existing fermentation process operating variables such as air flow, over pressure, stirrer speed, nutrient and acid/ base addition, antifoam addition, coolant flow, temperature etc.

17.4 CORRELATIONS OF RHEOLOGICAL VARIABLES

A large number of industrial bioprocessing use filamentous microorganism in which highly viscous broth with non - Newtonian rheological behaviour are produced. A great number of factors influence broth rheological behaviour including cell mass concentration and morphological characteristics. Besides correlating broth viscosity with cell concentration newer correlation could also seen between rheological parameters and physiological state of the cells in the culture. In dimensionless form it could be expressed by the following correlations in which C_1, C_2, C_3, C_4 and are dimensionless paramenters. Thus relative specific growth rates μ_x/μ_m C'_1, C'_2, C'_3, C'_4 relative, specific enzyme activity (A), relative production rate (E_A/E_{Am}) and relative specific consistency index (K/K_m) could be correlated. These conditions fitted very well in rheological changes of *Aspergillus awamori* NRRL, 3112 fermentation broth during amyloglucosidase production from starch. Above models known as modified Andrews model could show good fit with experimental results (Table 17.1).

$$\frac{\mu_x}{\mu_{xm}} = \frac{C_1}{\left[1+\frac{C_2}{(k/k_m)}\right]+\left[\frac{k/k_m}{C_3}\right]^{C_4}} \tag{17.2}$$

$$\frac{E_A}{E_{Am}} = \frac{C_1'}{\left[1+\left(\frac{C_2'}{K/K_m}\right)\right]+\left[\frac{K/K_m}{C_3'}\right]^{C_4'}} \quad (17.3)$$

Table 17.1 : Best fit of modified Andrews model (Equations 17.2 & 17.3)

Parameters	(μ_x/μ_{xm}) $R_1 - R_6$	(μ_A/μ_{Am}) $R_2 - R_6$	(μ_A/μ_{Am}) R_1
C_1 or C'_1	1.1 (fixed)	1.1	1.1
C_2 or C'_2	0.97×10^{-3}	1.08×10^{-3}	2.41×10^{-3}
C_3 or C'_3	0.226	0.681	0.262
C_4 or C'_4	1.241	0.849	1.680

$R_1 - R_6$ refers to viscometer rotors.

17.5 ADVANCEMENT IN VISCOSITY MEASUREMENT

Viscosity is an important parameter in industrial fermentation broth rheology. Viscosity measurement is widely used in industrial, laboratory and other research applications. In principle viscosity is the measurement of fluid friction, and it is defined as the ratio of shear stress to shear rate. Fluids that have a linear shear stress to shear rate relationship are termed Newtonian fluids. Accurate viscosity measurements of these fluids can be made over a wide range of shear rates. Most biological fluids, including fermentation fluids, have shear stresses that do not have a linear relationship to shear rates and they are termed non-Newtonian fluids.

Reproducible viscosity measurements is very important. In biological fluids in recent years viscosity measurements are often made in a low shear regime using falling— rolling ball viscometry in glass capillary pipette. In more recent years development of a computer interfaced falling ball viscometer has been reported. For measurement of viscosity this viscometer is calibrated at a given angle of fall of the ball. The inverse velocity of the steel ball is linear with respect to viscosity for Newtonian liquids. This relation is given by $1/V = m\,\mu + b$, where V is the velocity of the ball and μ is the viscosity value, m and b are constants. The calibration process defines this linear relation between 1/V and μ for several angles of the capillary tube with liquids of known viscosity.

❑❑❑

18

FOAM FORMATION AND ITS CONTROL

18.1 INTRODUCTION

The foam production in a submerged cultivation of microorganism is an integral part of fermentation and is caused by the introduction of gas as well as by the presence of growing cells in a medium containing sugars, salts, metabolites and surface active agents stabilizing the foam bubbles.

During growth, on the one hand, the occurrence of foam may imply a large phase contact surface leading to increasing oxygen transfer rates. On the other hand, intensive foam formation frequently may reduce the actual volume of the liquid; it may result in product loss and finally chances of contamination increase as a consequence of overflowing foam.

The objective of the present chapter is to understand the cause and mechanism for the foam production and describe the methods of its prevention/destruction in the fermenters.

In the present context the interest in foam formation during fermentation is purely negative as the main concern is to discover how to prevent its appearance or if this is impossible or impractiable how to destroy it. However, a review of the mechanism of foam formation is a necessary introduction as it serves to indicate the different types of action that may be taken to prevent foaming or to destroy foam once it has been formed.

18.2 MECHANISM OF FOAM FORMATION

Many fermentations require high rates of aeration and vigorous agitation if they are to yield their best. These conditions, imposed on the types of nutrient solutions commonly employed, provide an ideal situation for foam formation. The classical definition of a foam is that it is a dispersion of gas in liquid in the form of thin films separating gas bubbles. In most of the bacterial fermentations and some tissue cultures foam of this nature can be seen on the surface of the liquid. There is a clear division between the foam and aerated liquid underneath it. However, in some thick mycelial fermentations no such clear distinction exists: there is rather a gradual increase in the proportion of gas dispersed in the liquid, increasing from the bottom of the fermenter upwards. It is only near the top that the proportion of gas to liquid is such that the dispersion is recognized as foam. Two kinds of foam exist: there are unstable foams continuously approach the equilibrium state, and metastable foams in which the progress to equilibrium is arrested.

It is generally accepted that the pure liquids do not foam i.e. the stability of foam is very-very low. So the property which distinguishes a foaming solution from a non-foaming solution is its ability to resist local thinning of a liquid film whilst allowing general thinning to proceed. In a foaming solution the liquid films must possess elasticity. According to Gibb's theory of surface elasticity, If a film of liquid containing a surface-active agent is locally stretched, the surface tension of that part increases because the solute is positively adsorbed on the surface. The stretching leads to decrease of solute concentration within the film, and therefore, a rise in equilibrium surface tension. This force tends to resist local thinning and to restore the thin area.

In addition to the surface layer moving to restore local thinning, it is suggested that this movement drags the bulk liquid underneath it, and so enhances the purely surface effect.

Plateau's theory of foaming involves surface viscosity; this viscosity is similar to conventional viscosity except that it is in two dimensions only. The main function of surface viscosity is to reduce the rate of loss of liquid from films as they thin due to drainage. This effect is most strong in thin films, but rapid drainage is possible in relatively thick films where surface viscosity has less effect. Some protein solutions have high viscosity which can almost amount to surface rigidity. Rigidity itself does not necessarily favour foaming because a rigid structure may be brittle and lack elasticity. The theories so far mentioned can account for the formation of unstable foams.

The occurrence of metastable structures is explained by electrical double—layer repulsion. Thin films derive their stability from repulsion between ionic double layers formed by adsorption on the two sides of the film. This effect is particularly important in black films, that is, the thinnest possible films obtainable with a given solution.

18.3 EFFECTS AND PATTERNS OF FOAMING

In most microbiological processes, foaming is a problem. It may be due to a component in the medium or some factor produced by the microorganism. The most common cause of foaming is due to proteins in the medium, such as corn-steep liquor, Pharmamedia, peanut meal, soybean meal, yeast extract or meat extract (Schugerl, 1985). These proteins may denature at the air-broth interface and form a skin which does not rupture readily. The foaming can cause removal of cells from the medium which will lead to autolysis and the further release of microbial cell proteins will probably increase the stability of the foam. If uncontrolled then numerous changes may occur and physical and biological problems may be created. These include reduction in the working volume of the fermenter due to oxygen-exhausted gas bubbles circulating in the system (Lee and Tyman, 1988), changes in bubble size, lower mass and heat transfer rates, invalid process data due to interference at sensing electrodes and incorrect monitoring and control (Vardar-Sukan, 1992). The biological problems include deposition of cells in upper parts of the fermenter, problems of sterile operation with the air filter exits of the fermenter becoming wet, and there is danger of microbial infection and the possibility of siphoning leading to loss of product.

Hall *et al.*, (1973) have recognized five patterns of foaming in fermentations:

1. Foaming remains at a constant level throughout the fermentation. Initially it is due to the medium and later due to microbial activity.
2. A steady fall in foaming during the early part of the fermentation, after which it remains constant. Initially it is due to the medium but there are no later effects caused by the microorganism.
3. The foaming falls slightly in the early stages of the fermentation then rises. There are very slight effects caused by the medium but the major effects are due to microbial activity.
4. The fermentation has a low initial foaming capacity which rises. These effects are due solely to microbial activity.
5. A more complex foaming pattern during the fermentation which may be a combination of two or more of the previously described patterns.

18.4 FOAM BREAKING

The general principles of foam breaking by the following three methods will be discussed here: (A) Mechanical method of Foam Breaking (B) Physical Foam Breaking (C) Chemical Foam Breaking.

18.4.1 Mechanical Foam Breaking

Under this category there are many methods of foam breaking. Mechanical foam breakers may be divided into those types in which the foam to be broken is contained within the fermenter, and those in which the foam and exhaust air are separated outside the fermenter the liquid being returned.

The simplest form of internal foam breaker consists of rotating paddles or vanes attached to the stirrer shaft. Their action is most effective when the foam is well drained and has lost its self-healing properties. The working principle of these centrifugal foam breaker is simple. When the working part comes into contact with the foam the centrifugal forces act on bubbles and destroy them. The design of rotating paddles or vanes is very important in determining the efficiency of breakage.

Other method could be the use of cyclone where the foam entering it flow along a labyrinth and the liquid under the influence of centrifugal force is thrown at the labyrinth wall while the gas is forced to the cyclone centre and discharged through the outlet pipe. Successful application has found a bioreactor design with a built in cyclone.

Foam may also be broken by accelerating it through a nozzle, in which the kinetic energy of jet is responsible for the destruction of the liquid film of foam bubbles.

18.4.2 Physical Methods of Foam Breaking

Among the physical methods applied the ultrasonic method is the most popular one, but others occasiona1ly used are thermal and electrical treatment of foam. Ultrasonic whistles have been used to contain foam within a fermenter. The effectiveness of this method is attributed to four factors, these are: acoustic pressure, the unidirectional radiation pressure, the induced resonant vibrations in the bubbles, and the turbulence produced by the sonic wind.

Some times thermal method like bringing the foam in contact with a heated surface has been employed where foam bubbles are broken due to intensive expansion of bubbles and evaporation of moisture. The passage of electric current through foam can create forces which act differently on liquid and gas, thus breaking the foam. The exact mechanism of this phenomenon has not been investigated and not much profound literature is available about the efficiency of this method.

One of the main advantage of mechanical/physical foam destruction is that no extraneous agents need to be added to the system, thus do not create any further downstream processing problems. However, the disadvantage of mechanical/physical foam breakers are that they are often of complicated design and expensive to operate. This later aspect is particularly true of ultrasonic foam breakers, as not only do they need relatively large volumes of air for their whistles, but the pressure required is probably higher than the normal supply pressure of air used for sparging. Simple rotating vanes are cheap to fit and cost little to operate, and they may help to conserve chemical anti-foam agents as they are unlikely to control foam by themselves.

18.4.3 Chemical Methods of Foam Breaking

The use of chemical agents to control foam during fermentation is well established. Their action is fundamentally different from that of mechanical/physical foam breakers in that they can prevent foam formation as well as destroy existing foam; mechanical/physical devices destroy foam after it has been formed.

The properties of an anti-foam agent used in a fermenter fall into three attributes. There are those properties needed for it to function as an antifoam agent, those dictated by the fact that it is to be added to a living organism and finally it must not interfere with other analytical systems such as pH probes or dissolved-oxygen concentration detectors. Because the agent is to be added to the fermentation it must neither be toxic nor inhibit the process of the fermentation. Toxicity tests are conveniently carried out in shakeflasks, not only at the effective concentration of a single addition of the agent, but also at the concentration likely to have accumulated by the end of the fermentation.

The presence of any anti-foam agent should not interfere with any subsequent processing of the broth during extraction. For instance, it should not gum up ion-exchange-resin particles, and it should not promote emulsion formation during solvent extraction or broth concentration. As one must add the agent in a sterile manner, the act of sterilization by steam or dry heat must not worsen its properties. It is also worth finding out if the agent is corrosive at sterilization temperatures, or if it forms corrosive products.

Anti-foam agents is usually added to a fermenter on demand that is when the batch is foaming heavily. In this case, the first requirement is that it should act quickly and suppress the foam. Other desirable properties of the agent are that it should be effective in low concentrations, and that its action should last for as long as possible. It is a common practice to emulsify the agent with a carrier to increase its ability to spread. The ability of an agent to spread is given by the spreading coefficient S.

$$S = Y_F - Y_A - Y_{AF}$$

Where Y_F and Y_A are the surface tensions of foaming liquid and the antifoam agent and Y_{AF} is the interfacial tension. The greater the value of S the greater is the spreading power of the agent. The main function of an anti-foam agent is to eliminate surface elasticity, but it can also act by reducing surface viscosity and thus promoting more rapid drainage. The following section primarily deals with the different types of anti-foam agents.

18.5 ANTI-FOAM AGENTS

18.5.1 Properties of an Ideal Antifoam Agent

Antifoams are surface active agents, reducing the surface tension in the foams and destabilizing protein films by (a) hydrophobic bridges between two surfaces, (b) displacement of the absorbed protein, and (c) rapid spreading on the surface of the film (Van't Riet and Van Sonsbeck, 1992). Other possible mechanisms have been discussed by Ghildyal *et al.* (1988), Let and Ty-man (1988) and Vardar-Sukan (1992).

An ideal antifoam should have the following properties:

1. Should disperse readily and have fast action on an existing foam.
2. Should be active at low concentrations.
3. Should be long acting in preventing new foam formation.
4. Should not be metabolized by the microorganism.
5. Should be non-toxic to the microorganism.
6. Should be non-toxic to humans and animals.
7. Should not cause any problems in the extraction and purification of the product.
8. Should not cause any handling hazards.
9. Should be cheap.
10. Should have no effect on oxygen transfer.
11. Should be heat sterilizable.

The following compounds which meet most of these requirements have been found to be most suitable in different fermentation processes (Solomons, 1969; Ghildyal *et al*, 1988).

18.5.2 Examples of Anti-foam Agents

The chemical anti-foam agents suitable for use in fermentations can be classified as follows:

(i) Alcohols; stearyl and octyl decanol
(ii) Esters
(iii) Fatty acids
(iv) Fatty acid derivatives
(v) Silicones

(vi) Sulphites and Sulphonates

(vii) Miscellaneous; Alkaterge C, Oxazaline, Poly-propylene glycol

In addition to this wide range of chemical types there are other factors to consider. The first is the possible use of a carrier, secondly there is the problem of synergistic action. A carrier in generally an inert liquid, with respect to anti-foam action is which the anti-foam agent is dissolved or with which it is emulsified. The simplest view of the action of a carrier is that it serves as a reservoir from which the active anti-foam agent is slowly released. However, this does not fully explain its action as it has been shown that the use of a carrier accelerates and prolongs the effectiveness of the action of the anti-foam agent. Quite apart from this incompletely understood function a carrier may bring a further practical advantage by reducing the viscosity of the agent and so make it easier to add to a fermenter. Lard oil, liquid paraffin or castor oil are used as carriers.

The synergestic effect may be loosely described as the mutual enhancement of action when two different agents are mixed. Their effectiveness when combined is greater than the effectiveness of either one used separately. This effect may be exploited in two ways. The first is to mix the two agents prior to addition to the fermenter. The second is to use two different antifoam agents alternatively during the course of a fermentation.

(A) Alcohols

The alcohols are one group of compounds finding wide use as antifoam agents. Octadecanol is perhaps the most frequently used, either alone or with the addition of a carrier. It has been used with cold pressed lard oil to control foam during penicillin fermentation, while 3% alcohol in mineral oil has proved to be effective for the citric acid fermentation. The alcohols derived from fatty acids are known to be most effective anti-foam agents.

(B) Vegetable Oils

Vegetable oils have also long been used in fermentation to suppress foam formation and in addition they serve to control pH because they are metabolised. Some of these vegetable oils are cotton seed, linseed, soybean, olive oils as well as castor oil, sunflower oil, rapeseed oil and cod liver oil.

(C) Silicones

In addition to alcohols and vegetable oils there are, however, compounds now available specifically designed to destroy foam, the most notable class, being the silicone oils. There are two forms of silicone fluids available, the pure polysiloxanes in a carrier, or aqueous emulsions containing about 10% silicone.

(D) Polyglycols

Polyglycols have similarly found effective antifoam agents for mould fermentations. With a molecular weight of about 2000 the glycol is a clear slightly viscous fluid, and no carrier is required. Of the other compounds specially produced as anti-foam agents there are alkonylated fatty bases and alkylene glycols.

Enough has been written in the literature to show that the choice of anti-foam agents is extremely diverse. Because of this diversity of chemical types it is not possible to state that any one agent is ideal for a particular fermentation. The aim has been to show the choice and then one needs to experiment and find the best for a particular system of interest.

(E) Limitation of anti-foam agents

Unfortunately, the concentrations of many anti-foams which are necessary to control fermentations will reduce the oxygen-transfer rate by as much as 50%; therefore anti-foam additions must be kept to an absolute minimum. There are also other antifoams which will increase the oxygen-transfer rate (Ghildyal *et al.*, 1988). If the oxygen-transfer rate is severely affected by antifoam addition then mechanical foam breakers may have to be considered as a possible alternative.

18.6 FOAM CONTROL

The practical implementation of foam destruction in large scale is still far from ideal. Fermenters without mechanical/physical devices are still used and anti-foam agents are poured manually. The most common form of addition is on demand, when the batch is foaming heavily and some corrective action is needed immediately. In its simplest form the, addition of agent is controlled by the operator and its success depends entirely on the operator's vigilance and care. The disadvantage of this form of addition is that it is non-routine jobs and it is apt to lead to loss of control simultaneously in many tanks and clash with routine jobs.

For automatic control, the fermenter is fitted with a high level probe near its lid, and when foam reaches it a circuit is closed, allowing anti-foam to enter the fermenter. In its crudest form the closing of the probe circuit opens a solenoid valve, and anti-foam flows into the fermenter until the foam collapses, and probe circuit is opened again. This type of control system wastes anti-foam agents. Another disadvantage is that the agent is admitted when the probe is splashed or touched by transient foam.

In order to prevent waste it is usual to fit some delay into electrical circuit. In such a case when the probe circuit is closed a shot of anti-foam agent is added and, if the foam persists, there is a delay of 1 min. (say) before another addition is made. The period during which the valve is to be opened can be varied to allow different volumes of anti-foam agent to be added.

An important factor is design, construction and maintenance of all anti-foam addition kits is reliability. Too much antifoam harms the fermentation and subsequent extraction or treatment of the broth and too little anti-foam or none at all is equally harmful.

In control systems where the addition is made on demand, there results a wide variation of aeration and agitation conditions between addition of anti-foam. So sometimes antifoam is added at regular intervals to prevent the formation of foam. This method has been able to reduce the quantity of anti-foam and also periods of uncontrollable foaming were eliminated, these periods had occurred when anti-foam was added on demand.

The aim in the present chapter has been to describe the general principles of foam destruction and prevention rather than particular applications. The reason for this is that foam control like so much else in fermentation is an empirical art. The best method for a particular fermentation is not necessarily the best for the different fermentation system. The design and operating parameters of a fermenter may affect the properties and quantity of foam formed.

❑❑❑

19

COMPUTER APPLICATIONS IN FERMENTATION TECHNOLOGY

19.1 INTRODUCTION

Life is probably the most complex of processes; hence a new family of computerized instruments is needed to analyze and understand its manifestations. Molecular biology has a language, a grammar, logic and a mathematics of its own, and biological and computer languages have much in common. Fermentation biotechnology is a way of dealing with large numbers related to cells, amino acids, nucleotide bases and molecules. Fermentation biotechnology is not only a science, but also a business; therefore it must scale–up its production facilities and control the quality of its products.

For all these reasons and more, the marriage between fermentation biotechnology and computer science is a fortuitous one, likely to stimulate, nurture and expand both areas of human knowledge. Without computers, fermentation biotechnology would be progressing slowly and painfully, rather than exploding as it is today.

Computer applications in the fermentation industry began in the late 1950s. By the late 1960s, publications on computer applications in batch process control were fairly common. However, it was the widespread availability of minicomputers in the 1970s that led to an abundance of process–control applications. Since the beginning of the seventies, which marked the onset of automated regulation system in microbial industry, a considerable body of data has accumulated in favour of the use of computers for these purposes. Due to corporate secrecy policies, it is not clear which companies were the first to utilize computers in fermentation processes.

Computer control and computer applications in fermentation cover a broad spectrum of uses depending upon the needs of a particular process. The requirements for a pilot plant are quite different from those for a production plant. The degree of sophistication of the computer software depends upon the use of the system. Moving from data logging to data analysis, process modelling, process control, and finally to process optimisation requires an increasingly more complex system in terms of both the hardware and the software. All of these considerations along with future trends for computer applications are discussed in this chapter.

19.2 NEED OF COMPUTER–CONTROLLED SYSTEM

Recent advances in biochemistry and microbiology have resulted in a more complete understanding of fermentations, particularly with respect to substrate metabolism, growth requirements and enzyme regulation. These developments are enabling the fermentation technologist to describe the dynamic behaviour of fermentation processes in terms of kinetic models.

With this understanding, it is now possible to expand the control of fermentation processes beyond independent closed-loop feedback control of culture conditions, such as temperature, pH, and dissolved oxygen, into the more sophisticated strategies of adaptive or interactive control. Eventually, on-line optimization of a fermentation based upon model reference control of a process will be possible. This accomplishment will mean that a feed-forward, interactive control strategy will be utilized. In order to reach this capability, it is necessary to develop a system which encompasses both hardware and software needed to generate the raw data, analyze the resulting information, determine an optimal solution, and implement a control decision. A highly instrumented computer-coupled fermenter is capable of meeting this task.

The accomplishment of on-line optimization of a fermentation is an evolutionary process. The process began with the achievement of fermentation data acquisition by a computer system. Most fermentation companies now routinely utilize computers in data acquisition. This development involved on-line process monitoring and scanning, data logging, and generation of alarm signals. In many cases, this information was collected in a form that was transferred to a computer system for off-line analysis. With the advent of inexpensive and reliable minicomputers, it has become economical to utilize computer system on-line. With this capability, it has been possible to move rapidly to a computer applications level which has involved partial process control based upon using predetermined set points for controlling primary fermentation parameters, such as temperature, pH, and dissolved oxygen. It has also been possible to utilize the information generated on-line to study the process and to elaborate the process kinetics. Today, on-line computer systems are being utilized for reducing the time required for determining the dynamics of fermentation processes, for testing control strategies and for modeling the process.

With the development of more powerful minicomputer operating system software, it is possible to operate the computer system using both foreground/background tasks, i.e., simultaneously performing the on-line data acquisition and analysis tasks and simultaneously doing on-line process simulation. This allows the fermentation technologist to quickly develop and refine mathematical models and to identify the model parameters so that methods can be developed for predicting the interactive relationships between primary fermentation parameters. Once significant operating confidence is obtained with process kinetic models, computer systems can advance to the final stage of automatic process control, i.e., on-line optimization based upon model reference control.

19.3 ADVANTAGES OF COMPUTER–CONTROLLED SYSTEM

The advantages of using computer–controlled system in fermentation industry are the following—

(i) Reduction of human labour in the operation of fermenters.

(ii) Acquisition of precise information and a better control of the process.

On the other hand, the sophistication of measuring and control equipment poses higher demands on the personnel operating these devices, especially in the acquisition and evaluation of all possible data from the equipment, arriving at the correct conclusions and making the correct decision about the future course of the process.

The evaluation of the processes, i.e. the calculation of kinetic constants and other parameters, also becomes more laborious with an increasing number of parameters studied. Conventional means of data acquisition and process control are often unsatisfactory and such control may therefore be accomplished by means of computers. The widely known advantages of computer-controlled systems reflect three characteristic features of these systems:

a) fast and accurate processing of data,

b) storage of a large number of data in memory,

c) decisions are based on all the data available at the moment.

The advantages can be briefly characterised as follows:

a) process control is much more efficient than with conventional systems (operator + conventional equipment); the computer provides precise reproduction of the process under defined conditions, the control is continuous and encompasses all regulation circuits at the same time. The subjective element is excluded from process control;

b) computerisation reduces considerably the amount of labour necessary for fermenter operation and for data evaluation. Data appraisal performed by the computer during the process may be used practically immediately for process regulation and optimisation (Harrison and Harmes, 1972);

c) efficiency of the whole process increases due to a better utilisation of raw materials and energy and a reduction in product losses.

19.4 GENERAL APPLICATIONS

The application of computers to fermentation technology is very dependent upon the end use of the computer system. For example, requirements for a pilot-plant system are quite different from those for a production plant. In a pilot plant, the computer is used primarily as a research tool. It is necessary that the system be designed to be as flexible as possible. It should be particularly useful in obtaining kinetic data necessary for modeling the process and for developing control strategies aimed at process optimization. Pilot-plant computer-coupled systems should be structured to accommodate software development and software testing.

19.4.1 At Pilot-plant

There are two levels in which a computer system can be utilized in a pilot-plant. The first level involves operator-oriented programs which can be utilized by a technician. To run these programs, it is necessary for a technician to know only the computer command necessary to start them. The programs are designed to prompt the operator with the responses necessary for implementation. These programs include such functions as data logging, data analysis, and process control. By using operator-oriented routines, the technician can generate the necessary documentation and procedures for system operation and measurements. It is also helpful if the system contains programs for generating standard reports and graphical output.

The second level of computer utilization requires an engineer with programming ability to develop and insert into the operator-oriented programs new routines to be run by the technician. Also, on this level the engineer interacts with the computer to develop control strategies and process optimization routines. The engineer can interact with the system to quickly access large volumes of data. The easy access to this information frees him or her to develop and test control strategies and process optimization routines. Thus, the computer-coupled system is a research tool which allows the engineer to spend most of his or her time interpreting data rather than merely collecting and plotting it.

19.4.2 At Production-plant

For a full-scale production plant, the requirements for a computer-coupled system are quite different. The primary objective is to produce a product in an optimal manner. As a result, the computer is used for process control and optimization aimed at maintaining quality control and product uniformity. Here again, it is necessary to interact with the computer system on two levels. The first level involves an operator-oriented command language allowing easy access to the entire process. The command language will contain routines for generating process status information, calculating process variables, and scheduling process operations.

Plant status routines are used to send alarm messages to the operator and to supply him or her with information on the overall operation of the process. Raw material inputs, flow rates, and utility requirements in terms of steam, cooling water, and electric power demands are just a few of the process parameters and variables that are routinely monitored. Scheduling routines provide the operator with information for sequencing operations. This is particularly valuable for a batch process where there are strict protocols for cleaning, media sterilization, vessels charging, and product harvesting. The sequencing routines provide information to step the operator through the process in order to start up or to shut down the plant. With a sequencing routine, it is possible for the computer to verify a command by the operator and to prevent him or her from moving to the next sequence of events until the previous request has been satisfied. In this way, the computer system can be utilized to ensure that the same procedures are followed for each batch run which helps to assure product quality and product uniformity. These routines also monitor raw material inventories in order to maintain proper deliveries, and they assist the operator in planning maintenance schedules for major equipment. Finally, the command language generates specific process information regarding the status of each vessel and automatically produces reports and documentation at the end of each run.

The second level on which the plant-scale computer system can be utilized is similar to that in the pilot-plant. At this level, an engineer is required for making changes in the process-control or monitoring routines. These changes are required as process improvements are developed in the pilot-plant and transferred to plant-scale operations.

19.5 SPECIFIC APPLICATIONS

19.5.1 Data Logging

Table : 19.1 lists variables that are commonly monitored during a fermentation. In data logging applications the raw data are fed into a computer system where they are compared with calibration curves and correction factors in order to obtain values in standard engineering units. The data are usually printed out on a hard-copy device and written onto a mass storage device for a more detailed analysis by a larger computer system at a later time.

Table 19.1 : Variables Logged and Controlled

(A)	**Measured parameters**	
	1.	Inlet gas flow rate
	2.	Inlet gas pressure
	3.	Inlet gas temperature
	4.	Inlet gas carbon dioxide
	5.	Inlet gas oxygen
	6.	Cooling jacket flow rate
	7	Cooling jacket temperature differences
	8.	Agitator speed
	9.	Power to shaft
	10.	Vessel pressure
	11.	Glucose concentration
	12.	Culture turbidity
	13.	Culture fluorescence
	14.	Exit gas carbon dioxide
	15.	Exit gas oxygen
(B)	**Controlled variables**	
	1.	Dissolved oxygen
	2.	Vessel temperature
	3.	pH
	4.	Foam
	5.	Substrate addition rate
	6.	Acid addition rate
	7.	Base addition rate

19.5.2 Data Analysis

Numerous papers have been published dealing with the analysis of fermentation data (Yamashita and Murao, 1967; Yamashita *et al.*, 1969; Humphrey, 1971; Nyiri, 1971; Jefferis *et al.*, 1972; Wilson, 1975; Wilson *et al.*, 1975; Wang, 1976; Zabriskie, 1976). Many of the analytical techniques depend upon the particular process for which they were developed. However, there are certain calculations that are common for most fermentations. These include computations for the oxygen uptake rate, carbon dioxide evolution rate, respiratory quotient, overall mass transfer coefficient, heat balance, power requirements, cell mass, and cell growth rate. Table : 19.2 is a summary of some of these calculations.

Table 19.2 : Summary of Calculations

1.	Oxygen uptake rate
2.	Carbon dioxide evolution rate
3.	Respiratory quotient
4.	Total oxygen utilized
5.	Total carbon dioxide evolved
6.	Overall oxygen mass transfer rate
7.	Heat balance
8.	Metabolic heat evolution rate
9.	Cell mass concentration
10.	Cell mass growth rate
11.	Specific growth rate
12.	Cell yield
13.	Nitrogen utilization rate
14.	Glucose utilization rate
15.	Carbon balance
16.	Nitrogen balance
17.	Oxygen balance
18.	Apparent viscosity

These calculations require combining inputs from a number of measured variables. As a result, a computer system is a valuable tool for obtaining information on-line concerning the status of the fermentation process. Often, it is not practical to obtain these parameters by any other means. A good example of this is the measurement of biomass concentration and growth rate. Several papers have appeared in the literature concerning the indirect measurement of cell mass concentration, cell growth rate, and substrate concentration (Jefferis *et al.*, 1972; Zabriskie 1976; Wang, 1976; Armiger, 1977; Harima, 1977). Utilization of computers for indirect measurement to give on-line biomass and growth rate data will have a major impact on the fermentation industry.

Two of these techniques have been derived by performing material balances around the fermenter. The first approach selects a specific molecular component and uses a kinetic model relating cell growth with the material balance. Figure 19.1 shows a balance around a fermenter for component X. Zabriskie (1976) estimated biomass concentration by a material balance for oxygen. He corrected the balance for secondary metabolite formation by also monitoring the carbon dioxide evaluation rate. The basic relationship is shown as Eq. (1) (see Fig. 19.2), where the rate of change of component X with time is equal to the difference in input and output calculations plus an accumulation term.

$$\text{Gas phase balance: } \frac{d}{dt}\left(V_g C\right)dt = F_{in}C_{in} - F_{ex}C_{ex} - J_{g\to a} \tag{1}$$

Utilizing molecular oxygen for the material balance has several advantages over other components. First, there is a close relationship between oxygen and the energetics of growth. Second, oxygen concentration can be measured continuously in both the gaseous and the aqueous phase. Finally the solubility of oxygen in the aqueous phase is so low that a quasi-steady state may be assumed for the concentration of oxygen in the liquid phase. This simplifies the material balance equation. Thus, for oxygen, the uptake rate can be simply expressed as the difference between the inlet and outlet streams [See Eq. (2)].

$$F_{in}C_{in} - F_{ex}C_{ex} - R_a \cong 0 \tag{2}$$

If one assumes the fermenter is essentially in steady state conditions with respect to oxygen utilization, then the oxygen uptake rate is equal to the oxygen consumption rate. A kinetic model for oxygen consumption, first proposed by Marr, *et al.*, and by Luedeking and Piret (1959) is shown in Eq. (3).

$$\text{Model: } R_a = \left(M_{i/X}X + \frac{1}{Y_{X/i}}\frac{dX}{dt} \right) V_a \tag{3}$$

This is a two-parameter model which assumes that the oxygen is utilized for two basic purposes within the organism. A portion of the oxygen is consumed for cellular maintenance functions which are necessary to sustain cell viability and must be satisfied at all times. The balance of the oxygen is utilized for cell growth. Thus, the two parameters in the model are termed the growth yield and cell maintenance constants.

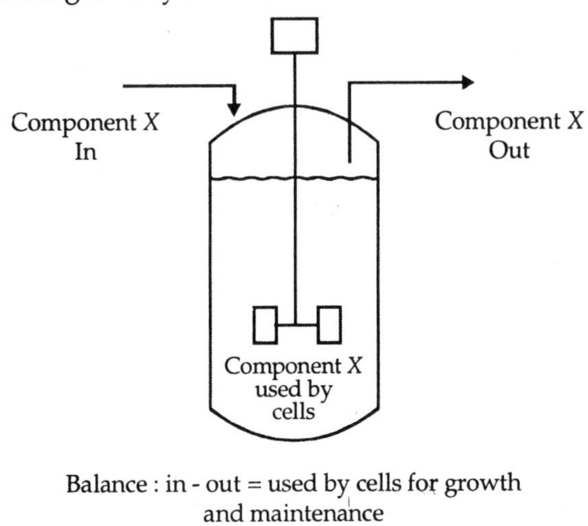

Fig. 19.1 : Material balance around a fermenter.

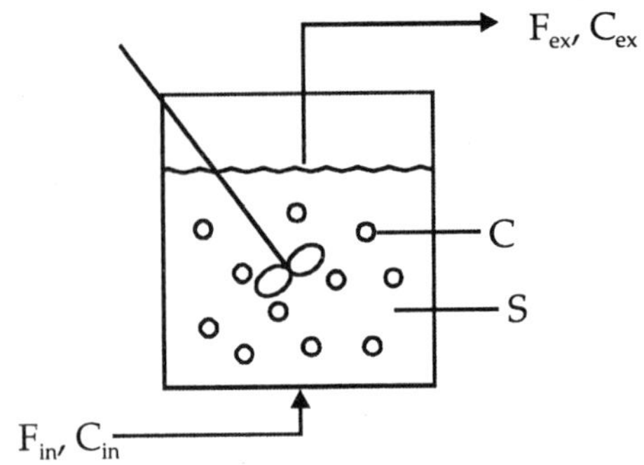

Fig. 19.2 : Oxygen balance around a fermenter. (From Zabriskie, 1976).

19.5.3 Process Modelling

Describing microbial processes by mathematical models is of increasing interest to microbiologists and biochemical engineers. Models are valuable for understanding processes and interactions among variables. In addition, models are necessary to provide the control engineer with insight into possible control strategies. Further, if the model can predict the dynamic behaviour of a process, it can be useful for process optimization. Dynamic models are valuable for simulating experiments and testing control strategies. They can be utilized to eliminate many expensive and time-consuming experiments and to design key experiments to test a hypothesis. A large number of models for microbial processes have been proposed in the literature. These range from very simple to very complex systems. Most of the models involve multivariable, nonlinear differential equations which must be solved by numerical techniques. Consequently, the utilization of a computer is absolutely essential in the solution of these fermentation models.

The objective of most fermentation models is to describe the formation of a primary product, such as biomass, or a secondary product, such as an antibiotic. These two situations have been modelled for batch, fed-batch, and continuous fermentations. There are two fundamental approaches to modelling systems. These involve the use of structured and unstructured models. Most fermentation models described unstructured systems. They do not account for individual differences between cells but examine variables that are characteristics of the macroscopic environment. Structured models do describe differences between cells and cell components; for example, one might describe the gross cell biomass, an enzyme component of that biomass, and the genetic mechanism activity. Unfortunately, the components of the structured model are difficult to measure; hence, they have not had the utility of the unstructured model. Tsuchiya *et al.*, (1966), Fredrickson *et al.*, (1970), and Yoshida and Taguchi (1977) all provide reviews of these approaches.

19.5.4 Process Control and Optimization

The application of computer control to the fermentation industry has lagged behind similar advances in the chemical process industry by about 10 years. When the complexity of the kinetics of microbial systems is compared to the kinetics of chemical systems, the reasons for this lag become obvious. In spite of these difficulties, process control of fermentations can be implemented. The ways in which control has been implemented include regulation, sequencing, and optimization methods.

Regulation is concerned with maintaining individual process loops at present values. Sequencing simply involves stepping through a predetermined program of events. Optimization is responsible for determining the best overall operating conditions for the process as a function of time. (All three of these functions are being utilized to some extent in fermentation production plants today.)

There are two basic types of application of a computer in process control. In the first case an operator acts as an intermediary between the computer and the process. The computer is used only for processing acquired data and the resulting information is used by the operator for process control. This type of application is called off-line (Fig. 19.3a). The advantages of a computer are utilised to the full extent in the second type of application—on-line—when the computer is connected directly into the regulation loops of the system, i.e. into the sensor—regulator—effector chain (Fig. 19.3b). There are two arrangements for this system. The computer may be used either as a controlling element for actual (usually analogue) regulators merely changing the adjustment of the value towards which the quantity under control tends (set-point)—digital set point control (DSC, Fig. 19.4) or it may replace the regulator entirely—direct digital control (DDC— Fig. 19.5). A comparison of these two arrangements shows that the properties of the DSC are limited by the properties of the regulators used (regulator type—P, I, D, delay, etc.) while DDC is limited only by the requirements of the process. On the other hand a computer failure in DDC completely abolishes control of the process while in DSC individual control loops may still operate. Since the DSC system is more expensive, actual systems are usually built as a compromise; the DSC system is used only in loops the failure of which would be critical for the whole process.

(a) Off-Line

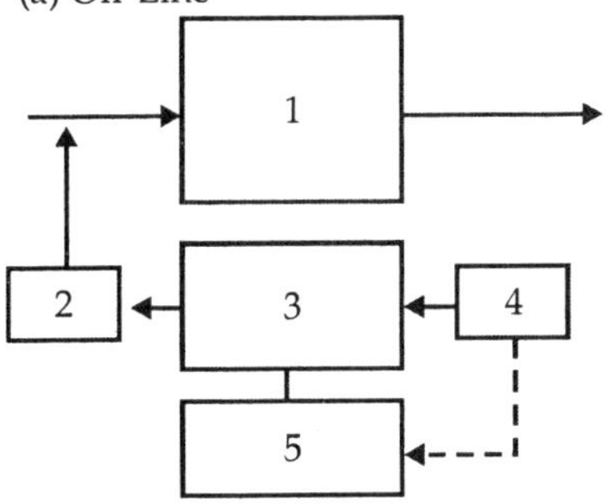

(a) On-Line

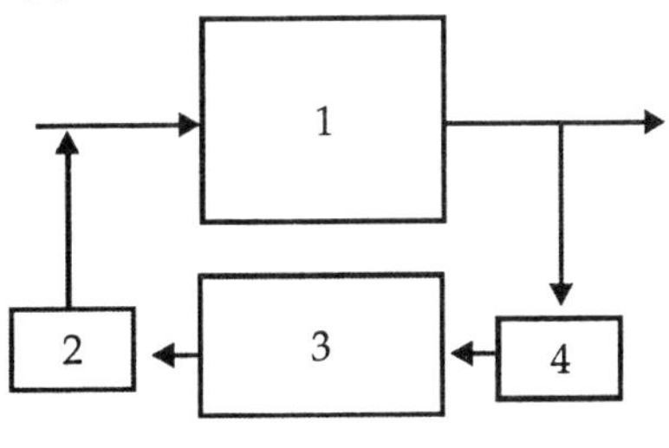

Fig. 19.3 : Basic connection of a computer with a process a) off-line: 1. system, 2. action components 3. regulators, 4. sensors, 5. controlling computer; b) on-line: 1. system, 2. action components, 3. controlling computer, 4. sensors

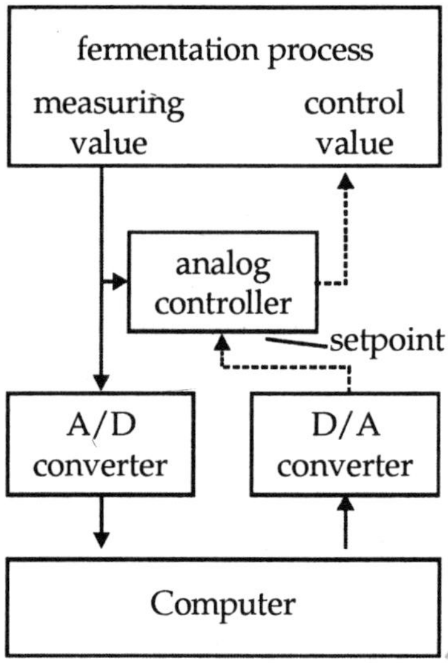

Fig. 19.4 : Scheme for DSC Control.

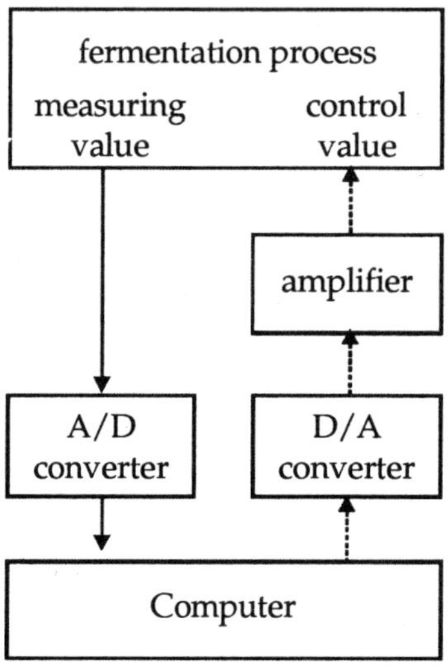

Fig. 19.5 : Scheme for DDC Contorl.

The function of an on-line computer includes several successive steps:

1. Data acquisition
2. Data proccssing
3. Process control.

Higher levels of process control which are currently being developed include:

4. Optimisation
5. System adaptation.

Data acquisition includes the gathering of values of different parameters from individual sensors, their transformation to a common digital form and their storage in memory. At the same time the data are being processed. Disturbing signals are filtered out and the resulting values are compared with preset parameters of the system at a given moment to assess the difference between set-point and observed values; they also serve for the calculation of further quantities and constants. The above difference is then used for actual process control either directly (DDC) or via analogue regulators (DSC). The computer program makes it possible, in addition, to perform changes in process sequence and adjust an optimum sequence of operations during the onset, course and termination of the process.

The computer has a particular role in process optimisation, i.e. in the search for a state in which the process exhibits the best properties determined by a certain criterion. The choice of optimisation criteria may differ widely; it is usual to use the maximum yield of a given product. Other criteria may involve the process in a wider context such as maximum profit. Optimisation also promotes the full use of the computer features, especially its operation and speed. Complex optimisation calculations, on the other hand, impose considerable demands on the hardware facilities of the control computer and on the level and extent of its software. This aspect, together with the characteristics of the controlled process (number and kind of measured and controlled quantities, etc.) are naturally decisive in the selection of the computer.

Numerous papers have been devoted to the computer control of microbial processes (Dobry and Jost, 1977, Arminger and Humphrey, 1979, Unden *et al.*, 1979). They reflect the general tendency to replace minicomputers by programmable calculators and a gradual introduction of specialised control units fitted with microprocessors. Changes in computer software reflect the transition from universal programming languages to specialised languages more suitable for process control.

19.6 SYSTEM CONFIGURATION

A highly instrumented fermentation system coupled to a digital computer is capable of generating on-line information for conceptualizing kinetic models and implementing strategies for process control and optimization. The process of utilizing the signal produced by an instrument and integrating it into a computer-controlled system involves both hardware and software development. With the advent of inexpensive but very powerful minicomputers, the problem of hardware development is basically one of selecting electronically compatible modules. The development of techniques and software for data analysis and process control is the area in which further research is needed. Configuration of system hardware and software must be structured to operate as an integrated unit. It must also be flexible enough to accommodate growth and expansion of both hardware and software as new products and techniques become available.

19.6.1 System Hardware Configuration

The system hardware consists of three main segments: (1) sensors and instrumentation, (2) computer interface, and (3) computer system. In general, the weakest segment in any system involves the sensors. As shown in Table : 19.1 the number of measurements is very limited. Most of the sensors are measuring variables in the macroscopic environment of the fermentation broth, while the key factors controlling growth and product formation involve the microscopic environment internal to the cell. The development of sensors and techniques for monitoring levels and activities of cellular constituents is only beginning. The physical problem of measuring fermentation parameters is further complicated by the sterilization requirements which severely restrict the choice of materials of construction of the sensors and often limit their sensitivity and accuracy.

The hardware for the computer interface is readily available. With sufficient attention to detail, it will reliably function. However, the architecture of the interface depends upon the application. Analogue scanning is the most economical method of interfacing if there is a large number (over 25) of channels involved. However, digital multiplexing, which is cost competitive on a small system (under 25 channels), offers many advantages. First, it allows each channel to function as a separate unit, and it provides the operator with a visual display of all the channels at one time. Second, it allows for all of the data to be logged simultaneously rather than sequentially so that calculations can be made on data from a specific instant in time. For large systems, the advantages of digital multiplexing are difficult to justify because of cost. Therefore, in most plant or pilot-plant applications involving many vessels, an analogue scanning interface is selected.

Computer interfacing involves a choice between increased flexibility and increased software requirements. This includes trade-offs such as random access scanning and increased hardware costs with less flexibility but lower software cost. Again, the number of channels involved in the interface will determine if the cost savings are to be gained by using more sophisticated hardware or software.

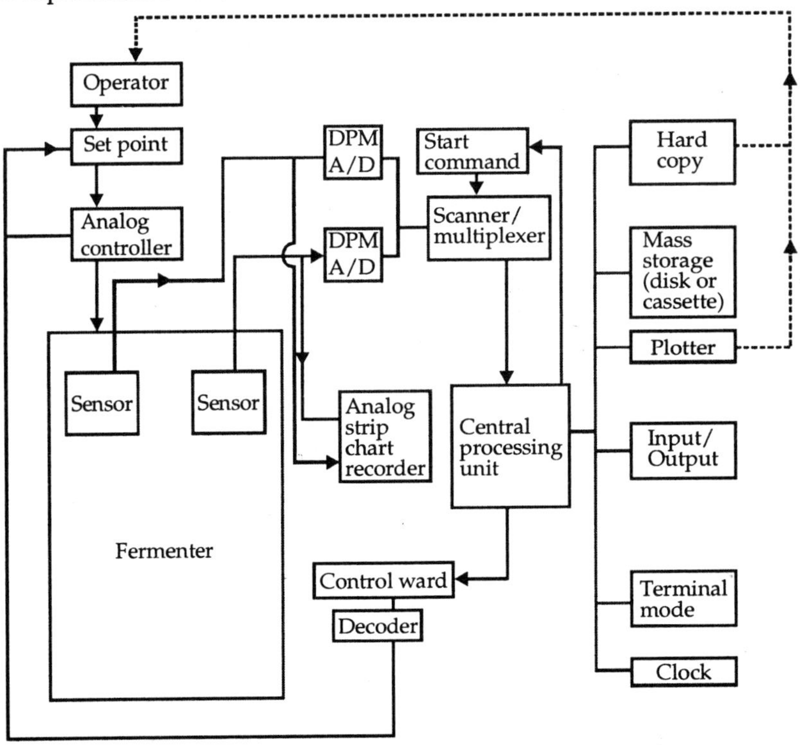

Fig. 19.6 : Schematic of physical configuration of system.

The selection of a computer system usually starts with the choice between a dedicated computer and a time-sharing system. With the availability of inexpensive but very powerful minicomputers, a dedicated system can usually be justified very easily on the basis of lower operating costs and increased system reliability. The details of selecting the size of computer memory and the peripheral devices are dependent upon the specific application. A disk-based system with sufficient memory to allow for foreground/background operation and some graphics capability is a good starting point. In many cases, communications with a large computer installation can be very beneficial for accessing special software or performing complex calculations.

Figure 19.6 is a hardware schematic and signal flow diagram for a typical installation. The analogue signals from the various sensors are fed to the interface where the analogue voltage is converted to a binary coded decimal (BCD) word. The BCD output of the interface is transmitted to the computer under control of the computer software which specifies the scanning rate and sequence. Once the data reach the computer, they are stored on the system disk and analyzed by the system programs. If the results of the analysis are to be used for process control, signals from the computer are then transmitted back to the fermenter to manipulate set points on analogue controllers (USC) or to manipulate variables directly (DOG). The results generated can also be output on a hard-copy device, stored on the disk or tapes, plotted on a graphical device, or transmitted to another computer system.

19.6.2 System Software Configuration

The first step in system software configuration is the selection of the proper computer operating system. An operating system with foreground/background, multiprogramming, and multitasking capabilities and a high-level language (i.e., FORTRAN) is preferred. The man-machine interface should be as simple as possible in order to make the computer operator as effective and reliable as possible.

An overview of the system software organization is illustrated in Fig. 19.7. The computer system can be conceptualized as two parts, the central processor, which contains the core memory; and the system disk. All other computer equipment are really peripheral devices to support these two systems. The central processor contains the computer's instructional logic and a relatively small region of core memory which is allocated into three sections. The first part contains the operating system resident monitor, which is the software enabling the operator to issue commands from the terminal to load programs into core, execute them, and control and monitor information transmitted to or from all system devices. The second portion contains the foreground program, which is responsible for sending the start command at the proper time to the scanner and logging the data on the disk. The third section is for the background program which is divided into many smaller regions. The first region contains the executive routine for managing the sequence in which different program modules enter, execute their function, and leave the overlay regions.

Since computer core is limited in minicomputers, only data and program modules required at a particular point in time reside in an overlay region. The bulk of the software remains on the disk. As time progresses and certain tasks are completed, data are transferred to a file on the disk and a new program module is read from the disk and overlaid on top of a module whose function is complete. The system disk has no computational power; it is a storage device only. It contains vast amounts of information consisting of the operating system modules required by the resident monitor, data storage files, temporary buffer files, FORTRAN programs, and assembly language routines.

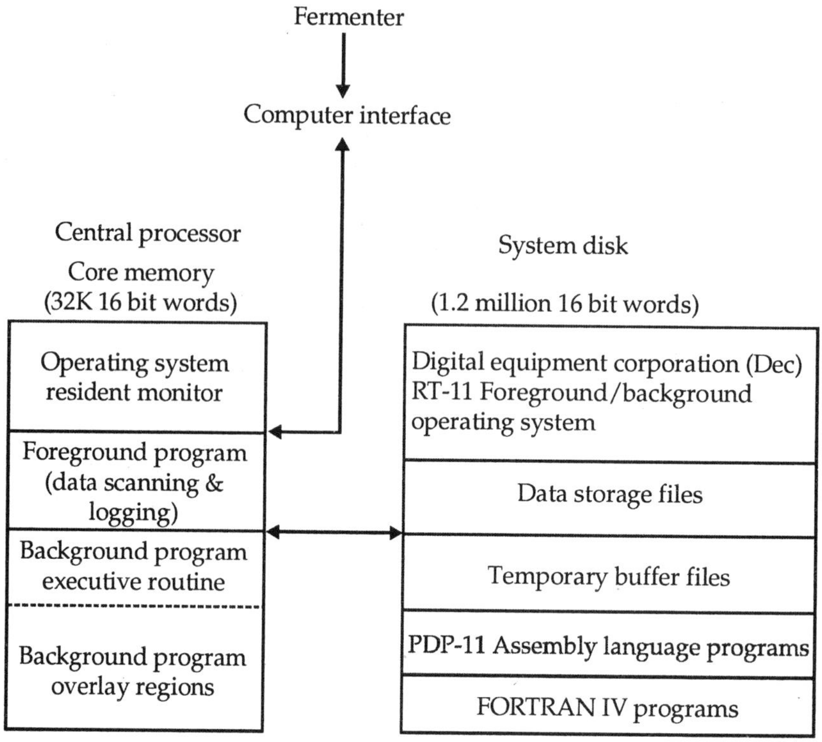

Fig. 19.7 : Overview of system software.

In a foreground/background operating system, tasks in the foreground have system priority. When the foreground has nothing to perform, control is automatically switched to the background to execute its task. Therefore, software must be divided into foreground and background tasks. The foreground functions include starting the scanner, logging data on the disk, and generating alarm messages. The benefits to this arrangement are manifest. The most time-dependent functions are performed first.

Clearly, it would be unacceptable if data acquisition had to be delayed because the computer was occupied by producing a report which could take as long as an hour. Typically, the tasks performed in the foreground can be executed very rapidly (seconds). Thus, without foreground/background capability, all of the time between the end of the foreground function and the generation of a successive data set by the scanner (1-10 minutes) would be wasted. However, with this arrangement, the system can be utilized automatically between scans for many different functions from analyzing laboratory data, simulating the fermentation, producing graphs, or developing new programs.

19.7 FUTURE TRENDS

Computer applications in fermentation technology will increase. The availability of microprocessors and software to support these systems will change the hardware configuration of a system and greatly reduce the costs. Soon the computer associated costs will be much less than those related to the sensor systems. However, the real future lies in developing improved mathematical models and strategies for process control and optimization. At this time, the capabilities of computer hardware greatly exceed the capabilities of the software that has been developed. Therefore, the greatest challenges for future application of computers to fermentation are involved with interpreting results rather than physically implementing control decisions. The time has arrived for the biochemical and control engineers to utilize their skills to bring computer applications in fermentation technology up to the level of the available computer technology.

❑❑❑

20

SCALE-UP

20.1 INTRODUCTION

Scale–up is the process of increasing the size of a fermentation system to arrive at a full–scale plant fermentation in the most efficient and economic manner. Problems will be encountered and resolved at the small–scale level so that only minor adjustments are necessary at the production–scale level. Obviously, mistakes made at the larger level are more expensive.

Also, additional information is gathered during scale–up so that the role of each factor affecting the fermentation is clearly understood and more amenable to control. Factors to be considered during scale–up include those of the small fermenter stage plus new problems that must often be resolved by a chemical engineer rather than the fermentation technologist.

Development of bioprocesses is usually carried out in three steps—

(i) Bench scale or flask scale where basic screening is carried out,

(ii) Pilot scale where optimal operating conditions are established,

(iii) Plant scale where the process is brought to an economically favourable level.

The fundamental idea of scale–up is that the optimal physiological conditions that are obtained in small–scale studies should be maintained on the large scale by controlling environmental conditions. These include not only physical factors (mass transfer ability, mixing ability, shear field, power consumption, etc.) and chemical factors (medium composition and concentration) but also process factors (number of precultures / inoculum, sterilization conditions, etc.). Many physical and chemical parameters that influence the behaviour of cultured organisms change as the scale of process becomes larger. For conditions that are chosen in small–scale fermenters to be reproduced in large fermenters, the factors significant to the cultivation should be determined and investigated in small–scale cultivation.

In this chapter, we discuss the factors to be considered in the scale–up procedure, such as construction of a new large–scale facility, translation of a new bioprocess with a new microorganism, and bioprocess improvement from the pilot plant into an existing fermentation facility.

20.2 SIGNIFICANCE OF SCALE–UP

The conversion of laboratory procedure to an industrial process is termed scale up.

It is well established in the field of industrial microbiology that a process which works well at the laboratory scale may work poorly or not at all when first attempted at large industrial scale. It is generally not possible to take fermentation conditions that have worked in the laboratory and blindly apply them to industrial–scale equipment. All of the skills of the biotechnologist must be brought into play in order to develop a successful large–scale process.

Scale–up is necessary in following circumstances—

(1) A new process is implemented in the plant.

(2) Mutants with 10–20% greater yield are to be introduced into large–scale production as soon as possible.

(3) Construction of a completely new fermentation plant, a rare occurrence

While laboratory microbiologists are primarily interested in yield based on weight of biomass, units per ml broth, or maximal yield per unit time, success in scale–up is evaluated on the basis of maximal yield for minimal operating cost and time.

Flow chart of basic fermentation process scale up is illustrated in figure 20.1

20.3 CRITERIA USED FOR SCALE–UP

Several criteria used for scale–up that focus on mass transfer and on the characteristics of a microorganism have been proposed.

The criteria are as follows—

(1) Constant agitation power per unit volume (P/V)

(2) Constant oxygen transfer coefficient (K_La)

(3) Constant mixing time

(4) Constant impeller tip speed.

(5) Feedback controlling the values of key environmental factors as nearly as possible between two scales of cultivation.

When a new fermenter is to be designed, criteria 1 and 2 are useful for comparing oxygen transfer ability between two different sizes of fermentation equipment. In these, the agitation power per unit volume (P/V) is the most frequently used parameter for scale–up.

As sensor technologies developed for dissolved oxygen (DO) concentration, scale–up methods to maintain DO at a constant level became popular. The scale–up method related to oxygen transfer is established by controlling the DO concentration at a suitable level. This is not difficult to control by changing agitation speed and the shape of impellers.

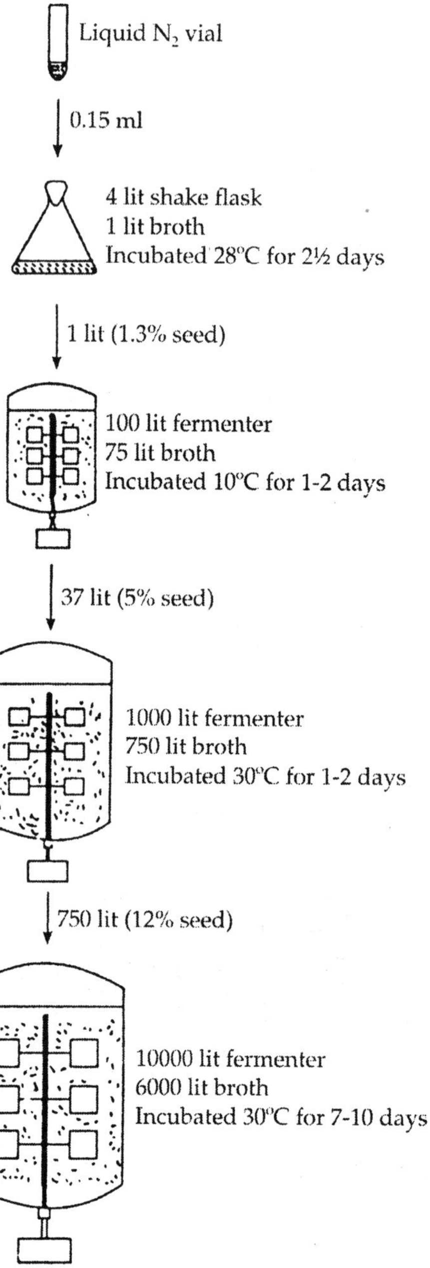

Fig : 20.1 Flowchart of basic fermentation process scale up.

On the other hand, adjusting factors that are important to the physiological condition of the microbial cell is very difficult. Many fermentation–based companies already possess large fermenters for multipurpose use. When a new fermentative process is ready for scale–up, the new process should be adjusted to the existing fermenter by changing a limited number of parameters.

In cultivation of filamentous microorganisms, such as fungi or actinomycetes, the viscosity of culture broth changes with culture time and sometimes increases to a very high level. In such cases, the characteristics of the fermentation broth are those of a non–Newtonian fluid. Addition of water to the culture broth occasionally is effective in decreasing the viscosity.

20.4 PHYSICAL FACTORS

20.4.1 Mass Transfer

For almost all microorganisms, oxygen is very important factor for growth and production. Many useful microbes that produce important secondary metabolites require a high level of oxygen for production. A lack of oxygen often causes a reduction of the yield or death of the microorganism.

The dissolved oxygen (DO) is the most commonly used indicator for monitoring conditions in a fermenter. To control DO in a fermenter, one changes agitation speed, aeration level, or pressure. Oxygen mass transfer coefficient and agitation power per unit volume are very important parameters for scaling up. However, as it is rather dangerous to rely solely on the calculation of these parameters, these values are often used as reference values for experiments to be carried out before designing a new large fermenter.

20.4.2 Oxygen Mass Transfer Coefficient [K_La]

K_La is an excellent parameter for comparing oxygen transfer ability in different size fermenters. K_La is an index of oxygen transfer ability from the gas phase to the liquid phase. It is defined as follows—

$$dc/dt = K_La\,(C^* – CL)$$

where C^* is oxygen concentration, t is time, and K_La is an oxygen transfer coefficient.

The value of K_La is affected by the size of fermenter, the shape and number of turbine impellers and baffles, agitation speed, aeration rate, pressure, and so on.

One major problem is that the K_La value should be measured using cultured broth for accurate estimation, since the rheological characteristics of cultured broth significantly affect the K_La value.

20.4.3 Agitation Power per Unit Volume [P/V]

A scale–up method using P/V has often been applied to industrial fermentations. The benefit of this method is that the cultivation data obtained from small–scale fermentation are often available. Factors that have great influence, such as viscosity or non–Newtonian fluid characteristics, are already included in the experimental value of P/V. When the motor power of the exiting fermenter is not strong enough to increase DO for cultivation of a new microorganism, a decrease in volume of fermentation broth is sometimes effective as a means to maintain DO.

20.4.4 Theoretical Calculation of Agitation Process Parameters

The calculation of some parameters of the agitation process (e.g. the impeller power in aerated and unaerated systems and partially also the homogenisation time) during the scale-up of a process presents no serious difficulties as these parameters can be described (for geometrically similar systems) by numerical values of dimensionless criteria. However the scale-up of a process does not preserve hydrodynamic similarity analogous to the geometrical similarly since the values of all dimensionless criteria cannot be kept constant. The effect of scale-up is most pronounced in aerobic microbial processes as the impeller serves two functions—a pump for culture fluid circulation and gas dispersion—and its working regime is restricted by numerous limitations (Einsele, 1978). Some formulae (e.g. for the power input in an aerated system or for choking) exist for the region close to the impeller but not for the whole fermenter volume. These problems are usually solved with the aid of local values of the parameters throughout the culture fluid volume, obtained by the study of suitable mathematical models (Miura, 1976).

Some of the criteria suggested for the scale-up of microbial processes include the constant value of the specific power input, P/V, the constant value of the Reynolds Number, Re_M, a constant circumferential speed of the impeller, $\pi\, dn$, a constant value of the expression $ndD^{-0.5}$ and others. None of the criteria is universal and the application of any one of them depends on the type of the scaled-up process. Scale-up often includes a change in the geometry of the system which causes additional complications. Also, most valuable data were obtained in laboratory devices; data from industrial devices are rare and so are studies on their relations to laboratory systems pertaining to the same process. The scale-up of microbial processes is therefore still more an art than an exact procedure; the designer has to work with a wide margin and has to take into account the possibility of changes in some parameters.

20.4.5 Shear Stress

When filamentous microorganisms, such as fungi or actinomycetes, are cultured in a large fermenter, cells are damaged by shear stress, and yield of product is reduced.

Shear stress is considered to be proportional to the tip speed to the agitation impeller. Oldshue showed the relationship between the tip speed of the impeller and the amount of nucleic acid–related compound leaked from the cells.

A microorganism that is sensitive to shear stress should be cultured with large–diameter impellers or multiple impellers, to maintain a suitable DO level with a low agitation speed.

20.4.6 Effect of Time Delay in a Large Fermenter

Mixing time is often a key factor in the bioprocesses in which pH control or substrate feeding is needed. When concentrated alkali or acids are fed and mixing time is long, a pH gradient is formed in the fermenter. Especially when a fed substrate inhibits cell growth or when substrate concentration has to be controlled at a very low level, long mixing time in a large fermenter results in low production yield. To shorten mixing time, one can use large–diameter or sloped impellers or feeding from many points in the fermenter.

As the size of the fermenter is increased, fermenter depth also increases, and DO at the bottom of a fermenter is higher than that at the surface. Microorganisms in a large fermenter repeatedly travel up and down. When the fermenter depth is 10 m, the pressure and the DO difference are about 1 atmosphere and about 8 ppm, respectively.

For microorganisms that show low productivity at high DO levels, production yield sometimes decrease in a large fermenter because of a high Do level at the bottom. In Bialaphos production by *Streptomyces hygroscopicus*, production yield under a high DO condition is low. When the process was carried out and the DO was measured in the middle of a fermenter (which was considered to represent the DO in the fermenter), the yield of Bialaphos was about 85% of that in small–scale fermentation. To improve the fermentation yield, the DO was measured at the bottom of the fermenter and controlled at 0.5 ppm. As a result Bialaphos production yield increased to 96% of that in small–scale culture.

20.5 CHEMICAL FACTORS

For economic reasons, low–grade or cheap natural ingredients are often used for medium preparation in large–scale fermentations instead of special–grade ingredients. The products obtained with these low–grade ingredients may be of lower purity than those obtained with high–grade media or may contain harmful by–products. When medium composition or grades are changed, several small–scale studies are necessary to ensure the purity of the final product.

When the nutrient requirements of a large–scale fermentation are different from those at the small–scale, nutrient feeding is effective. Amounts of nutrients in a culture broth can be measured by the combination system of a continuous filtration module and high–performance liquid chromatography.

When a main medium component that affects the culture conditions is known, controlling the nutrient concentration to a suitable level is an effective way to scale up the chemical condition of culture broth.

20.6 MICROBIAL ASPECTS OF SCALE–UP

A wide variety of minor problems may arise during scale–up. Some microbial aspects of frequently occurring problems are discussed in following section.

20.6.1 Initiation of Inoculum Build–Up

The transfer of microbial growth from stock cultures to liquid media, and the initiation of growth in the liquid media at times present special problems as regards the initiation of inoculum build-up. Vegetative cells and spores of bacteria are suspended in a sterile diluent, such as tap water, for addition to the broth medium, in addition, the bacterial spores, and, particularly, those from the genus *Clostridium,* often require a heat treatment, or *heat shock,* to allow germination of a high percentage of spores. For example, the highly heat resistant spores of *Clostridium acetobutylicum* for the acetone fermentation are placed in boiling water for approximately 90 seconds before delivery to the bottom of freshly steamed liquid medium. On the other hand, spores of actinomycetes and fungi are not particularly heat resistant and will not withstand heat shock. Thus it may be necessary to *pregerminate* spores of these organisms in a special medium if there is some question as to whether the spore will germinate in the normal inoculum medium. Usually, however, pregermination is not required. Spore suspensions of fungi or actinomycetes are prepared by adding a suitable diluent, such as sterile tap water, to sporulated agar growth. The spores are then scraped loose with a sterile loop so that they become suspended in the diluent. With certain organisms, the spores are not wetted by water and tend to remain as a film on the surface of the hyphae, or they float on liquid surface and creep up the inner wall of the growth vessel without suspending in the diluent. This condition is corrected by adding a small amount of a non-toxic surface active agent, such as sodium lauryl sufonate, to the dilution liquid. In contrast, certain stationary fungal fermentations, such as the citric acid fermentation, using the surface culture of *Aspergillus niger* require that the fungal spores be floated on the surface of the liquid medium so that germination and growth of the spores will yield a floating mycelial mat. This is accomplished by carefully draining the spore suspension down the wall of the culture vessel so that the spores float out across the surface of the liquid medium, or by spraying the spore suspension across the surface of the medium with an atomizer. Non-sporulating or poorly sporulating actinomycetes or fungi are handled in a different manner so that fragmented hyphae are transferred to the broth inoculum medium. The hyphae are broken up by vigorously scraping the agar surface with a loop to provide an aqueous hyphal suspension, or the hyphae can be torn apart in a waring blender or a similar macerating device.

20.6.2 Size of Inoculum

All steps of be fermentation process, except that of the initial inoculation from a stock culture, require the transfer of approximately 0.5 to 10 percent inoculum by volume from the preceding step in the sequence. At each step, the organisms should grow quickly and in high numbers so that the period of incubation required is relatively short. Little if any fermentation product should accumulate during the inoculum stages, because the cells should be transferred to larger batches of media while they are still in their logarithmic growth phase and before accumulation of product would normally occur. In fact, inoculum media are usually balanced for rapid cell growth and not for product formation.

The inoculum level introduced into a production tank is also usually in the 0.5 to 10 percent range. It is common practice to employ large volumes of inoculum on the production scale (for example 10%) in order to shorten the initial growth phase of the production-stage fermentation and thus make better use of available plant capacity. At times it may be as high as 20 percent or greater. These high levels of inocula are employed under conditions in which the presence of growth inhibitors or of poor energy recovery of the cells do not allow extensive growth of the microorganism in the fermentation production medium (for instance in the bisulfite-yeast-glycerol fermentation). A similar condition exists when pregrown inoculum is utilized for enzymatic conversions. These various additions of high levels of inocula to fermentations may require that the inoculum cells be sedimented, centrifuged, or in some other manner separated from the inoculumn growth medium so that the addition of a high level of cells does not concurrently cause excessive dilution of the production medium, unwanted pH changes, or carry-over of unwanted nutrients or metabolic waste products.

The quality and reproducibility of the inoculum are critical factors in reproducing high yields from one production to another and in obtaining valid experimental results in smaller tanks used in research studies. For these studies, enough inoculum should be available in a single batch so that it can be split uniformly and allow each tank to receive similar inoculum.

20.6.3 Stability of Inoculum

Many generations of microbial cells must occur during the production of inoculum between the starting point of a stock culture and the termination point of a suitable inoculum for a large production tank. We assume that, during this cellular multiplication, the microorganism does not change physiologically or genetically. However, mutation does occur among these cells, although the occurrence of a particular mutation is usually infrequent and, when it does occur, it is usually lethal to the cell or provides no selective advantage to the cell in its growth competition with non-mutated cells. Thus, the occurrence of a mutation is not serious in inoculum build-up unless it provides the microorganism with a distinct growth advantage and, at the same time, with an altered ability to produce the fermentation product. A serious problem can exist, however, if the organism used for the fermentation is in itself a mutant strain. Mutant cells are not always stable, and the frequency of occurrence of *back mutation*, or loss of a particular mutation, can be rather high; thus inoculum for production tanks can contain a considerable portion of back mutants. Obviously, fermentation and media conditions tending to select for growth of the mutant fermentation strain or against growth of the back mutant are of definite benefit in inoculum production for this type of organism. For example, a *Corynebacterium glutamicum* mutant strain requiring L-threonine for growth should not be allowed to exhaust the supply of this amino acid in the medium during inoculum production. Should this occur, the mutant could not compete in growth with the non-threonine-requiring back mutant, and this back mutant would become dominant in the inoculum. Stock-culture maintenance of mutated fermentation strains should be rigorously controlled so that the least possible numbers of back mutants will find their way into inoculum production.

Enzymatic adaptation in microorganisms resembles a mutational change, but it is not an inbherited characteristic. A microorganism does not continuously produce high levels of enzymes to attack all of the possible substrates for which it has the inherent capability. It produces in quantity only those enzymes that it really needs. Some of these enzymes may be *adaptive enzymes* produced in response to the presence of a substrate that the organism can utilize for energy and growth. If such a substrate is deleted from the medium or becomes exhausted, the microorganism in succeeding generations decreases the formation of the respective adaptive enzyme. Industrial fermentations often utilize the activities of the adaptive enzymes of microorganisms and, hence these enzymes should be present in the highest possible quantities during production. Therefore, provision should be made that the *inducing* substrate be present either in all the steps of inoculum build-up or, at least, in the final stages. If this consideration is not met, a distinct lag in growth may be observed in the production tank immediately after inoculation while the microorganisms are adaptively forming enzymes to allow utilization of the inducing substrate.

To avoid or minimize culture degeneration problems the following steps should be taken: 1) Start with a pure culture, obtained from a single-cell isolate of a high-yielding strain. 2) Preserve the strain in a form in which growth, and hence DNA replication, cannot occur. The preserved master culture should be prepared from a pure culture of the productive organism as quickly as possible, with a minimum number of sub-cultures. Various techniques are available for culture preservation including lyophilisation, freezing under liquid nitrogen, storage under mineral oil, dessication and so on; these have been described in chapter-3. Serial sub-culture, as a means of preserving highly mutated strains, should be avoided at all costs.

20.6.4 Prevention of Contamination

Contamination is an ever-present risk in inoculum production. Contamination may be present but undetected, or it may be so gross as to be easily observed. Every attempt should be made to detect either level of contamination so that inoculum for production tanks is completely devoid of contaminants.

20.6.4.1 Sterility Testing in Fermentation Processes

With fermentation employing the hyphae of actinomycetes or fungi, high-level bacterial contamination is sometimes easily observed as general turbidity and even foaming. Usually, however, contamination is detected by resorting to microscopic observation of wet mounts or stained preparation and by culturing samples of the inoculum on a medium that allows growth of suspected contaminating microorganisms. Inoculum for anaerobic fermentations should be tested for contamination by culturing under both anaerobic and aerobic incubation conditions so as to detect the presence of facultative anaerobes. Contamination by Gram positive bacteria usually originates from insufficient tightness of the fermentation equipment and from insufficient sterilization of air used in aeration. Gram negative bacteria usually come from leaking cooling coils or heat exchangers. Fungi are less likely contaminants with the exception of yeast-like organisms which may originate from a medium that has not been sufficiently sterilized, for instance from corn steep liquor.

For some fermentations, such as the *Clostridium acetobutylicum* acetone-butanol fermentation, there are characteristic fermentation patterns which provide a clue to the presence of contamination during growth of this organism, there is an initial production of acetic and butyric acids accompanied by adaptive enzyme formation for the conversion of these acids to acetone and butanol. Concurrent with the production of the latter solvents, the levels of acetic and butyric acids decrease markedly—the *acid-break*; this phenomenon can be observed by titrating successive samples of inoculum or production culture with dilute alkali. Inoculum of this organism should be used only after a decisive acid-break has occurred, since contaminating microorganisms prevent the detection of a sharp acid-break.

20.6.4.2 Precautions Against Attack by Phages

The actual amount of damage to a fermentation caused by a phage attack depends on the particular fermentation and on the phase of the fermentation at the time at which the attack occurs. Thus, a phage infection of a fermentation is not a problem if the phage happens to be one that can neither lyse nor establish lysogeny in the fermentation organism. However, if the phage finds rapidly growing sensitive host microorganisms mediating an industrial fermentation, it will multiply in and lyse those organisms so that, within a relatively short time, a high percentage of the cells will have disappeared. The surviving cells are those few that were naturally resistant to the phage, and these cells will slowly multiply and reproduce the fermentation medium. This occurrence markedly extends the incubation period for the fermentation. In contrast, an infection occurring late in the fermentation will probably hardly be noticed, since phage multiplication and cell lysis require young, actively growing cells. Therefore, the most critical time for a phage attack is during logarithmic growth of the sensitive fermentation organism.

Every attempt should be made to prevent contamination of the stock cultures and inoculum, and to prevent contamination of a fermentation with dust from the air which may be carrying phages. Nevertheless if it is suspected that a phage attack occurred, the presence of the phage should be demonstrated by filtration of a sample of the culture broth followed by plating the culture filtrate for plaque formation with the uninfected host fermentation organism. If this procedure reveals the presence of phage, a sample of the culture filtrate is frozen as a source of the phage for later testing, or the phage is recovered by inoculation from the plaques into an actively growing culture of the host. Also, the fermentation microorganism is reisolated from the fermentation broth after regrowth of resistant cells has occurred, and the isolates thus obtained are purified and tested against the phage to confirm the resistance. The fermentation microorganism can also be mutated, using the phage filtrate to select the phage-resistant mutants. Regardless of which procedure is utilized, the resulting phage-resistant strains must also be retested for production capacity for the fermentation product. It is possible that, in the process of obtaining a phage-resistant strain, we may have selected a normally poor-yielding strain from the natural population, or that phage resistance, for some mutants, may be genetically tied to low-yield capacity.

A fermentation microorganism selected as above for resistance to a phage will not be resistant to subsequent attacks by other phages. Should another phage attack occur, we must start over again to obtain resistance to the new phage. There are too many different phage-strain possibilities to make the fermentation culture resistant to all the possible phages that could conceivably attack it during industrial production. Thus, we cannot incorporate phage resistance into fermentation microorganisms by mutation and selection during process development. We must constantly be on the watch for a phage attack and be ready to recover a high-yielding, phage-resistant strain of the fermentation organism. Some reports indicate that some media possibly allow increased protection for a fermentation organism against phages. Also, the incorporation of low concentrations of antibiotics into fermentation media has been reported to have this effect (Hongo, 1972).

A phage that establishes lysogeny in a fermentation microorganisms presents a different type of problems for industrial fermentations. Cell lysis is negligible and, thus, the presence of the phage is not detected unless culture filtrate is combined for plaque formation with a sensitive indicator organism. This procedure usually is not followed because these phages are considered to have little influence on an industrial fermentation and are not determined during routine observation of the progress of an industrial fermentation. However, this is a faulty assumption. The nucleic acids of the phages carry genetic material that is incorporated into the genetic material of the host microorganism and, as a result, new enzymes may be formed or old enzymes deleted from the normal enzymatic component of the cell. This can have a profound effect on the growth and yield potential of the fermentation organism. The genetic changes induced by a phage during lysogeny resemble a mutation of the host organism, but as one regards an industrial fermentation, they differ in one important aspect. Although the frequency of a particular mutation in a normal population of microorganisms is very low, and rigorous selection is required for the mutant to gain ascendancy in the population, a phage that established lysogeny in the fermentation organism in an inoculum or production tank presents a different picture. The phage multiply rapidly in the few sensitive organism present, librating further phages so that most of the microbial population becomes lysogenic. The genetic changes attributable to phage in the host organism, however, are incorporated rapidly into the entire microbial population, and selection for the genetic variant is not required. Thus, if the genetic change affects growth or yield, it will become apparent relatively quickly.

There is little that can be done about a phage attack that includes lysogeny in a fermentation microorganism. All fermentation equipment should be carefully cleaned and sterilized, but this may have little effect if the phage is still present in high numbers in the air of the plant. We should be extremely careful that such a phage does not get into stock cultures. However, from a different viewpoint, this type of phage attack on a fermentation could be a boon, for example, when the genetic changes associated with lysogeny cause increased yields instead of the more likely decreased yields. In this case, it would be advantageous to reisolate and purify the fermentation organism from this rare high-yielding tank.

20.6.4.3 *Medium Sterilization*

The basic nature of the fermentation medium will presumably be the same both on the small and the large scale, that is, it will remain unchanged during scale-up. However, crude, technical-grade medium constituents will almost certainly be employed on the production scale for obvious economic reasons, whereas laboratory-grade reagents may well be employed on the small scale. The impurities introduced via these crude materials may give rise to scale-up problems, either directly or indirectly by participating in chemical reactions during sterilization of the medium. Medium sterilization itself poses a major problem. The fermentation medium is subjected to a more severe heat treatment on the large scale during batch sterilization. This is unavoidable because extended periods of time are necessary to heat the large volumes of medium up to the required sterilization temperature and to cool them down to normal operating temperature when the sterilization cycle has been completed. Heating and cooling periods of 3-6 hours are, for example, quite common in large production fermenters. It is also necessary to employ a more severe heat treatment on the large scale from the point of view of sterility. The larger volumes of medium contain greater numbers of contaminating microorganisms, which require longer exposure to high temperatures for their complete destruction. The greater severity of the heat treatment employed on the larger scale may have an adverse effect on the nutritive quality of the medium, due to thermal degradation of essential nutrients or to the formation of toxic compounds by reaction between medium constituents at elevated temperatures. The result of the reduced nutritive quality of the medium will inevitably be lower yields of the desired product. The requirement of minimum destruction of proteins and minimal formation of toxic substances by reactions of carbohydrates with amino acids lead to the fact that sterilization of the nutrient medium on an industrial scale is usually performed in such a way that the medium after sterilization is not strictly sterile, but *production sterile.*

20.6.5 Foam Control in Fermentation Processes

Foam formation during microbial fermentations is usually considered an undesirable but generally unavoidable evil. At worst it may threaten an entire operation with contamination and loss of product. Such foaming is a familiar problem to most fermentation technologists, but the treatments adopted to combat it are as variable as the processes themselves. They range from the purely empirical approach of adding the nearest available antifoam to a full-scale investigation of the underlying causes and a complete and systematic evaluation of the available control methods.

It a chemical method is to be adopted, then the choice of product available is considerable, ranging from more recently developed silicones and polyoxyalkylene ethers of traditional formulations based on raw or partially oxidized animal and vegetable oil and other natural products. One great advantage of the newer silicones and polyoxyalkylene glycols is their high activity at very low concentrations, where additions even at a later stage of a fermentation are not likely to prove troublesome from the standpoint of product recovery and toxicology.

Numerous mechanical methods for foam control are also available, but many suffer from the disadvantages that they entail extensive modification to an existing plant, have high running costs and are often only partially successful in controlling foam. Further, many of the patented designs have only been developed as far as the experimental stage and are not commercially available. Foam and foam control have been reviewed by Hall *et al.* (1973).

20.6.6 Experimental Approach for Scale-up of Fermentations

It would seem that the best way to obtain fermentation information for production tanks would be to carry out fermentation studies directly in the large tanks. However, this is not usually possible for a new fermentation, and it is not particularly feasible for the study of variables of a fermentation already in production. Production tanks are in continuous use for commercial production of fermentation products, and the utilization of a production tank for an experimental determination removes this tank from the normal production capacity of a company. Moreover, valid experiments cannot be carried out with only a single tank; one or more tanks are required as experimental controls. Aside from these considerations, the various costs, including media, associated with the use of these large production tanks are far too great for their extensive utilization for experimental studies. The commonly accepted alternative is to obtain the best possible information during scale-up studies in the hope that this information will be valid at the production level.

Erlenmeyer flasks are the conventional vessels for fermentation studies during secondary screening or laboratory process development. However, these flasks, even when shaken on a reciprocating or rotary mechanical shaker, provide a poor estimate of the fermentation potential for a microorganism and its medium because of the relatively poor aeration characteristics associated with these vessels. Similar flasks with baffles provide better aeration but do not provide the favourable aeration conditions obtainable in a fermentation tank. Small laboratory fermentation tanks of 1 to 10 or 12 litre sizes are more suited for these studies, since their aeration and agitation conditions can be varied, and since the overall fermentation conditions of these tanks more closely resemble those of the larger production tanks. Although these laboratory-scale tanks do provide considerable information, it is still necessary to study a fermentation in larger experimental tanks (80 to 300 litres or slightly larger) such as are found in a *pilot plant*. These tanks allow fermentation studies on a scale that is meaningful in relation to production-tank conditions, but without too great expense for media, labour, power input, and so forth. Also at times yet larger tanks intermediate in size between the latter tanks and production tanks are employed to gain additional information on a particular fermentation before it is carried out in production fermenters.

Various attempts have been made to mathematically predict the fermentation conditions for a large-scale production tank based on information obtained from laboratory or small pilot-plant tanks. But experience with particular fermentation equipment and previous fermentations are the only real guide in translating scale-up information to production tanks. Experimentation in production tanks is not completely ruled out. All fermentations require some production-scale experimentation at the time of process development and during later production. Various schemes have been proposed for applying experimental variations to one or two production tanks with the remaining production tanks serving equally as experimental controls and for commercial production of the fermentation product. These procedures usually allow the recovery of at least minimal amounts of valuable information, but the success of this approach again depends, to a large extent, on previous experience with the particular fermentation being studied, with other fermentations, and with the fermentation equipment itself.

20.7 SCALE-DOWN

Scale-down is the situation where laboratory- or pilot-scale experiments are conducted under conditions which mimic the industrial-scale conditions. This approach is important in both the development of a new product and the improvement of an existing full-scale fermentation. The procedure has been reviewed by Jem (1989). Frequently, conditions achievable on a laboratory scale are impractical on an industrial scale, which means that if inappropriate conditions have been used in the laboratory unrealistic yield objectives may be set for the scaled-up process. The aspects to consider in the design of laboratory- or pilot-plant experiments in the context of scale-down may be summarized as follows:

(i) **Medium design.** Media relevant to the industrial situation should be used in development experiments.

(ii) **Medium sterilization.** If the medium is to be batch sterilized on the large scale its exposure time at a high temperature will be much greater than that experienced in the laboratory or pilot plant. Thus, the sterilization times on the smaller scales should be increased to mimic the industrial situation. Alternatively, medium sterilized in the production fermenter may be used in the laboratory and pilot plant. This highlights the advantage of continuous sterilization where little loss of medium quality occurs. Furthermore, the same continuous sterilizer may be used for both full-scale and pilot scale vessels.

(iii) **Inoculation procedures.** Due to a range of circumstances, it may not always be possible to inoculate every production fermentation with inoculum in optimum condition. The scale-down approach can be used to predict the consequences of such events by mimicking these situations in the laboratory, for example by storing inoculum or using inocula of different ages.

(iv) **Number of generations.** An industrial scale fermentation requires a greater number of generations than does a laboratory one; this may place more severe stability criteria on the process strain than may have been appreciated on the small scale. The industrial situation may be modelled in the laboratory by using serial sub-culture to ensure that the strain is sufficiently stable. This approach is particularly pertinent in the development of recombinant fermentations.

(v) **Mixing.** As indicated in the previous section it is almost inevitable that the degree of mixing will decrease with an increase in scale. Thus, it is possible to model inadequate mixing in the laboratory by subjecting the organism to pulse medium feeds or fluctuating process conditions such as oxygen concentration, pH and temperature. Such scaled-down experiments then allow predictions to be made about the suitability of new strains for industrial exploitation.

(vi) **Oxygen transfer rate.** Far higher oxygen transfer rates can be achieved in laboratory fermenters than in industrial-scale ones. Thus, unrealistic demands may be made of a fermentation plant if the development work has been done at very high oxygen-transfer rates. Therefore, the laboratory and pilot fermenters should reflect the oxygen transfer rates achievable in the full-scale fermenters.

The adoption of these simple approaches to small scale experimentation can prevent many scale-up problems before they even occur.

20.8 SUMMARY

Before designing and planning any fermentation process, it is necessary to scale-up experimentation required for fermentation process. Small to big differences exist in design and efficiency of bioreactor. There is a vast difference between the laboratory fermentation process/conditions and pilot plant. Therefore, detailed study is required to know the conditions, which virtually exists during production. Small-scale experiments are the starting point for scale-up procedure. Few changes are required in small scale-plan before employing it for large-scale production. For development of bioreactor for bioprocess development many factors like geometric of reactors, kinematics and dynamic similarities, thermodynamic, micro kinetic, transport, flow and diffusion (i.e., shear, mixing, mass transfer, and heat transfer etc.) should be studied and planned properly. Other criteria are for required production at large scale are efficient mixing in large vessel, aeration, improved strains, type of nutrient, proper sterilization conditions, separation of growth inhibitory substances, regulation of secondary metabolite biosynthesis, carbon and nitrogen source control, etc.

❑❑❑

SELECTED BIBLIOGRAPHY

Aiba, S., Humphrey, A. E. and Mills, N. F. 1965. *"Biochemical Engineering"*, Academic Press, New York.

Aiba, S. (Ed.) 1987. *"Horizons of Biochemical Engineering"*, Uni. Tokyo Press, Tokyo.

Aida, K. *et.al.* 1986. *"Progress in Industrial Microbiology"*, Kodansha Ltd., Tokyo.

Alderman, N. J. and Heywood, N. I. 2004. *"Improving slurry viscosity and flow curve measurements"*, *CEP*, **4**: 27-32

Andrews, B. A. and Asenjo, J. A. 1987. *"Enzymatic lysis and Disruption of Microbial Cells"*, Trends. Biotechnol., **5**: 273-277.

Anonymous, 1986. *"NIH Guidelines for Recombinant DNA-Technology Research"*, N. I. H., Bethesda, Maryland.

Anonymous, 1993. *"Biosafety in Microbiological and Biomedical Laboratories"*, 3rd ed., U. S. DHHS, Atlanta, Georgia and N. I. H., Bethesda, Maryland.

Arora, D. K. *et al.* (Eds.) 1992. *"Handbook of Applied Mycology"*, Marcel Dekker, N. Y.

Askew, M. W. 1975. *"Fermentation : Water, Waste and Money"*, *Proc. Biochem*, **10**: 5-7.

Atherton, K. T. *et al.* 1979. *"Genetic Manipulation for Industrial Processes"*, *Soc. Gen. Micro. Symp.*, **29**: 379-406.

Atkinson, B. 1974. *"Biochemical Reactors"*, Pion Ltd., London.

Atkinson, B. and Daoud, I. S. 1976. *Microbial flocs and flocculation in fermentation process engineering"*, *Adv. Biochem., Eng.*, **4**: 41-124.

Atkinson, B. and Mavituna, F. 1983. *"Biochemical Engineering and Biotechnology Hand book"*, Macmillan Pub. Ltd., New York.

Atkinson, B. and Sainter, P. 1982. *"Development of downstream processing"*, *J. Chem. Tech. Biotech No.* **32**: 100-108.

Ausubel, F. M., *et al.* 2001. *"Current Protocols in Molecular Biology"*, John Wiley & Sons, N. Y.

Bailey, J. E. and Ollis, D. F. 1977. *"Biochemical Engineering Fundamentals"*, McGraw-Hill book Co., London.

Baltz, R. H. and Matsuhima, P. 1980. *"Applications of protoplast fusion, site directed mutagenesis and gene amplification to antibiotic yield improvement in* ***Streptomyces****"*, *Actinomycetes*, **15:** 95-120.

Baumberg, S., *et al.*(Eds.), 1989. *"Microbial Products : New Approaches"*, *Soc. Gen. Microbiol. Symp.*, **44**: 95-120.

Birch, G. G. and others (Eds.), 1981. *"Enzymes and Food processing"*, Applied Science Ltd., London.

Birge, E. A. 1988. *"Biodiversity as a source of innovation in Biotechnology"*, *Ann, Rev. Microbiol.*, **46**: 219-252.

Blakebrough, N. (Ed.), 1967-68. *"Biochemical and Biological Engineering Science"*, *Vols. 1 and* 2, Academic Press, London.

Blanch, H. W. and Clark, D. S. 1966. *"Biochemical Engineering"*, Marcel Dekker, N. Y.

Board, R. G. and Lovelock, D. W. (Eds.) 1975. *"Some Methods for Microbiological Assay"*, Academic Press, London.

Brenner, S., Hartley, B. S. and Rodgers, P. J. (Eds.), 1980. *"New Horizons in Industrial Microbiology"*, The Royal Society, London.

Bryant, R. 1984. *"The Pharmaceutical Quality Control Handbook"*, Aster Pub. Corp., Springfield, Oregon, U.S.A.

Bu`Lock, J. D. and Kristiansen, B. (Eds.) 1987. *"Basic Biotechnology"*, Academic Press, London.

Bu`Lock, J. D., Nisbet, L. J. and Winstanely, D. J. (Eds.) 1982. *"Bioactive Microbial Products : Search and Discovery"*, Academic Press, London.

Bull, A. T. and Medow, P. M. (Eds.) 1978. *"Companion to Microbiology"*, Longman Group Ltd., London.

Bull, A. T. and others, (Eds) 1979. *"Microbial Technology"*, University of Cambridge, Cambridge.

Bushell, M. E. and Slater, J. H., (Eds.) 1981. *"Mixed Culture Fermentations"*, Academic Press, London.

Calam, C. T. 1967. *"Media for industrial fermentations"*, Proc. Biochem., **2**: 19-22.

Calam, C. T. 1976. *"Starting investigational and production cultures"*, Proc. Biochem., **11**: 7-28.

Casida, L. E. 1968. *"Industrial Microbiology"*, John Wiley & Sons Ltd., New York.

Charles, M. 1978. *"Technical aspects of the rheological properties of microbial cultures"*, Adv. Biochem. Eng., **8**: 1-62.

Chater, K. F. *et al.*1982. *"Gene cloning in Streptomyces"*, Curr. Top. Microbiol. Immunol., **96**: 69-95.

Cheetham, P. S. J. 1987. *"Screening for novel biocatalysts"*, *Enz. Microb Technol.*, **9**: 194-213.

Chen, N. Y. 1990. *"The design of airlift fermenters for use in biotechnology"*, Biotechnol. Gen. Eng. Rev., **8**: 379-396.

Cherry, G. B. *et al.* 1963. *"Sterilization of air"*, *Prog. Ind. Microbiol*, **4**: 37-60.

Cheryan, M. 1986. *"Ultrafiltration Handbook"*, Technomic Publ., Lancaster.

Claremont, D. J. 1987. *"Biosensors : Clinical requirements and Scientific Promise"*, J. Med. Engg. Tech. **11**: 51-56.

Cohen, B. L. 1973. *"The neutral and alkaline Proteases of* ***Aspergillus nidulans****"*, J. Gen. Microbiol., **77**, 21-28.

Collins, C. H. and Beale, A. J. (Eds.) 1992. *"Safety in Microbiology and Biotechnology"*, Butterworth-Heinemann, Oxford.

Collins, C. H. and Lyne, P. M. 1979. *"Microbiological Methods"*, Uni. Press, Baltimore.

Cooney, C. L. 1979. *"Computer Applications in fermentation technology"*, Biotechnology and Bioengineering Symposium, **9**: 1-11.

Corum, C. J. (Eds.) 1970. *"Developments in Industrial Microbiology"*, American Institute of Biological Sciences, Washington, U.S.A.

Curran, J. S. and Smith, J. 1989. *"Heat and Power in Industrial fermentation"*, Appl. Energy, **34**: 9-16.

Daies, R. (Eds.), 1982. *"Developments in Food Microbiology"*, Applied Science Pub. Ltd., London.

Datar, R. 1993. *"Economics of primary separation steps in relation to fermentation and genetic engineering"*, *Proc. Biochem.*, **21**: 19-26.

Davies, C. N. 1973. *"Air Filtration"*, Academic Press, London.

Dawson, P. S. S. 1974. *"The cyclone column fermenter"*, *Biotech. Bioeng. Symp.*, **4**: 809-819.

DBT 1984. *"The Indian Recombinant DNA safety Guidelines and Regulations"*, DBT, New Delhi.

Dean, A. C. R. and others (Eds.), 1976. *"Continuous Culture : Applications and New Fields"*, Ellis Horwood Ltd., Chichester.

Dellwey, K. (Eds.), 1983. *"Biotechnology"*, Verlag Chemie, Florida.

Demain, A. L. 1973. *"Mutation and production of secondary metabolites"*, *Adv. Appl. Micro.*, **16**: 177-202.

Demain, A. L. 1983. *"New applications of microbial products"*, Science, **219**: 709-719.

Demain, A. L. and Solomon, N. A. 1986. *"Industrial Microbiology and Biotechnology"*, American Microbiological Society, Washington.

Denyer, S. and Baird, R. (Eds.), 1990. *"Guide to Microbiological Control in Pharmaceuticals"*, Ellis Horwood, New York.

Diaz, M. 1988. *"Three-phase extractive fermentation"*, TIBTECH, **6**: 126-130.

Dier, O. D. *et al.* 2000. *"Safe Biotechnology"*, TIBTECH, **18**: 141.

Dunnill, P. and Lilly, M. D. 1972. *"Continuous enzyme isolation"*, *Biotech. Bioeng. Symp.*, **3**: 97-113.

East, D. *et al.* 1984. *"Reduction of biological risk in fermentation processes by physical containment"*, *Dev. Ind. Micro.*, **25**: 89-105.

Edwards, V. H. 1969. *"The Recovery and Purification of Biochemicals"*, *Adv. Appl. Microbiol.*, **11**: 159-210.

Einsele, A. 1978. *"Scaling-up of Bioreactors"*, *Proc. Biochem.* **July**: 1-14.

Elander, R. P. 1980., *"New Genetic Approaches to industrially important fungi"*, *Biotech. Bioeng.* **22**: 49-61.

Elander, R. P. 1985. *"Present and future roles by biotechnology in the pharmaceutical industry"*, *Drug Dev. Ind. Pharm.*, **11**: 965-999.

Elander, R. P. *et al.* 1977. *"Genetics of Industrial Microorganisms"*, *Ann. Rep. Ferment. Proc.*, **1**: 1-40.

Ettler, P. 1992. *"The determination of optimum inoculum quality for submersed fermentation process"*, *Collect. Czech. Chem. Commun.* **57**: 303-308.

European Federation of Biochemistry. 1983. *"Guidelines for the characterization of Immobilized Biocatalysts"*, *Enz. Microb. Tech.*, **5**: 304-307.

Fallick, G. F. 1969. *"Industrial Ultrafiltration"*, *Process Biochem.*, **4**: 29-34.

Fantini, A. A. 1966. *"Experimental approaches to strain improvement"*, *Dev. Ind. Micro.*, **7**: 79-87.

Ferenczy, L. *et al.* 1974. *"Fusion of fungal protoplasts"*, *Nature*, **248**: 793-794.

Finkelstein, D. B. and Ball, C. (Eds.) 1992. *"Biotechnology of Filamentous Fungi"*, Butterworth-Heinemann, Stoneham.

Flannery, R. J. and Steinschneider, A. 1983. *"Fermentation process economics in relation to Genetic Engineering"*, *Biotechnology*, **1**: 773-776.

Fogarty, W. M. (Ed.) 1983. *"Microbial Enzymes and Biotechnology"*, Applied Science Publishers, London.

Freitas, Y. M. and Fernandes, F. (Eds.) 1971. *"Global Impacts of Applied Microbiology"*, Uni. of Bombay, Bombay.

Fritz, J. S. *et al.* 1982. *"Ion Chromatography"*, Huthing Verlag, Heidelberg.

Gavora, J. *et al.* (Eds.) 1988. *"Biotechnology Research and Applications"*, Elsevier, London.

Geysen, H. M. and Gray, P. P. 1972. *"A graphical method for optimizing fermentations"*, *Biotech. Bioeng.*, **14**: 857-860.

Ghildyal, N. P. *et al.* 1988. *"Foam control in submerged fermentation"*, *Adv. Appl. Microbiol.*, **33**: 173-221.

Ghose, T. K. (Ed.) 1985. *"Biotechnology and Bioprocess Engineering"*, United India Press, New Delhi.

Giorgio, R. J. and Wu, J. J. 1986. *"Design of large scale containment facilities for recombinant DNA fermentations"*, *Trends. Biotech.*, **4**: 60-65.

Glasby, J. S. 1976. *"Encyclopaedia of Antibiotics"*, J. Wiley & Sons, London.

Glazer, A. N. and Nikaido, H. 1995. *"Microbial Biotechnology: Fundamentals of Applied Microbiology"*, W.H. Freeman, New York.

Godfrey, T. and Richelt, J. 1983. *"Industrial Enzymology"*, Macmillan Pub. Ltd., New York.

Gottschalk, G. 1986. *"Bacterial Metabolism"*, Springer-Verlag, Heidelberg.

Gutcho, S. J. 1973. *"Chemicals by Fermentation"*, Noyes Data Corporation, London.

Halme, A. 1983. *"Modelling and control of fermentation processes"*, Pregamon Press, Oxford.

Hambleton, P. *et al.* 1992. *"Biosafety monitoring devices for biotechnology processes"*, *Trends. Biotechnol.*, **10**: 192-199.

Harris, T. J. R. (Ed.) 1990. *"Protein Production by Biotechnology"*, Elsevier, London.

Harrison, D. E. and Others, (Eds.) 1980. *"Hydrocarbons in Biotechnology"*, Heyden & Sons. Ltd., London.

Heijnen, J. J. *et al.* 1992. *"Fundamental bottlenecks in the application of continuous bioprocesses"*, *J. Biotechnol.* **22**: 3-20.

Hellman, A. *et al.* (Eds.) 1973. *"Biohazards in Biological Research"*, Cold Spring Harbor Laboratory, N.Y.

Hewitt, W. 1977. *"Microbiological Assay"*, Academic Press, London.

Hockenhull, D. J. D. (Ed.) 1972. *"Progress in Industrial Microbiology"*, Churchill Livingstone, London.

Hopwood, D. A. *et al.* 1985. *"Genetic manipulation of Streptomyces"* (A Lab. Manual), The John Innes Foundation, Norwich.

Howard, J. R. (Ed.) 1983. *"Fluidised beds"*, Applied Sci. Pub., London.

Howells, E. R. 1982. *"Single-cell protein and related technology"*, *Chem. Indust.*, **7**: 508-511.

Hughes, M. N. and Poole, R. K. 1989. *"Metals and Microorganisms"*, Chapman and Hall, London.

Hugo, W. B. and Russell, A. D. 1980. *"Pharmaceutical Microbiology"*, Blackwell Scientific Publications, London.

Humphrey, A. E. 1960. *"Air Sterilization"*, *Adv. Appl. Microbiol.*, **2**: 302-312.

Hutner, S. H. 1972. *"Inorganic nutrition"*, *Ann. Rev. Microbiol.*, **26**: 313-346.

Hutt, R. 1967. *"Recovery of Fermentation Products"*, *Biochem. Biol. Engn. Sci.*, **1**: 245-264.

Hutter, R. *et al.* (Eds.) 1978. *Antibiotics and other secondary metabolites"*, Academic Press, London.

Huxsoll, J. F. 1994. *"Quality Assurance of Biopharmaceuticals"*, John Wiley & Sons, New York.

Imrie, F. K. E. 1969. *"Fermentation Media"*, *Proc. Biochem.*, **8**: 34-35.

Iwai, Y. and Omura, S. 1982. *"Culture conditions for screening of new antibiotics"*, *J. Antibiot.*, **35**: 123-141.

Jain, D. and Buckland, B. C. 1988. *"Scale-up of the efrotomycin fermentation using a computer-controlled plant"*, *Bioprocess Eng.*, **3**: 31-36.

Janson, J. C. 1984. *"Large Scale purification-state of the art and future prospects"*, *Trends. Biotechnol.*, **2**: 31-38.

Jem, K. J. 1989. *"Scale-down techniques in fermentation"*, *Biopharmacology*, **2**: 30-39.

Juran, J. M. *et al.* 1988. *"Quality Control Handbook"*, McGraw-Hill Book Co., New York.

Kale, G. P. and Others (Eds.) 1978. *"Perspectives in Industrial Microbiology"*, A.M.I., Bombay.

Katoh, S. 1987. *"Scaling-up affinity chromatography"*, *Trends Biotechnol.*, **5**: 328-331.

Keay, L. 1971. *"Microbial Proteases"*, *Proc. Biochem.* **6**: 17-21.

Keim, C. R. and Venkatasubramanian, K. 1989. *"Economics of current biotechnological methods of producing ethanol"*, *Trends. Biotechnol.*, **7**: 22-29.

King, P. P. 1982. *"Biotechnology: an industrial view"*, *J. Chem. Tech. Biotech.*, **32**: 2-8.

Kirsop, B. E. and Doyle, A. (Eds.) 1991. *"Maintenance of Microorganisms and cultured cells"*, Academic Press, London.

Kjaergaard, L. 1977. *"The redox potential: Its use and control in biotechnology"*, *Adv. Biochem. Eng.*, **7**: 131.

Kloyosov, A. A. 1986. *"Enzymic conversion of cellulosic materials to sugars and alcohols"*, *Appl. Biochem. Biotechnol.*, **12**: 249-300.

Knecht, R. *et al.* 1977. *"Microbiology and biotechnology of SCP produced from n-parrafin"*, *Proc. Biochem.* **12**: 11-14.

Kobayashi, T. *et al.* 1973. *"Oxygen transfer into mycelial pellets"*, *Biotech. Bioeng.*, **15**: 27-45.

Koryta, J. 1972. *"Theory and application of ion-selective electrodes"*, *Anal. Chem.. Acta.*, **61**: 329.

Kroner, K. H. *et al.* 1984. *"Cross-flow filtration in the down stream processing of enzymes"*, *Process Biochem.*, **April**: 67-74.

Krumphanzi, V. *et al.* (Eds.) 1982. *"Overproduction of Microbial Products"*, Academic Press, London.

Kula, M. R. *et al.* 1982. *"Purification of enzymes by liquid-liquid extraction"*, *Adv. Biochem. Eng. Biotechnol.*, **24**: 73-118.

Kybal, J. and Vlcek, V. 1976. *"Simple device for stationary cultivation of microorganisms"*, *Biotechnol. Bioeng.*, **18**: 1713-1718.

Lacey, R. E. and Loeb, S. 1972. *"Industrial Processing with Membranes"* Wiley-Interscience, New York.

Ladisch, M. R. 2001. *"Bioseparations/Engineering Principles, Practice and Economics"*, Wiley-VCH, Germany.

Ladisch, M. R. and Bose, A. 1992. *"Harnessing Biotechnology for 21st century"*, American Chemical Society, Washington, D.C.

Lafferty, R. M. (Eds.) 1983. *"Enzyme Technology"*, Springer Verlag, Berlin.

Laine, J. and Kuoppamaki, R. 1979. *"Development of the design of large-scale fermenters"*, *Ind. Eng. Chem. Prod. Res. Dev.*, **18**: 501-506.

Le, M. S. and Atkinson, T. 1985. *"Crossflow microfiltration for recovery of intracellular products"*, *Proc. Biochem.*, **20**: 26-31.

Lee, Y. H. and Tsao, G. T. 1979. *"Dissolved oxygen electrodes"*, *Adv. Biochem. Eng.* **13**: 35.

Li, N. N. and Strathmann, H. (Eds.) 1988. *"Separation Technology"*, Engineering Foundations. N.Y.

Luckner, M. *et al.* 1977. *"Secondary Metabolism and Cell Differentiation"*, Springer-Verlag, Berlin.

Luong, J. H. *et al.* 1987. *"Recent developments in downstream processing based on affinity interactions"*, *Trends Biotechnol.*, **5**: 281.

Macdonald, K. D. (Ed.) 1976. *"Genetic of Industrial Microorganisms"*, Academic Press, London.

Maclean, G. T. 1977. *"Oxygen transfer in aerated systems with two liquid phases"*, *Proc. Biochem.*, **12**: 22.

Malik, V. S. 1979. *"Genetics of Applied Microbiology"*, *Adv. Genet.*, **20**: 37-126.

Mann, I. 1977. *"Secondary Metabolism"*, Oxford Uni. Press, Oxford.

Martini, L. J. 1984. *"Practical Seal Design"*, Marcel Dekker, N.Y.

Mellor, J. D. 1978. *"Fundamentals of Freeze Drying"*, Academic Press, N.Y.

Mezger, T. G. 2002. *"The Rheology Handbook"*, Vinemtz Verlag, Hannover, Germany.

Miller, B. M. and Litsky, W. 1976. *"Industrial Microbiology"*, McGraw-Hill Book Co., New York.

Miller, D. N. 1974. *"Scale-up of agitated vessel gas-liquid mass transfer"*, *A.I.Ch.E.J.*, **20**: 445.

Minoda, Y. 1986. *"Raw materials for amino acid fermentation"*, *Dev. Indust. Microbiol.*, **24**: 51-66.

Miura, Y. 1976. *"Transfer of oxygen and scale-up in submerged aerobic fermentation"*, *Adv. Biochem. Eng.*, **4**: 3-40.

Modi, H. A. 1995, 1996. *"Elementary Microbiology"* Vols. 1 and 2, Akta Prakashan, Nadiad, India.

Modi, H. A. 2007. *"Introductory Food Microbiology"*, Aavishkar Publishers, Distributors, Jaipur, India.

Moo-Young, M. (Eds.) 1985. *"Comprehensive Biotechnology"*, Vol. 1 to 4, Pergamon Press, Oxford.

Moo-Young, M. 1977. *"Economics of SCP production"*, *Proc. Biochem.*, **12**: 6-10.

Moo-Young, M. and Blanch, H. W. 1981. *"Design of biochemical reactors"*, *Adv. Biochem. Eng.*, **19**: 1-69.

Moss, M. O. and Smith, J. E. (Eds.) 1977. *"Industrial Applications of Microbiology"*, Surrey Uni. Press, London.

Mukhopadhyay, S. N. and Das, D. K. 1994. *"Oxygen Responses, Reactivities and Measurements in Biosystems"*, CRC Press, Boca Raton, Florida.

Mullin, J. W. 1972. *"Crystallisation"* 2nd ed., Butterworth, London.

Nelder, J. A. and Mead, R. 1965. *"A simplex method for function minimization"*, *Computer Journal*, **7**: 308-313.

Nelson, L. S. 1982. *"Technical aids"*, *J. Qual. Tech.*, **14**: 99-100.

Neway, J. O. (Ed.) 1989. *"Fermentation Process Development of Industrial Organisms"*, Marcel Dekker, N. Y.

Nienow, A. W. 1990. *"Agitators for mycelial fermentations"*, *Trends. Biotechnol.*, **8**: 224-233.

Nisbet, L. J. 1982. *"Current strategies in the search for bioactive products"*, *J. Chem. Tech. Biotech.*, **32**: 251-270.

Norris, J. R. and Ribbons, D. W. (Eds.) 1970. *"Methods in Microbiology"*, Academic Press, New York.

Norris, J. R. and Richmond, M. H. (Eds.) 1978. *"Essays in Applied Microbiology"*, John Wiley & Sons, New York.

Nyiri, L. K. 1972. *"A philosophy of data acquisition, analysis and computer control in fermentation"*, *Dev. Ind. Microb.* **13**: 136-145.

Old, R. W. and Primrose, S. B. 1985. *"Principles of Gene Manipulation"*, 3rd ed., Blackwell Sci. Pub., Oxford.

Oldshue, J. Y. 1970. *"The case for deep aeration tanks"* *Chem. Eng. Proc.*, **66**: 73-78.

Omura, S. 1986. *"Philosophy of new drug discovery"*, *Microbiological Reviews*, **50**: 259-279.

Onions, A. H. S. and Others 1981. *"Smith's Introduction of Industrial Mycology"*, Edward Arnold, London.

Peberdy, J. F. 1979. *"Protoplasts-Application in Microbial Genetics"*, Uni. of Nottingham, U.K.

Peppler, H. J. and Others (Eds.) 1979. *"Microbial Technology"* Vols. 1 and 2, Academic Press, New York.

Perkowski, C. A. 1983. *"Fermentation process air filtration via cartridge filters"*, *Biotechnol. Bioeng.*, **25**: 1215-1221.

Perlman, D. (Ed.) 1969. *"Fermentation Advances"*, Academic Press, New York.

Perlman, D. (Ed.) 1977. *"The Fermentation Industry"*, Academic Press, New York.

Perlman, D. 1970. *"Some prospects for the fermentation industries"*, *Wallerstein Comm.*, **33**: 165-173.

Pirt, S. J. 1975. *"Principles of Microbe and Cell Cultivation"*, Blackwell, Oxford.

Pirt, S. J. 1979. *"The theory of fed-batch culture of microbes"*, *Ann. N.Y. Acad. Sci.*, **326**: 119-125.

Plackett, R.L. and Burman, J. P. 1946. *"The design of multifactorial experiments Biometrika*, **33**: 305-325.

Poole, R. K. *et al.* 1990. *"Microbial Growth Dynamics"*, IRL Press, Oxford.

Prasher, B. D. 1975. *"Mass transfer coefficients and interfacial areas in agitated dispersions"*, *A.I. Ch. E.J.*, **21**: 407.

Pratten, C. F. 1971. *"Economics of scale in manufacturing industry"*, Uni. Press, Cambridge.

Price, N. C. and Stevens, L. 2000. *"Fundamentals of Enzymology"*, 3rd ed. Oxford Uni. Press, Oxford.

Queener, S. and Swartz, R. W. 1979. *"Penicillins: biosynthetic and synthetic"*, In "Secondary Products of Metabolism, Economic Microbiology", Vol. 3, pp. 35-122 (Editor Rose, A. H.), Academic Press, London.

Radovich, J. M. 1985. *"Mass transfer effects in fermentations using immobilized whole cells"*, *Enz. Microb. Technol.*, **7**: 2-10.

Rainbow, C. and Rose, A. H. 1963. *"Biochemistry of Industrial Microorganisms"*, Academic Press, London.

Ratledge, C. 1977. *"Fermentation substrates"*, *Ann. Rep. Ferm. Proc.*, **1**: 49-71.

Reed, G. (Ed.) 1982. *"Prescott and Dunn's Industrial Microbiology"*, Macmillan Pub. Ltd., London.

Reed, G. and Nagodawithana, T. W. 1991. *"Yeast Technology"*, 2nd Ed., Van Norstrand Reinhold, N.Y.

Rehm, H. J. and Reed, G. (Eds.) 1981-1988. *"Biotechnology"*, Vols. 1 to 8, VCH Pub. (Verlag Chemie), Florida.

Reisman, H. B. 1988. *"Economic alanlysis of fermentation process"*, CRC Press, Baton Ronge, Florida.

Reisman, H. B. 1993. *"Problems in Scale-up of biotechnology production processes"*, *Crit. Rev. Biotechnol.*, **13**: 195-253.

Rhodes, A. and Fletcher, D. L. 1966. *"Principles of Industrial Microbiology"*, Pergamon Press Ltd., Oxford.

Richards, J. W. 1968. *"Introduction to Industrial Sterilization"*, Academic Press, London.

Riviere, J. 1977. *"Industrial Application of Microbiology"* (translated and edited by Moss, M. O. and Smith, J. E.), Suray Uni. Press, Guildford.

Roels, J. A. *et al.* 1974. *"Rheology of mycelial broths", Biotech. Bioeng.*, **16**: 181.

Roffler, S. R. *et al.* 1984. *"In situ recovery of fermentation products", Trends Biotechnol.*, **2**: 129-136.

Rose, A. H. (Ed.) 1979. *"Secondary Products of Metabolism"*, Academic Press, London.

Rose, A. H. (Ed.) 1982. *"Economic Microbiology"*, Vols. , Academic Press, London.

Sambrook, J. and Russell, D. W. 2001. *"Molecular Biology: A Laboratory Manual"*, Cold Spring Harbor Lab. Press, N. Y.

Schugerl, K. 1982. *"New bioreactors for aerobic processes", Int. Chem. Eng.*, **22**: 591.

Schugerl, K. 1985. *"Foam formation, foam suppression and the effect of foam on growth", Proc. Biochem.*, **2**: 122-123.

Schuler, M. L. and Kargi, F. 1992. *"Bioprocess Engineering"*, Prentice Hall Inc., New Jersy.

Scouten, W. H. (Ed.) 1981. *"Affinity Chromatography"*, J. Wiley & Sons, N.Y.

Scragg, A. H. (1991). *"Bioreactors in Biotechnology"*, Ellis Horwood, Chichester.

Sebek, O. K. and Laskin, A. I. (Eds.) 1979. *"Genetics of Industrial Microorganisms"*, ASM, Washington.

Sharp, D. H. 1989. *"Bioprotein Manufacture"*, Ellis Horwood, Chichester.

Shortle, D. *et al.* 1981. *"Directed mutagenesis", Ann. Rev. Gen.*, **15**: 265-294.

Smith, H. E. and Berry, D. R. (Eds.) 1971. *"An Introduction to Biochemistry of Fungal Development"*, Academic Press, London.

Smith, H. E. and Berry, D. R. (Eds.) 1975. *"Industrial Mycology"*, Edward Arnold Pub. Ltd., London.

Solomons, G. L. 1967. *"Antifoams", Proc. Biochem.*, **2**: 47-48.

Solomons, G. L. (1969). *"Materials and Methods in Fermentation"*, Academic Press, London.

Steel, R. and Miller, T. L. 1970. *"Fermenter Design", Adv. Appl. Microbiol.*, **12**: 153-188.

Stowell, J. D. *et al.* 1987. *"Carbon substrates in Biotechnology"* IRL Press, London.

Svensson, R. 1988. *"Continuous media sterilization in biotechnical fermentation", Dechema Monographien*, **113** 225-237.

Taguchi, H. 1971. *"The nature of fermentation fluids", Adv. Biochem. Eng.*, **1**:.

Tanny, G. B. *et al.* 1980. *"Improved filtration technique for concentrating and harvesting bacteria", Appl. Env. Microbiol.*, **40**: 267-273.

Terui, G. (Ed.), 1972. *"Fermentation Technology Today"*, Yamada-Kami, Osaka.

Thoma, R. W. (Ed.) 1977. *"Industrial Microbiology"*, Dowden Hutchinson & Ross Inc., Pennsylvania.

Topiwala, H. H. and Khosrovi, B. 1978. *"Water recycle in biomass production processes", Biochem. Bioeng.*, **20**: 73-85.

Trinci, A. P. J. 1992. *"Mycoportein: A twenty year overnight success story", Mycol. Res.*, **96**: 1-13.

Tsao, G. T. (Ed.) 1982. *"Annual Reports on Fermentation Processes"*, Academic Press, London.

Tuffile, C. M. and Pinho, F. 1970. *"Determination of oxygen-transfer coefficients in viscous stretomycete fermentation", Biotech. Bioeng.*, **12**: 849-871.

Turkova, J. 1978. *"Affinity Chromatography"*, Elsevier, Amsterdum.

Turner, W. B. 1971. *"Fungal Metabolites*, Academic Press, London.

Umezawa, H. 1972. *"Enzyme inhibitors of Microbial Origin"*, Uni. of Tokyo Press, Tokyo.

Underkofler, L. (Ed.) 1975. *"Developments in Industrial Microbiology"*, American Institute of Biological Sciences, Washington.

Van't Riet, K. and Tramper, J. 1991. *"Basic Bioreactor Design"*, Marcel Dekker, N.Y.

Vandamme, E. J. (Ed.) 1984. *"Biotechnology of Industrial Antibiotics"*, Marcel Dekker, Inc., New York.

Venn, R. F. (Ed.) 2000. *"Principles and Practice of Bioanalysis"*, Taylor and Francis, London.

Verrall, M. S. (Ed.) 1985. *"Discovery and Isolation of Microbial Products"*, Ellis Horwood Pub., Chichester.

Walker, J. M. and Gingold, E. B. (Eds.) 1993. *"Molecular Biology and Biotechnology"*, Royal Society of Chemistry, Cambridge.

Wang, D. I. C. *et al.* 1979. *"Fermentation and Enzyme Technology"*, John Wiley & Sons, N.Y.

Whitaker, A. 1973. *"Fermentation Economics"*, *Proc. Biochem.*, **8**: 23-26.

Wise, D. L. (Ed.) 1991. *"Bioinstrumentation and Biosensors"*, Marcel Dekker, N.Y.

Wiseman, A. (Ed.) 1975. *"Handbook of Enzyme Biotechnology"*, Ellis Horwood Ltd., Chichester.

Wiseman, A. (Ed.) 1983. *"Principles of Biotechnology"*, Surrey Uni. Press, London.

Wiseman, A. 1969. *"Enzyme for breakage of microorganisms"*, *Process Biochem.*, **4**: 63-65.

Wolf, F. J. 1974. *"Outline for Fermentation Technology"*, MIT, Cambridge.

Yamane, T. and Shimizu, S. 1984. *"Fed-batch techniques in microbial processes"*, *Adv. Biochem. Eng./Biotechnol.*, **30**: 147-184.

Yamashita, S. and Murao, C. 1978. *"Fermentation process"*, *J. Soc. Instr. Control. Eng.*, **6**: 735-740.

Yanofsky, C. 1981. *"Attenuation in the control of expression of bacterial operons"*, *Nature*, **289**: 751-758.

Yoshida, F. *et al.* 1973. *"Fed-batch hydrocarbon fermentations with colloidal emulsion feed"*, *Biotech. Bioeng.*, **15**: 257-270.

Zahner, H. 1979. *"What are secondary metabolites?"* *Folia Microbiol.*, **24**: 435-443.

Zetelaki, K. 1969. *"Distruption of mycelia for enzymes"*, *Proc. Biochem.*, **4**: 19-27.

Zlokarnik, M. 1978. *"Sorption characteristics for gas-liquid contacting in mixing vessels"*, *Adv. Biochem. Eng.*, **8**: 133-151.

USEFUL WORLD WIDE WEB SITES

(1) About Biotechnology
http://www.outcast.gene.com/ae/ab

(2) Access Excellence
http://www.gene.com.ae

(3) Biotechnology Industry Organisation
http://www.accessexcellence.com/ab/bc/bioprocesstechnology.html

(4) Biotechnology Information Centre
http://www.inform.umed.edu:8080/edres/topic/agrenv/biotech

(5) Encyclopaedia of Bioprocess Technology
http://www.cbs.umn.edu/bpti/mcf/ebt.html
(6) Genetic Engineering News
http://www.genengnews.com
(7) Microbiology Laboratory Simulation Programme
http://www.leeds.ac.uk/bionet/compend/bnt05pst.html
(8) About Enzymes
http://www.expasy.org/enzyme
(9) www.emc.maricopa.edu
(10) www.biochem.unizh.ch.
(11) www.bio.miami.edu/dana
(12) www.basic.northwestern

❑❑❑

Index

O

P

❑❑❑

Other Books by Dr. H.A. Modi

INTRODUCTORY FOOD MICROBIOLOGY

This book should prove to be an useful source of information for anyone with an interest in food microbiology for both undergraduate as well as post-graduate courses of microbiology. The contents of this book will be also useful to the students of food technology, biotechnology, public health, veterinary science, food and nutrition, environmental studies, hotel and catering management and other food-related courses. U.G.C. syllabus for food microbiology suggested for vocational Industrial Microbiology course is also adopted in framing the chapterization of the book. This book will be proved as an useful reference to get first hand information of food microbiology for the people who are working in food-processing industries, state and local Government agencies and in academic institutions.

The book includes forty chapters and four appendices. The contents of the book have been divided thematically in seven units. The theme of Unit-I is "Microorganisms and Food" covering chapters 1 to 8, elaborating interactions between microorganisms and foods leading to the development of Food Microbiology. The Unit-II has ten chapters (No. 9 to 18) dealing with microbial spoilage of different types of foods. The Unit-III describes "Food-borne Illness" covering four chapters (No. 19 to 22). The theme of Unit-IV is "Food Preservation" covering total 12 chapters (No. 23 to 34), explaining various methods of food preservation. "Beneficial activities of Microbes in Foods" theme is described in Unit-V with three chapters (No. 35 to 37). The theme of Unit-VI is "Microbial Quality Control and Safety of Foods" covered in three chapters (No. 38 to 40), discussing various aspects as well as recent issues of food quality control. Unit-VII includes four appendices which can be used as additional reading materials.

ISBN 978-81-7190-221-3 **Rs. 2000.00**

FOOD BORNE ILLNESSES

The aim of this book is to review food-borne hazards and illnesses to protect the people from the victimization by the food-borne pathogens.

The first chapter elaborates interactions between microorganisms and foods leading to the development of Food Microbiology. The second chapter describes all the nutrients that we must obtain from food. The basic principles of food-borne diseases are elaborately explained in chapter-3, which also helps the readers in understanding the control of food-borne illnesses. The various features of major bacterial food-borne infections and intoxications are summarized in chapter-4. Various types of mycotoxins are described in chapter-5. Other food-borne hazards—viral infections, animal toxins, parasitic infections, mushroom and chemical poisoning etc. are discussed in chapter-6. The basic principles of microbial control are briefed in chapter-7. The basic principles and practice of cleaning and sanitation involved in food industry are described in chapter-8 and so this chapter is very important for the students of food science and food technology. Similarly, chapter-9 "Microbiological Examination of Food" describes sampling, various test procedures used for detecting food-borne pathogens, food-spoilage organisms etc.

The book will prove to be an useful source of information for anyone with an interest in food microbiology especially in food-borne illnesses for both undergraduate as well as postgraduate courses of Microbiology. It will also be useful to the students of Food Technology, Biotechnology, Medicine, Public Health and Sanitary Courses, Home Science, Hotel and Catering Management and for the people who are working in food-processing industries and Government organizations involved in public health.

ISBN 978-81-7190-256-5

Rs. 650.00

FOOD MICROORGANISMS

The food-processing industry is the oldest and larger user of microorganisms and biotechnological processes. The first chapter elaborates interactions between microorganisms and foods. The second chapter describes all the nutrients that we must obtain from food and significance of human nutrition. Important characteristics of food-associated microorganisms are summarized in third chapter. Various factors influencing microbial growth in foods are discussed in chapter 4. The sources of microorganisms are summarized in chapter 5. Incidence and significance of microorganisms in foods are mentioned in chapter 6. Among the food components, microbial metabolism of carbohydrates, proteins and lipids are of major importance for many reasons which are elaborately discussed in chapter 7. The detailed description on "Microbiological examination of food" covered in chapter 8 will be proved very useful to the students of food science and technology The principles of microbial spoilage of foods are discussed in chapter 9. Many hundreds of different types of fermented foods are found worldwide and these are briefly summarized in chapter 10. Health benefits of beneficial bacteria in foods are highlighted in chapter 11.

The contents of this book will be useful to the students of Microbiology, Food Science and Technology, Biotechnology, Biochemistry, Home Science, Food and Nutrition, Veterinary Science, Hotel and Catering Management and other food-related courses. The book will serve as an useful first-hand information of food microorganisms for the people who are working in food-processing industries.

ISBN 978-81-7190-246-6

Rs. 650.00